PCBs and the Environment

Volume I

Editor

John S. Waid, D. Phil.(Oxford), F.I.Biol., M.A.S.M.

Chairman and Head
Department of Microbiology
La Trobe University
Bundoora, Victoria, Australia

CRC Press, Inc.
Boca Raton, Florida

Library of Congress Cataloging-in-Publication Data

PCBs and the environment.

Includes bibliographies and indexes.
1. Polychlorinated biphenyls--Environmental aspects.
I. Waid, John S.
QH545.P6P236 1986 363.1'79 86-17549
ISBN 0-8493-5923-6 (v. 1)
0-8493-5924-4 (v. 2)
0-8493-5925-2 (v. 3)

This book represents information obtained from authentic and highly regarded sources. Reprinted material is quoted with permission, and sources are indicated. A wide variety of references are listed. Every reasonable effort has been made to give reliable data and information, but the author and the publisher cannot assume responsibility for the validity of all materials or for the consequences of their use.

Direct all inquiries to CRC Press, Inc., 2000 Corporate Blvd., N.W., Boca Raton, Florida, 33431.

International Standard Book Number 0-8493-5923-6 (v. 1)
International Standard Book Number 0-8493-5924-4 (v. 2)
International Standard Book Number 0-8493-5925-2 (v. 3)
International Standard Book Number 0-8493-5929-5 (set)

Library of Congress Card Number 86-17549
Printed in the United States

THE EDITOR

John S. Waid, D.Phil.(Oxford), F.I.Biol., M.A.S.M., is Chairman and Head, Department of Microbiology, La Trobe University, Bundoora, Victoria, Australia, a position he has held since 1974. He also served as Dean of the School of Biological Sciences from 1977-1979.

Prior to his offices at La Trobe University, Professor Waid served in various capacities at the Grassland Research Institute; the Nature Conservancy; Levington Research Station, Fissons, Ltd.; University of New England, New South Wales; Cornell University; Reading University; and the Univesity of Canterbury, New Zealand.

His experimental experience includes the post as joint editor of the proceedings and a convenor of an international symposium on the ecology of soil fungi held in Liverpool in 1968; initiator and Executive Editor of *Soil Biology and Biochemistry* from 1969 to the present; New Zealand delegate to the Committee on Microbial Ecology, International Association of Microbiology Societies (1972-1974); leader of the advisory team from New Zealand under the auspices of the Colombo Plan to review and report on science teaching at Khon Kaen University in northeast Thailand (1974); Victorian representative on the Standing Committee for Environmental Community Problems (SCENCOP) of the Australian Society for Microbiology (1975-1977); and the national representative for the Australian Society for Microbiology on the International Committee of Microbial Ecology (1979-present).

Professor Waid's research interests at present are root biology (in particular the decomposition of grass and legume roots and release of legume-nitrogen to the soil, a study which has been supported by ARGC grants since 1977) and polychlorinated biphenyls in the environment and their degradation. The PCB work has been supported by grants from the Rural Credits Development Fund (1976-1978), the Victoria Ministry for Conservation (1979-1989), and the Australian Marine Sciences and Technologies Advisory Committee (1980-1981). He has also had contracts with the Victoria Ministry for Conservation, and the Federal Department of Science and the Environment to review problems associated with PCBs and the monitoring of hazardous chemicals in the environment. He has made surveys of PCBs in soils and waterways on behalf of a government department, an urban authority, and a large industrial organization (all unnamed), and is consulted regularly on matters pertaining to PCBs by the Victoria Environment Protection Authority.

CONTRIBUTORS

Elliot Atlas, Ph.D.
Associate Research Scientist
Department of Oceanography
Texas A & M University
College Station, Texas

Terry Bidleman, Ph.D.
Associate Professor
Department of Chemistry
Marine Science Program/Belle W. Baruch Institute for Marine Biology and Coastal Research
University of South Carolina
Columbia, South Carolina

Thomas Cairns, Ph.D.
Department of Health and Human Services
Food and Drug Administration
Office of Regulatory Affairs
Los Angeles, California

Sheng-Fu Chou, Ph.D.
Associate Organic Chemist
Geochemistry Section
Illinois State Geological Survey
Champaign, Illinois

D. W. Connell, M.Sc., Ph.D.
Reader
School of Australian Environmental Studies
Griffith University
Nathan, Queensland, Australia

Gregory M. Doose
Department of Health and Human Services
Food and Drug Administration
Office of Regulatory Affairs
Los Angeles, California

Jerry E. Froberg
Department of Health and Human Services
Food and Drug Administration
Office of Regulatory Affairs
Los Angeles, California

C. S. Giam, Ph.D.
Professor
Department of Chemistry
Graduate School of Public Health
Department of Industrial Environmental Health Sciences
University of Pittsburgh
Pittsburgh, Pennsylvania

R. A. Griffin, Ph.D.
Geochemist and Head
Geochemistry Section
Illinois State Geological Survey
Champaign, Illinois

Alan V. Holden
Manse Road
Moulin, Pitlochry, Scotland

Robert A. Jacobson
Department of Health and Human Servies
Food and Drug Administration
Office of Regulatory Affairs
Los Angeles, California

John A. G. Roach
Mass Spectrometrist
CFSAN
Food and Drug Administration
Washington, D.C.

Brij L. Sawhney, Ph.D.
Soil Chemist
The Connecticut Agricultural Experiment Station
New Haven, Connecticut

Glen R. Shaw, B. App. Sci.
Senior Chemist
Government Chemical Laboratory
Brisbane, Queensland, Australia

Emil G. Siegmund
Department of Health and Human Services
Food and Drug Administration
Office of Regulatory Affairs
Los Angeles, California

Virginia F. Stout, Ph.D.
Research Chemist
Utilization Research Division
Northwest & Alaska Fisheries Center
National Oceanic and Atmospheric Administration
U.S. Department of Commerce
Seattle, Washington

Shinsuku Tanabe, Ph.D.(Agric.)
Assistant Professor
Department of Environment Conservation
Ehime University
Matsuyama, Japan

Ryo Tatsukawa, Ph.D.(Agric.)
Professor
Department of Environment Conservation
Ehime University
Matsuyama, Japan

TABLE OF CONTENTS

Volume I

Volume II

Volume III

Chapter 1

ANALYTICAL CHEMISTRY OF PCBs

Thomas Cairns, Gregory M. Doose, Jerry E. Froberg, Robert A. Jacobson, and Emil G. Siegmund

TABLE OF CONTENTS

I. GLOBAL USES AND CONTAMINATION

A. Industrial Applications

Although the synthesis of polychlorinated biphenyls (PCBs) was first described in 1881 by Schmidt and Schultz,[1] the potential industrial applications were not fully realized until about 1930. The increased use of PCBs as important industrial chemicals for use as nonflammable oils in a host of products gave birth to a series of commercially available raw materials marketed under various trade names; e.g., Aroclor® (Monsanto, U.S.), Clophen® (Bayer, West Germany), Phenoclor® (Caffaro, Italy), Kanechlor® (Kanegafuchi, Japan), Pyralene® (Prodelec, France), and Sovol® (U.S.S.R.) — to name a few. These mixtures of PCBs quickly gained wide acceptance in industrial products where nonflammability and heat-resistant properties were highly desired. For reference purposes, the Aroclor® series[2] manufactured in the U.S. can be taken as representative of the range of PCB mixtures available on the world marketplace. Foreign counterparts had different names but their PCB content was very similar to an existing member of the Aroclor® series. This series of Aroclors® was sold under various Arabic number designations which reflected their relative compositions and physical properties (Table 1). In theory, the total number of possible compounds resulting from chlorination of the biphenyl nucleus is 209 (Figure 1). Molecular structural type was defined by the first two digits: 12 for PCBs, 25 and 44 for blends of PCBs and polychlorinated terphenyls (PCTs), and 54 for PCTs. The last two digits were an approximate estimate for the wt% of chlorine. Popularity in industrial applications of such materials as constituents of heat transfer systems, hydraulics/lubricants, transformers, capacitors, plasticizer applications, and petroleum additives catapulted U.S. domestic sales[3] of Aroclors® (1221 to 1268) from 32 million lb (14,515 t) in 1957 to almost 80 million lb (36,287 t) in 1970. The most popular blend was Aroclor® 1242 (18 million lb [8,165 t] in 1957 to almost 50 million lb [22,680 t] in 1970). Industrial applications involving transformers and capacitors were the leaders in usage pattern (29 million lb [13,154 t] in 1957 to almost 40 million lb [18,182 t] in 1969). This quantitative historical data has been deliberately limited to pre-1970 to provide a suitable backdrop for the environmental consequences that were to follow this intense growth in production and uses of PCBs.

B. Need for Regulatory Controls

Grave concerns about the environmental fate of such vast quantities of PCBs and their resultant toxicological effects were alerted by the first reported findings of PCBs in fish and wildlife by Jensen[4] in 1966. Soon it became glaringly obvious that the widespread popularity of PCBs in a somewhat uncontrolled fashion had eventually led to their environmental incorporation as persistent and ubiquitous contaminants on a global scale.[5-8] In the ecosystem, PCBs had become the most abundant of the chlorinated aromatic pollutants,[9] rivaling DDE. Sadly, indirect contamination by PCBs had eventually led to their presence in the food chain.[10-12]

Shortly after Jensen's report of PCBs in 200 pike,[4] an unfortunate accident in Japan[13] in 1968 was a preview of similar events that would occur in the U.S. during the next decade. PCBs had accidentally leaked from a heat exchanger used in the production of rice oil. Resultant levels of PCBs in the rice oil, when ingested, produced a spectrum of adverse symptoms: chloracne, discoloration of the gums and nailbeds, swelling of the joints, waxy secretions of the glands in the eyelids, as well as more general manifestations such as lethargy and joint pain. Perhaps it was this single historical event rather than the widespread environmental contamination reported in the scientific literature that prompted regulatory intervention. The U.S. Food and Drug Administration (FDA) initiated a national survey to determine the exact extent and levels to which PCBs might have made their way into the food chain by indirect use of PCB contaminated animal feed, industrial and environmental

Table 1
GENERAL PHYSICAL AND TOXICOLOGICAL PROPERTIES OF VARIOUS AROCLORS® [2]

Aroclor® no.	Form	Sp. gravity	Distillation range (°C)	Rats oral LD_{50} (mg/kg)	Rabbits skin MLD (mg/kg)
1221	Clear, mobile oil	1.182—1.192	275—320	3,980	>2,000
1232	Clear, mobile oil	1.270—1.280	290—325	4,470	>1,260
1242	Clear, mobile oil	1.381—1.392	325—366	8,650	>794
1248	Clear, mobile oil	1.405—1.415	340—375	11,000	>794
1254	Light yellow viscous oil	1.495—1.505	365—390	11,900	>1,260
1260	Light yellow, soft sticky resin	1.555—1.566	385—420	10,000[a]	>1,260[a]
1262	Light yellow sticky viscous resin	1.572—1.583	395—425	11,300[a]	>1,260[a]
1268	White to off-white powder	1.804—1.811	435—450	10,900[a]	>2,510[b]

[a] Administered as 50% solution in corn oil.
[b] 10% solution in corn oil.

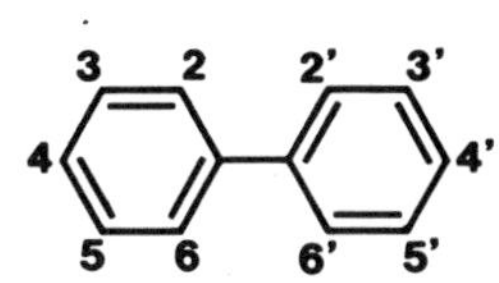

Chlorine Substitution	Number of Possible Isomers
Mono-	3
Di-	12
Tri-	24
Tetra-	42
Penta-	46
Hexa-	42
Hepta-	24
Octa-	12
Nona-	3
Deca-	1
Total:	209

FIGURE 1. Number of possible isomers of PCBs.

sources, and the use of PCB-containing paper food-packaging materials. This extensive survey finally resulted in a "Notice of Proposed Rule Making"[14] to limit the level of PCBs in foods containing unavoidable residues from environmental and industrial sources. It was interesting to note that of the animal feed samples analyzed less than 5% contained PCBs (0 to 0.6 ppm). Several accidents similar to the "Yusho" incident had, however, directly

Table 2
FDA TOLERANCES FOR PCBs IN SEVERAL CLASSES OF FOOD

Product	Concentration (ppm) 1973	1979—
Milk & dairy products (fat basis)	2.5	1.5
Poultry (fat basis)	5.0	3.0
Eggs	0.5	0.3
Fish & shellfish (edible portion)	5.0	5.0[a]
Infant & junior foods	0.2	0.2
Feed for food producing animals	0.5	0.2
Animal feed components including fishmeal & other by-products of marine origin	5.0	2.0
Paper food packaging materials intended for or used with human food	10.0	10.0

[a] A regulation establishing a new level of 2 ppm was promulgated but was stayed on October 5, 1979 until further notice.

contaminated animal feed and subsequently the poultry and eggs intended for human consumption. Other parts of the survey indicated that the use of PCB-containing coatings on the inner walls of grain silos had been responsible for PCB residues in milk[15] derived from dairy cows who fed on the grain stored in such silos. Results on food packagings were revealing; 67% of the samples tested contained PCBs with the highest value found being 338 ppm, and of these samples only 19% of the actual food contained in such packages contained PCBs at the maximum level of 0.1 ppm. A parallel survey of infant food packaged cereals found PCBs in 75% of the samples tested with an average concentration of 0.3 ppm and a maximum concentration of 1 ppm. In spite of the scanty quantitative knowledge of the toxicological effects of PCBs in 1973, the FDA concluded that it would be in the interest of the public health to reduce human low-level exposure to PCBs by limiting the ways in which PCBs might enter the food chain as well as limit the levels of PCBs in food containing unavoidable residues from environmental or industrial sources by the establishment of temporary tolerances (Table 2) until such time as additional toxicological data might cause reconsideration.[16] In this interim period, domestic production of PCBs in the U.S. declined drastically to below 10 million lb (4535 t). The age of unpopularity of PCBs as industrial chemicals had arrived. However, their popularity for scientific inquiry was spectacular. Seemingly, the concern about the ubiquitous distribution of PCBs in the ecosystem prompted a multitude of additional evidence to emphasize their prevalence as global contaminants.[17-26] Uppermost in this massive scientific probe was the need to increase awareness of the potential toxicity of PCBs. By 1975, the EPA responded to this new wealth of data by sponsoring a "National Conference on PCBs"[27] at which the FDA announced it had initiated a review of the appropriateness of the 1973 temporary tolerances. After such a review the FDA proposed to reduce the temporary tolerances[28] for unavoidable residues of PCBs in several classes of food. Interested parties likewise responded with over 100 comments being received for consideration before final disposition. The FDA Commissioner then issued a final order reducing PCB tolerances (Table 2) through publication of his responses and reasons for adoption of these new reduced tolerances.[29]

Since part of the problem with PCBs was the vulnerability of food and feed commodities to direct contamination through accidental causes, the EPA issued rules governing the continued deployment of PCBs in certain industrial applications. These regulatory controls were made under the Toxic Substances Control Act of 1976 and proposed the discontinued use of PCBs in heat transfer systems in plants manufacturing or processing food, drugs, and cosmetics. An interagency alert notice[30] was then issued by the EPA to urge voluntary compliance in removal of equipment containing PCBs and replacement with non-PCB units to prevent food contamination.[31]

C. Implications in Trace Analysis

In any attempt to extract, separate, and identify trace levels of PCBs from a wide variety of matrices several major problem areas must first be fully understood. The analytical problems encountered in dealing with PCBs have recently been reviewed by Cairns and Siegmund.[32]

1. Composition

The complications induced by analyzing mixtures of PCBs rather than any single specific isomer have probably contributed the most serious impediment to both identification and quantification. Although the theoretical calculations establish 209 possible congeners, the actual number of major components in Aroclor® 1254, for example, by capillary GC[33] was only 19 (69 eluting peaks, 50 represented minor constituents). Parallel studies using packed columns[34] have demonstrated elution profiles with less than 20 discernible peaks.[35] In analytical terms, this problem can be summed up as having to deal with a potential group of compounds within the mol wt range 188 to 494 daltons possessing vastly different chemical and physical properties.

2. Analytical Sensitivity

Halogen-sensitive chromatographic detectors such as the Hall electrolytic conductivity detector (HECD) and the electron capture detector (EC) are now capable of detection in the subnanogram range and offer the residue chemist a convenient analytical approach[36] when dealing with trace levels of PCBs (below 1 ppm). Added confirmation of residue levels by gas chromatography/mass spectrometry (GC/MS) often must employ multiple ion detection (MID) techniques[37] to achieve the necessary and parallel level of sensitivity of such GC detectors. Other analytical techniques such as nuclear magnetic resonance (NMR)[38] and liquid chromatography with UV detection[39] do offer alternative methods for analysis but can only be successfully applied where the suspected concentration level of PCBs is in excess of 1000 ppm.

3. Incurred Residues

Perhaps the most serious problem in identifying PCB residues from environmental samples is the inability to correlate the results with a known reference standard.[40] In spite of the availability of a wide range of reference materials, the EC/GC elution profiles obtained from samples containing PCBs do not always match directly. Although a number of reasons have been advanced (see Section III) for this phenomenon, the challenge experienced by the analytical chemist is often met by employing an auxiliary technique such as GC/MS or HE/CD which can assist in the judicious choice as to which reference standard most closely resembles the elution profile detected in the actual sample. The analyst in this arena must clearly develop a high degree of skill in the application of these types of methodologies by acquiring an ability to interpret elution profiles correctly, particularly if quantitation is required.

II. EXTRACTION AND SAMPLE CLEANUP

A. Separation

Jensen's historical breakthrough in 1966 of the first confirmed report[4] of PCBs in fish and wildlife was made after repeated and somewhat frequent encounters of similar GC elution patterns while routinely analyzing for DDT and other chlorinated pesticides. Indeed, the earlier failures to properly recognize this PCB presence must surely have contributed to the overestimation of DDT and TDE in the environment. The noticeable shift in emphasis to PCBs after 1966 reversed the roles almost overnight and PCB residues were then described to have interferences from a wide variety of organochlorine pesticides. Use of a halogen-sensitive GC detector such as EC in conjunction with a nonpolar stationary phase such as OV-101 or SE-30 had resulted in interferences with DDT, DDD, and DDE as well as early eluting pesticides such as BHC isomers, aldrin, heptachlor, and heptachlor epoxide because of similar retention times. Dependent upon the matrix selected for analysis, three distinct analytical protocols emerged involving either no prior separation, separation via column chromatography, or separation with destruction or conversion of interfering compounds.

In the case of fatty foods (milk, cheese, fish), the officially adopted method by the regulatory agencies[41] involved petroleum ether extraction of the fat (3 g) followed by residue partitioning into acetonitrile, dilution with water, re-extraction into petroleum ether, and finally column chromatography on Florisil® with 6% ethyl ether in petroleum ether. Additional cleanup procedures have also been devised to permit separation of DDT and its analogues from some of the PCBs.[42] Masumoto[43] has pointed out, however, that such procedures can be flawed by various experimental factors which cause variability in column preparation and hence recoveries. Additionally, those congeners with the lowest chlorination were held on the silicic acid column and could only be eluted by a more polar solvent than petroleum ether.

Lower recoveries with the less chlorinated residues were demonstrated. This methodology has also been miniaturized to analyze PCBs in fish.[44] Similar approaches have been developed to efficiently separate PCBs from such organochlorine pesticides as DDT and toxaphene,[45] heptachlor epoxide, lindane, and dieldrin.[46] In the analysis of paper and packaging, advantage of the chemical stability of PCBs to treatment with alkali[47] was optimized in devising a procedure based on extraction by refluxing with alcoholic KOH followed by an abbreviated cleanup. Sequential use of the alkaline hydrolysis and oxidation with chromic acid/acetic acid converted DDE and DDT to the respective dichlorobenzoquinones[48-50] and left the PCBs unchanged for analysis by EC/GC. Various other techniques have been described in the literature for the separation of PCBs from DDT and its analogues by similar approaches involving chemical derivatization and column chromatography.[51-54] Thin layer chromatographic procedures[55,56] were also developed to analyze PCBs in the presence of DDT, but lacked the specificity to distinguish the source of contamination.

B. Weathering and Metabolism

Previous mention has already been made to the fact that GC elution profiles observed experimentally by EC for PCB residues in biological samples[57-59] very often do not resemble in any recognizable way the profile for one of the many standard reference materials. This observation could be due to two main processes, metabolism and/or degradation. Dechlorination, arene oxide formation, and hydroxylation have already been identified as metabolic pathways for PCBs.[60,61] The main ramification of dechlorination to lesser chlorinated congeners is that it will tend to contribute to the production of an elution profile suggestive of a lower chlorinated reference standard and lead to confusion in the choice of the most suitable standard. However, the more serious consequence is the preferential removal from the PCB GC profile of certain congeners which have undergone metabolic hydroxylation. Such chlorobiphenylols do not exhibit the same chemical characteristics as their parent moieties and therefore do not co-extract and/or co-elute with the sample preparation and cleanup techniques adopted for routine analysis. Metabolic processes in biological samples can therefore complicate the assignment of PCB residues to contamination by a primary commercial product.

Additionally, the extraction and cleanup procedures might introduce a minor complication in that the attempt to solvent partition the various congeners as a single chemical entity is regarded as a compromise.

III. CHEMICAL CHARACTERIZATION

Because of the large number of standard reference materials available throughout the world, the following subsections are the first attempt in the scientific literature to provide a comprehensive library of elution profiles and patterns obtained by those analytical techniques both in common use as well as practiced by specialized laboratories. It must be stressed that not all these techniques can be employed for trace analysis work. However, the range and sensitivity of the analytical methods exhibited here should assist the reader in selecting a suitable reference standard and an analytical approach relative to their current problem. The wealth of data to be presented is somewhat overwhelming but serves to emphasize the magnitude of the problems of analyzing for PCB residues. Although the relative sensitivity of the various techniques are illustrated, the information offered is to be considered qualitative. A full discussion of the experimental approaches to quantitation will be presented in Section IV.

A. Gas Chromatography

Three GC detectors have been extensively employed in the analysis of PCBs in environmental and biological samples: namely, electron capture (EC), Hall electrolytic conductivity detector (HECD), and flame ionization detection (FID).[57] While the first two are sensitive to chlorine and are used mainly for trace work, the continued use of FID is still in practice where samples have a high concentration of PCBs. Clearly the advantages of both EC and HECD are that they are halogen sensitive and essentially ignore the majority of other sample components that have been co-extracted provided they are nonhalogen-containing entities. In the case of EC, sample cleanup is usually required to avoid detection of other electron capturing species present from the sample matrix. This is not the case with FID and sample extracts require extensive cleanup before analysis to avoid possible interferences from endogeneous components of the sample matrix. However, with EC as the detecting system for chlorinated aromatic samples such as PCBs, the main disadvantage is that of differences in response, i.e., highly dependent on the degree and location of chlorination. Detector responses have been shown to vary as much as two to four orders of magnitude between mono- and polychlorinated species.[58] The EC responses to 1 ng injected on the same GC column of five selected dichlorobiphenyls is illustrated in Figure 2. Dependent on the relative location of the two chlorine atoms on the biphenyl nucleus, a wide range of detector responses were obtained illustrating the marked differences within the same molecular weight group. This situation is disturbing because the reference standards are employed as standards for quantitation and the proper selection is dependent on the closest match to the incurred sample profile. Disproportionality of response with EC has been the main feature of criticism concerning analysis of several homologous series such as toxaphene,[59] strobane,[60] chlordane,[61] dibenzodioxins and dibenzofurans,[61] PCTs,[62] and PCBs.[63] Whereas EC responds to all electron capturing species, the HECD in the case of PCBs offers a direct measure of the HCl produced by the various congeners as they elute, i.e., proportional to Cl content. Therefore, it is not surprising that the elution profiles produced by EC and HECD for the same sample will be different. Both these detectors, however, offer high sensitivity and are capable of providing routine and reliable results. On the other hand, the responses of the FID are directly related to thermally excited ions produced in a hydrogen-air flame, i.e., a function of the number and type of burnable carbon atoms in the molecule. Response differences are most noticeable for the lowest members of a homologous series such as the

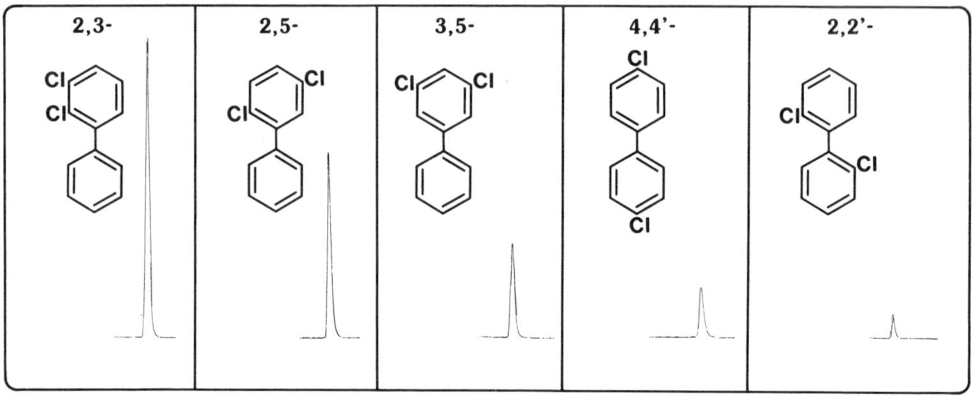

FIGURE 2. EC responses for 1 ng injected on column of selected PCBs: 2% OV-101, 125 cm × 2 mm I.D., 180°C, attenuation 16×.

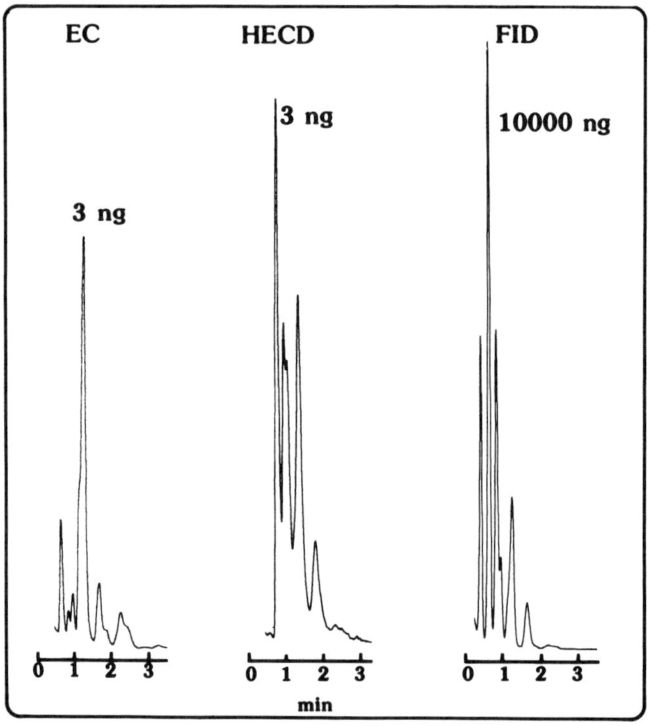

FIGURE 3. GC detector response profiles for Aroclor® 1221 together with amounts injected on-column.

PCBs. In terms of relative sensitivity, both EC and HECD demonstrate a level usually three orders of magnitude greater than that observed with FID.

With this informational background as to the utility of these various GC detectors, a tripartite graphical approach has been adopted to convey the various elution profiles obtained from 15 PCB standard reference materials and three PCTs (Figures 3 to 20). Standard operating conditions were employed throughout this series of experiments to provide comparative data: 2% OV-101, 120 cm × 2 mm I.D., 200°C isothermal, 40 mℓ/min argon/methane. To permit reproduction and/or extrapolation to other systems the elution times

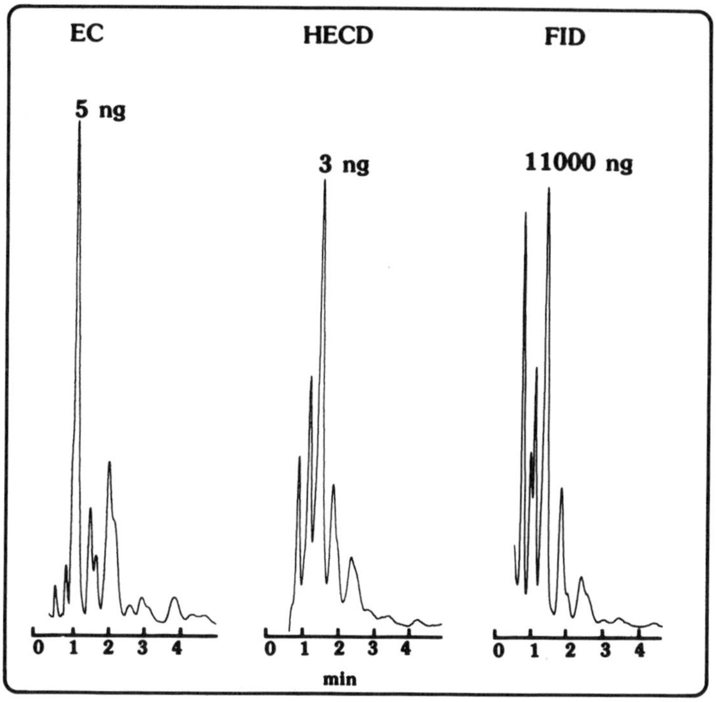

FIGURE 4. GC detector response profiles for Aroclor® 1232 together with amounts injected on-column.

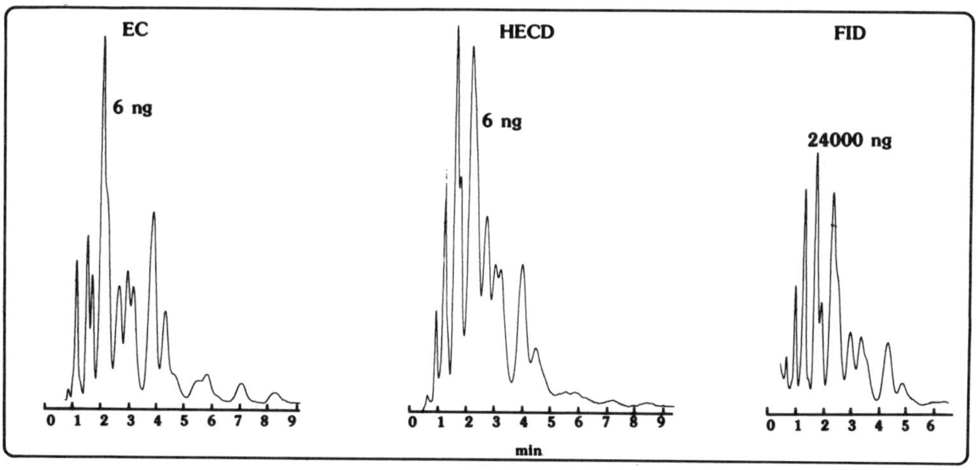

FIGURE 5. GC detector response profiles for Aroclor® 1242 together with amounts injected on-column.

under these selected conditions for aldrin, *p,p*-DDE, and *p,p*-DDT were 6.5, 8, and 13.5 min, respectively. Furthermore, standard reference materials from different manufacturing sources which resemble each other have been conveniently grouped together for discussion purposes. It is hoped that the illustration of all three elution profiles to describe each reference material (Table 3) will facilitate the choice of the most suitable one to describe the incurred residue.

In viewing this data base, it is clear that each detector has provided a response pattern entirely different from its two companion detectors. As previously stated, this experimental

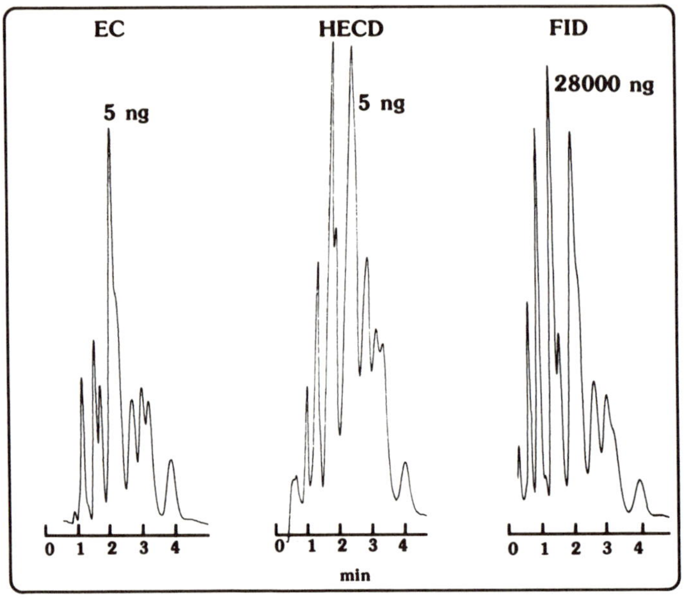

FIGURE 6. GC detector response profiles for Aroclor® 1016 together with amounts injected on-column.

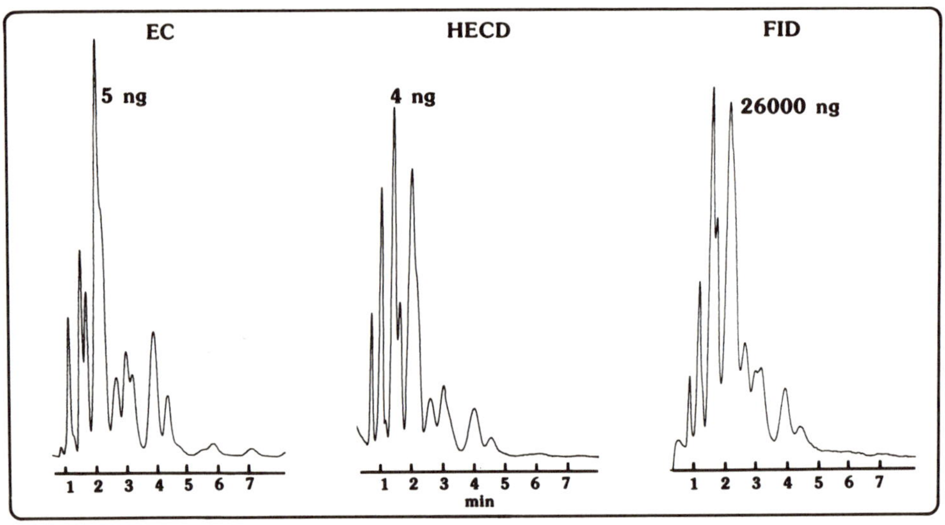

FIGURE 7. GC detector response profiles for Clophen® A-30 together with amounts injected on-column.

evidence is not that surprising since each detector has a different criterion for detection. However, it is reassuring to know that from an examination of these profiles we can conclude which reference materials are equivalents. While it might be difficult to select a suitable standard from a single detector elution profile, the additional information from a second and even third detector should make the task easier. Another factor in the postdetermination of the nature of the PCB residue is the retention time of the elution profile. As the retention time increases so does the degree of chlorination encountered.

A number of additional observations are worthy of mention. While the three GC detector profiles are distinctly different at low levels of %Cl, the responses recorded for HECD and

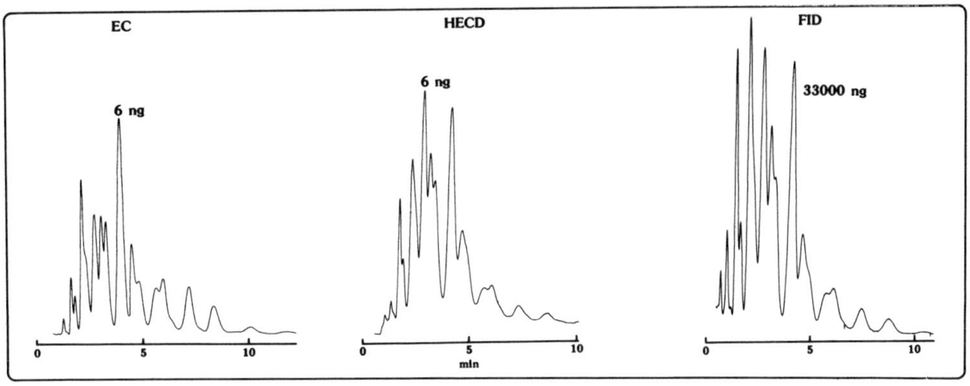

FIGURE 8. GC detector response profiles for Aroclor® 1248 together with amounts injected on-column.

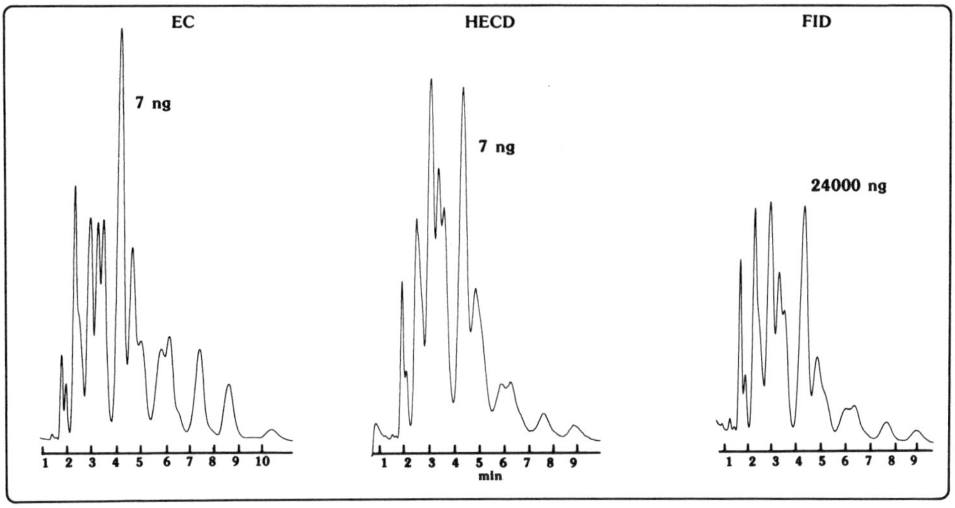

FIGURE 9. GC detector response profiles for Clophen® A-40 together with amounts injected on-column.

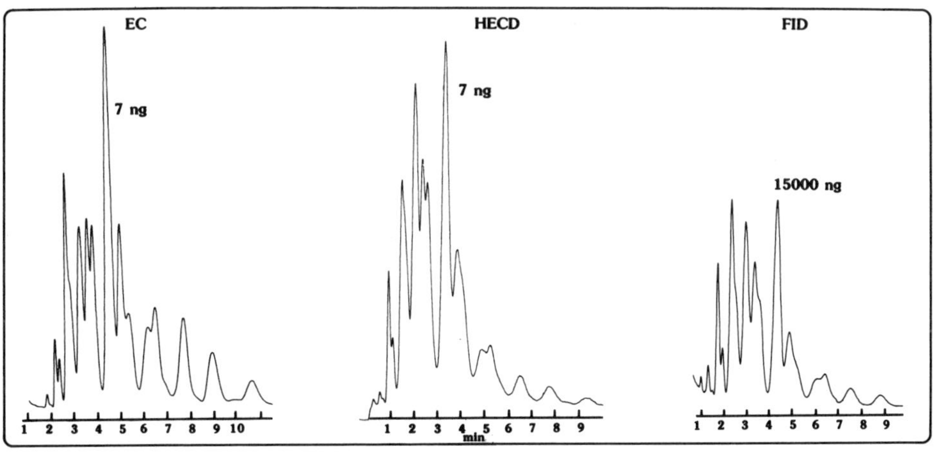

FIGURE 10. GC detector response profiles for Kanechlor® 400 together with amounts injected on-column.

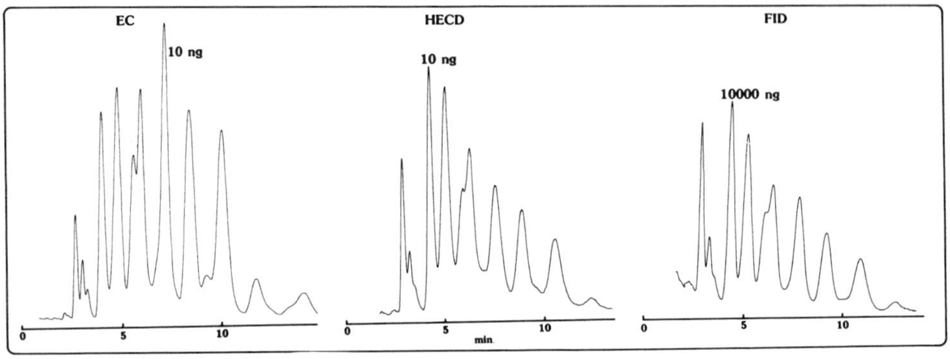

FIGURE 11. GC detector response profiles for Aroclor® 1254 together with amounts injected on-column.

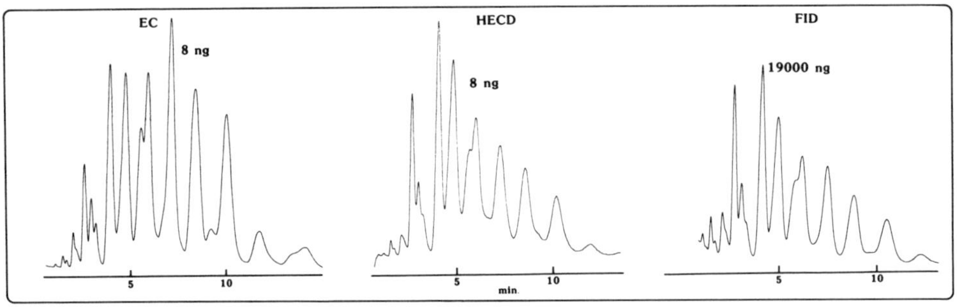

FIGURE 12. GC detector response profiles for Clophen® A-50 together with amounts injected on-column.

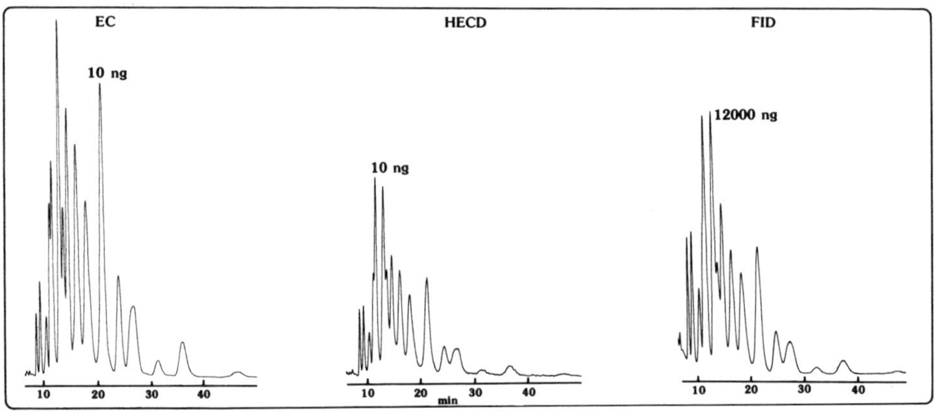

FIGURE 13. GC detector response profiles for Aroclor® 1260 together with amounts injected on-column.

FID begin to look similar at %Cl values greater than 48 (Figures 11 to 17). In only one case (Aroclor® 1268, Figure 17) did all three elution profiles match. This observation might not be that significant in that at this level of chlorination, only a few congeners are possible and with the high degree of chlorine substitution, disproportionality of response, etc., becomes a relatively unimportant matter to consider. The remaining Aroclors® (Figures 18 to 20) were PCTs and elution profiles varied little between detectors because of the high degree of chlorination. Perhaps the most enlightening value of this data base is the amount of material injected under FID conditions to match the responses recorded for both EC and HECD.

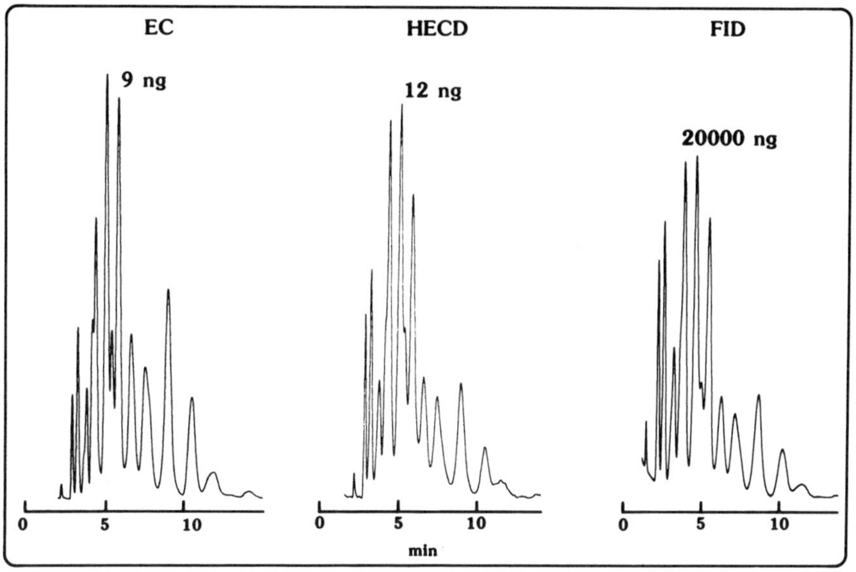

FIGURE 14. GC detector response profiles for Clophen® A-60 together with amounts injected on-column.

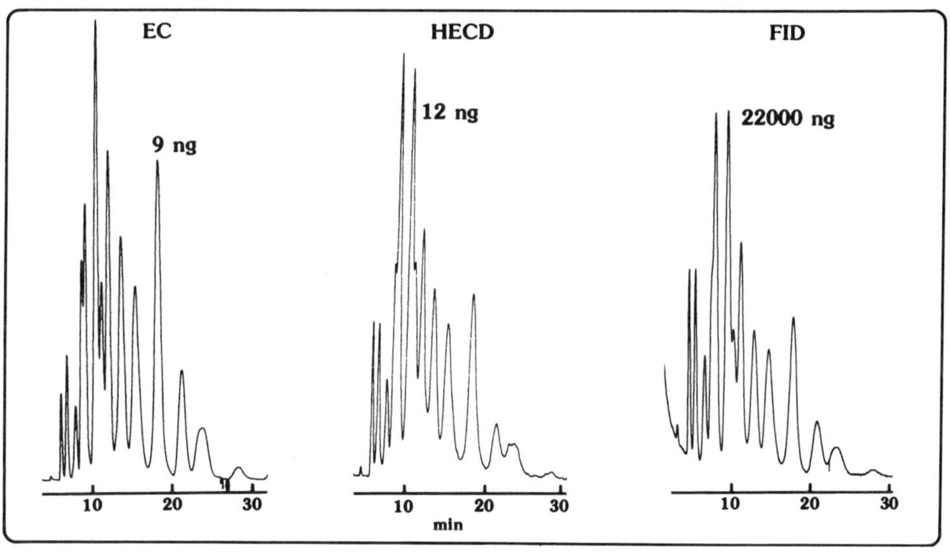

FIGURE 15. GC detector response profiles for Pyralene® 6000 together with amounts injected on-column.

In summary, the potential value of all three detectors for identifying PCBs in residue work as well as a description of the possible source of origin is now documented. This data base in graphical format should assist the analytical chemist to make the proper choice of a suitable standard reference material when quantitation is required. Where the incurred residue is in the low parts per million range, only EC and HECD can provide the necessary level of sensitivity required for detection. While the incurred residue profiles may not always bear an exact resemblance to one of the standard reference materials presented here, a choice can be made as to which is the most appropriate.

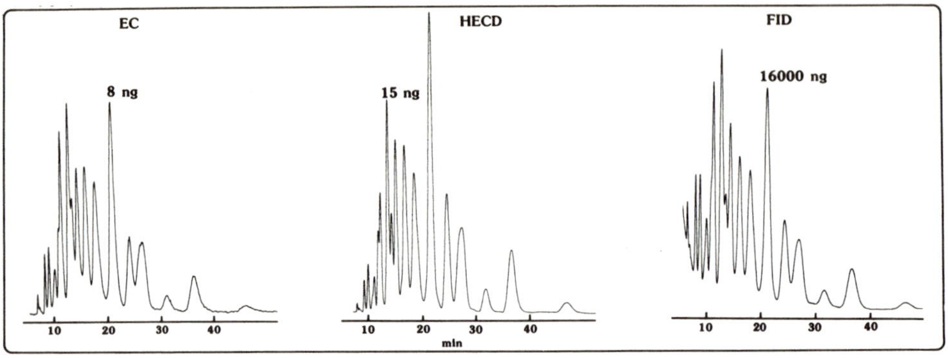

FIGURE 16. GC detector response profiles for Aroclor® 1262 together with amounts injected on-column.

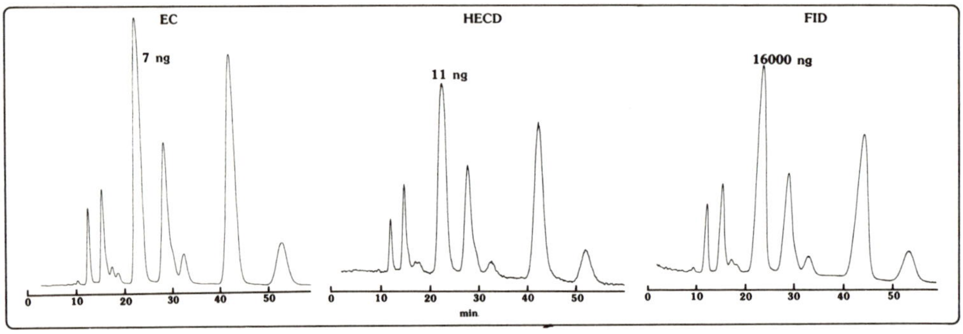

FIGURE 17. GC detector response profiles for Aroclor® 1268 together with amounts injected on-column.

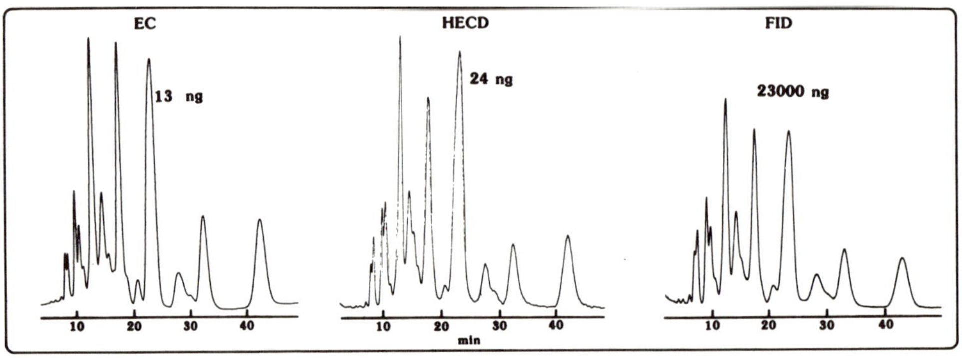

FIGURE 18. GC detector response profiles for Aroclor® 4465 together with amounts injected on-column.

B. Gas Chromatography/Mass Spectrometry

The analytical pitfalls or difficulties that can arise when using GC with one of the detectors mentioned above have led to the increased application of mass spectrometry to help resolve difficult identification problems. In particular, confirmation of incurred residues[63-69] has often been provided by GC/MS where full use is made of the isotopic distribution of chlorine (75.53% ^{35}Cl and 24.47% ^{37}Cl) in identifying ion clusters corresponding to certain degrees of chlorination.[70]

The electron impact (EI) spectra of PCBs have been previously studied in detail by Safe and Huntzinger[68,69] who concluded that the fragmentation scheme of the PCB congeners

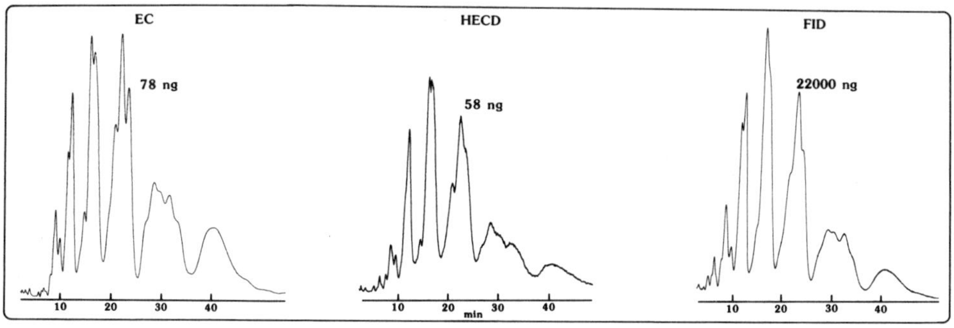

FIGURE 19. GC detector response profiles for Aroclor® 5432 together with amounts injected on-column.

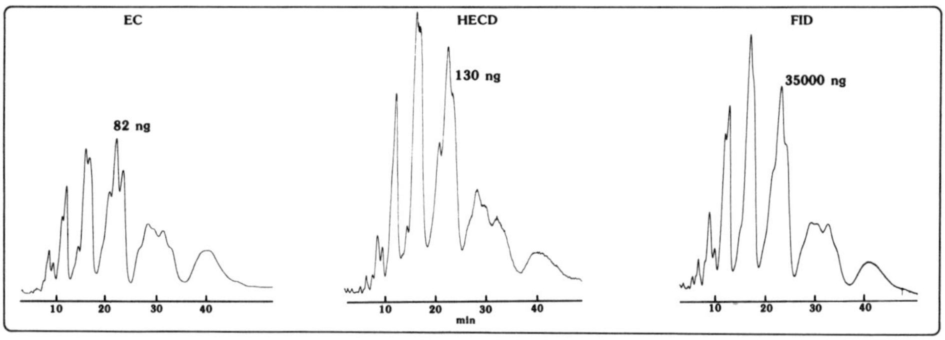

FIGURE 20. GC detector response profiles for Aroclor® 5442 together with amounts injected on-column.

was predominantly successive expulsion of chlorine atoms. This fragmentation pattern of PCBs made it difficult to distinguish the various molecular weight groups that might occur within an eluting peak. A recent study of the methane (CH_4) chemical ionization (CI) spectra of PCBs by Cairns and Siegmund[63] indicated that very little fragmentation took place (Figure 21). There was some evidence to suggest a minor population (less than 2%) of ions corresponding to the loss of HCl from the protonated molecular species which might be misinterpreted as belonging to the next lower homolog. Since the protonated molecular ions (MH^+) were the most abundant in the spectra whatever the degree of chlorination, a method of analysis involving the ions corresponding to chlorine content (in essence, molecular weight plus 1 dalton) was possible. It was found that responses from individual members of each molecular weight group were identical (within 5%) but decreased between molecular weight groups with increasing molecular weight, i.e., sensitivity of detection decreased with increasing molecular weight. This decrease in sensitivity of detection on going from mono- to hexachlorobiphenyl (1:0.4) was rationalized in terms of the availability of additional sites of protonation on the aromatic nucleus when only one or two chlorines were present. Since PCBs on CI produced almost entirely molecular ion clusters, an experimental approach was then devised involving monitoring only those ions corresponding to the different levels of chlorination. The resulting ion profiles obtained were found to be highly characteristic in providing differentiation for the various Aroclors®.

In keeping with the format established above for EC/HECD/FID (Figures 7 to 20), the necessary GC/MS information obtained under CI conditions has been compiled to add to that library of data. To provide a perspective view of the GC/MS data all the total ion current (TIC) profiles were obtained under identical conditions (Figures 22 to 24): 3% SP2100, 120 cm × 2 mm I.D., 190 to 250°C at 20°C/min, 30 mℓ/min methane. These elution profiles

Table 3
LOCATION OF GC ELUTION PROFILES ON REFERENCE STANDARDS

Reference standard	%Cl content	GC data[a] (EC/HECD/FID)
Aroclor® 1221	21	Figure 3
Aroclor® 1232	32	Figure 4
Aroclor® 1242	42	Figure 5
Aroclor® 1016		Figure 6
Clophen® A-30		Figure 7
Aroclor® 1248	48	Figure 8
Clophen® A-40		Figure 9
Kanechlor® 400		Figure 10
Aroclor® 1254	54	Figure 11
Clophen® A-50		Figure 12
Aroclor® 1260	60	Figure 13
Clophen® A-60		Figure 14
Pyralene® 6000		Figure 15
Aroclor® 1262	62	Figure 16
Aroclor® 1268	68	Figure 17
Aroclor® 4465 (PCBs + PCTs)	65	Figure 18
Aroclor® 5432 (PCTs only)	32	Figure 19
Aroclor® 5442 (PCTs only)	42	Figure 20

[a] All data generated under standard recording conditions to permit direct comparisons: electron capture (EC), Hall electroconductivity detector (HECD), and flame ionization detector (FID). The amount injected on column is displayed in each figure at the same attenuation factor of the recorder.

when compared to the GC data do not bear any strong resemblance to the previous elution patterns obtained by EC or HECD. However, the resemblance between FID and GC/MS-CI-CH_4 is established. Presumably the criterion for detection in both these cases runs parallel. Once again, the ability to group certain standards as equivalents (Table 3) has been demonstrated by observing similar elution profiles. The close similarity between Aroclor® 1262 and Aroclor® 1260 (Figure 22) is a reflection of the inability of GC/MS-CI-CH_4 to distinguish between such a small %Cl difference. As with the EC/HECD/FID data for Aroclor® 1268 (Figure 17) the GC/MS data (Figure 22) are a perfect profile match to all three detectors for the same reasons previously outlined. Besides this ability to distinguish standards via TIC profiles, the retention time ranges involved must not be totally forgotten as an additional criterion for identification. The higher chlorinated Aroclors® have larger ranges of retention time. While the GC data previously presented were based on a column temperature of 190°C, a temperature programming approach had to be adopted for GC/MS to provide the necessary peak widths (time for about 10 scans) for total ion detection (*m/z* 80 to 500) as well as completing the analysis within 10 min. To achieve these profiles illustrated in Figures 22 to 24, the amount injected on column was in the range of 3 to 7 μg, i.e., two orders of magnitude greater than that employed for EC or HECD. While it is not anticipated that this particular mode of detection (i.e., TIC over *m/z* 80 to 500) will be used in actual sample analysis, these basic profiles do serve as the primary data source in understanding the mass chromatograms that can be extracted from the data collected. In simple terms, a mass chromatogram is a computer-assisted technique that searches the total data collected and plots out only the responses recorded for the particular single *m/z* value requested. In the case of PCBs, this is a very useful technique since CI conditions produce only protonated molecular ions. Therefore, by specification of the *m/z* values corresponding to protonated molecular ions of biphenyl, monochlorobiphenyl, dichlorobiphenyl, trichlorobiphenyl, tetrachlorobiphenyl, pentachlorobiphenyl, hexachlorobiphenyl, heptachlorobiphenyl, and octachlorobiphenyl (155, 189, 223, 257, 293, 327, 361, 395, and 431, respectively) the total ion current can be subdivided to give maps of the composition of each eluting peak by molecular weight or chlorine content (Figures 25 to 33). This process can be considered as providing a second dimensional analysis of the TIC. Such mass chromatograms are very revealing in that they provide an instant visual representation of the composition of each Aroclor® by molecular weight. The ability to perform such mass chromatograms has increased the power of the GC/MS-CI-CH_4 data to the extent of providing unambiguous proof of identification as well as possible origin of contamination. In actual analysis of residue samples, however, the technique employed is to scan only for those ions of interest thereby achieving the level of sensitivity equivalent to that of EC or HECD.

C. Liquid Chromatography

While liquid chromatography (LC) has been demonstrated to be particularly useful in the analysis of thermally labile compounds, its potential utility in PCB analysis has been reported by only a few investigators.[39,71-77] At the present time, two LC techniques have been used to analyze PCB samples — adsorption liquid solid chromatography (LSC) and reversed-phase liquid liquid chromatography (RPLC). Aitzetmuller,[75] Brinkman et al.,[39] and Chiosi et al.[77] all favored the application of LSC since the observed elution patterns of PCBs were the reverse order of that observed both in GC with nonpolar stationary phases and RPLC, i.e., order of elution was according to decreasing chlorine content. With this technique, DDT and its analogues were separated from PCBs and hence removed the need for chemical separation before analysis. However, the resultant elution profiles illustrated poor separation of the congeners making assignment of the possible source of contamination difficult. Dry hexane[39] was used to increase resolution of the elution profile requiring additional analysis time and at the same time destroying the advantage of DDT separation from PCBs. On the other hand, Kirkland[78] favored RPLC for the nonpolar congeners. Sieber[73] has shown that RPLC can be used to assign individual chlorinated biphenyls to the certain reference standards.

In the continuing effort to provide extensive graphical documentation on elution profiles, the solvent system (5% water in methanol with ODS column) was deliberately chosen for direct transfer to liquid chromatography/mass spectrometry (LC/MS) using a moving belt interface.[79] Elution patterns obtained in this manner are illustrated in Figures 34 to 37 and can be considered similar to the GC elution profiles illustrated in the previous sections. Such elution patterns are in general agreement with those obtained by Sieber[73] and demonstrate decreased solubility in the mobile phase solvent for increasing molecular weight. As with GC/MS, deployment of SIM techniques (Figures 38 and 39) can be useful in determining the composition by degree of chlorination, thereby assisting in the proper assignment of source of contamination whenever possible. Additionally, the use of SIM is able to selectively ignore the presence of DDT and its analogues eliminating the disadvantage reported by those workers favoring LSC. The most widely used ultraviolet (UV) detector unfortunately requires extensive sample cleanup before analysis. Cairns et al.[79] have already demonstrated that LC/MS-SIM can eliminate the need for sample cleanup prior to quantitative analysis. However, one disadvantage continues to plague LC and LC/MS. Operating sensitivity is often two orders of magnitude lower than that experienced with EC or HECD. With the advent of modern detectors such as the photoconductivity detector (PCD), some of these sensitivity problems should begin to disappear.

In conclusion, LC has the potential to become an additional powerful analytical tool in the hands of the residue chemist. It is already established as a confirmatory technique in the absence of MS, but lacks sufficient sensitivity to compete with EC or HECD in residue analysis.

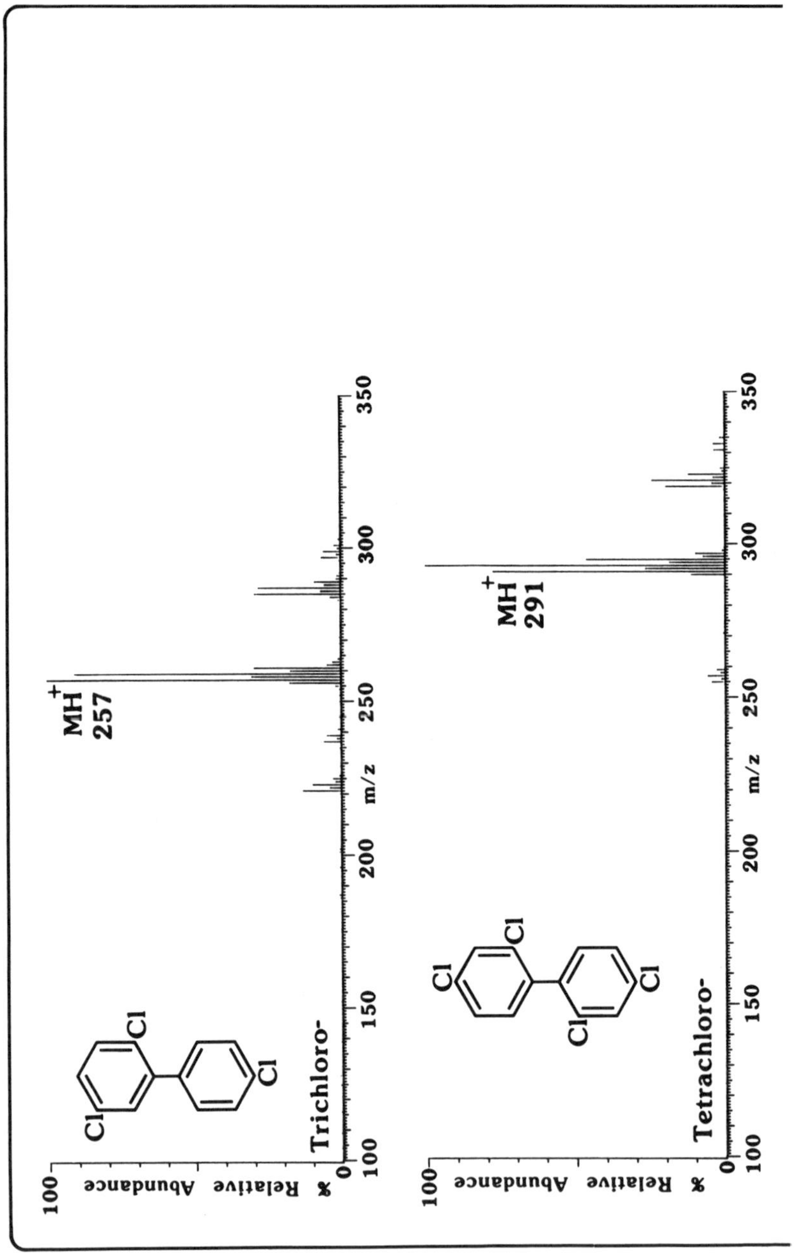
Trichloro-
MH+ 257
Tetrachloro-
MH+ 291
% Relative Abundance
m/z
Cl

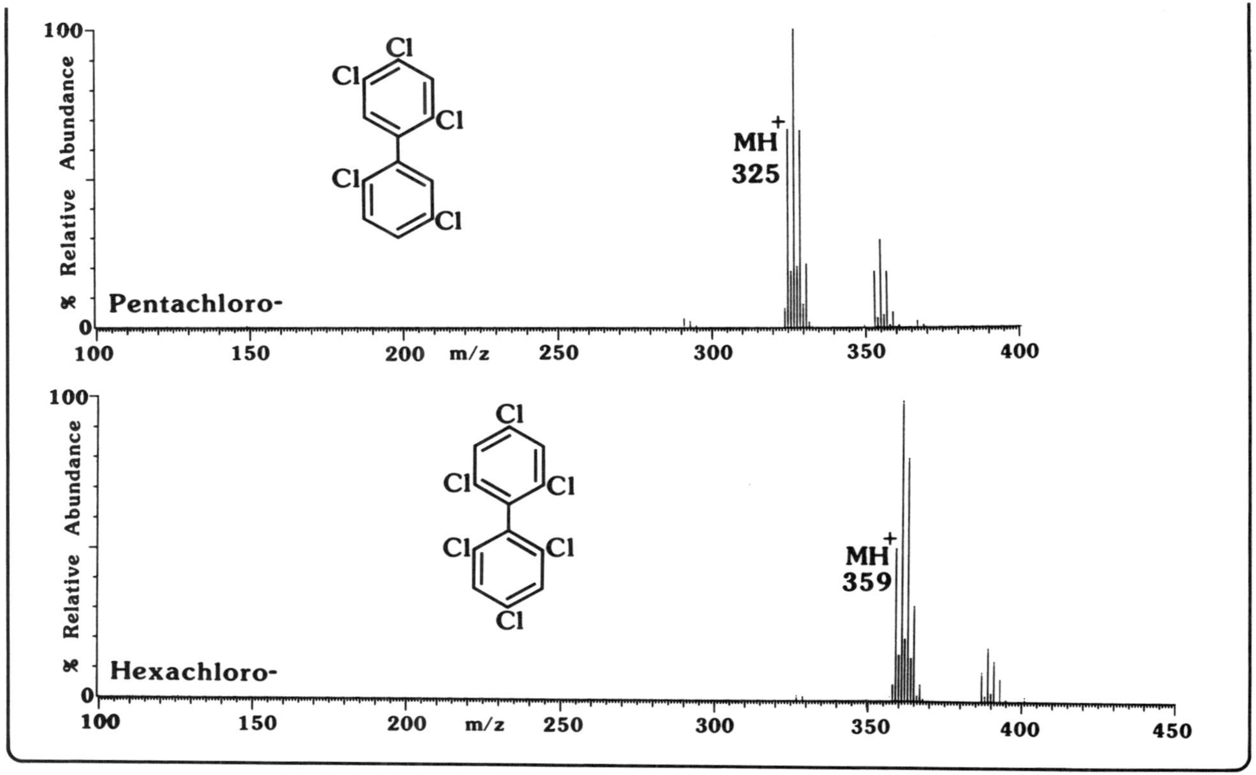

FIGURE 21. Chemical ionization mass spectra (CI/MS) of selected PCB isomers.

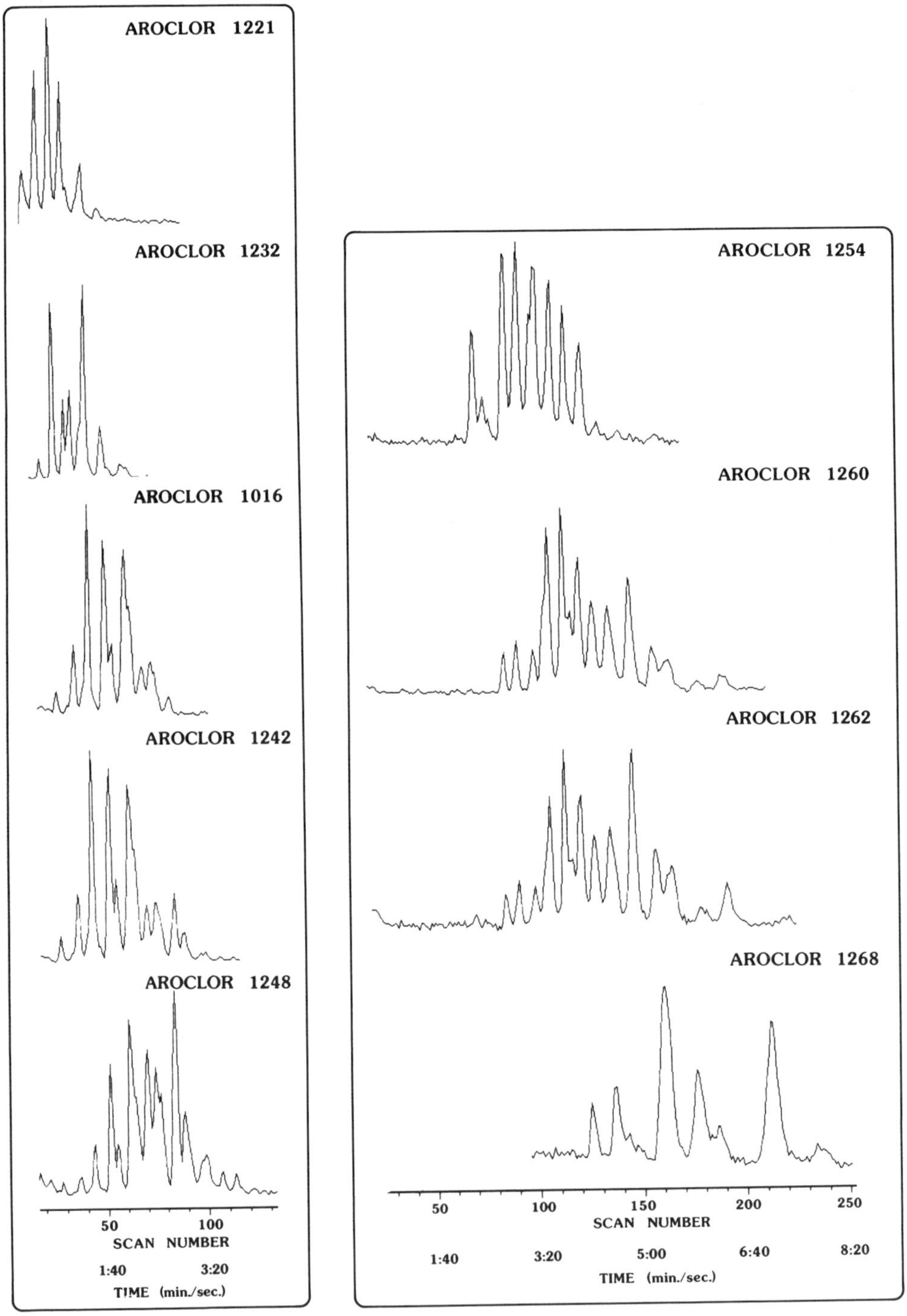

FIGURE 22. Total ion current profiles for Aroclors® under GC/MS-CI-CH_4.

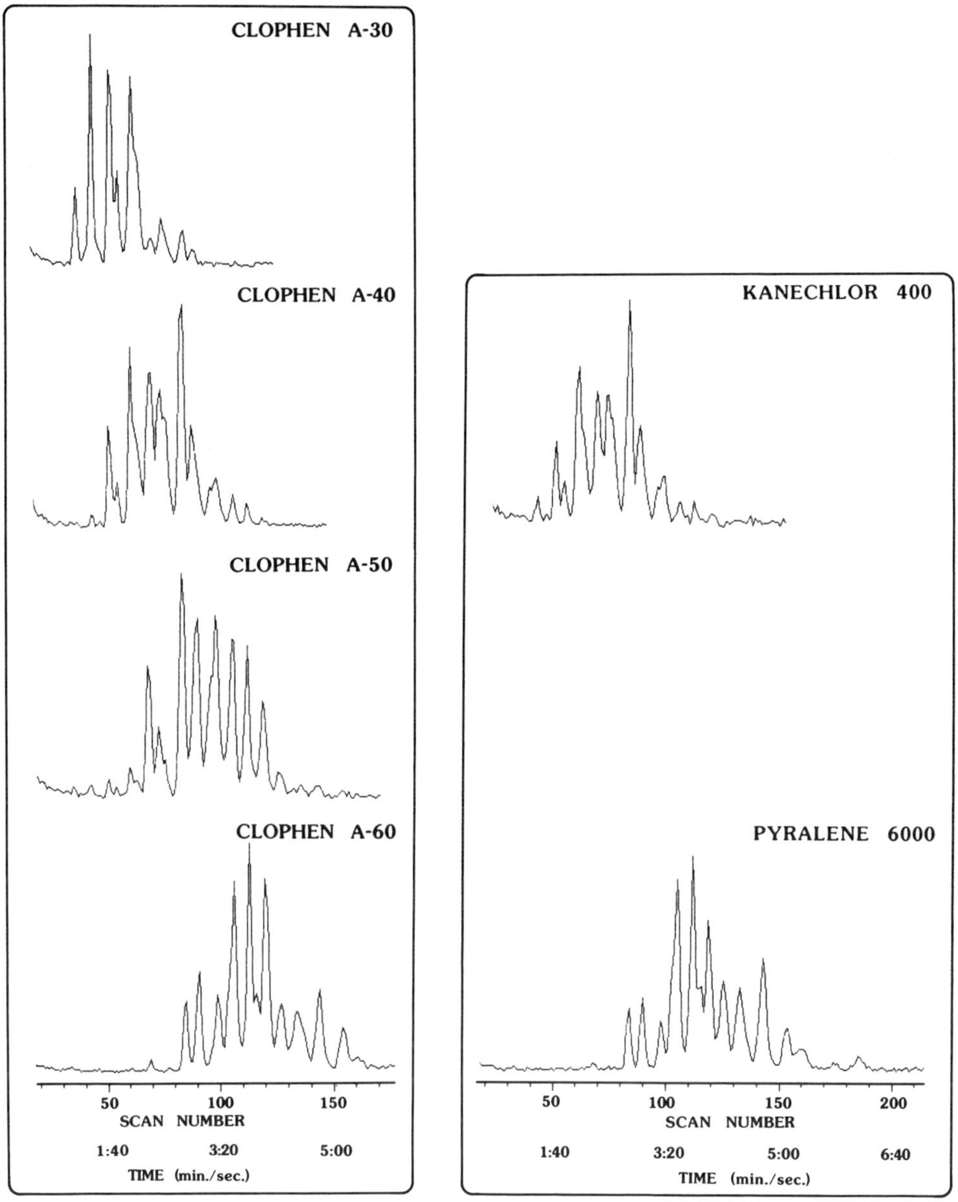

FIGURE 23. Total ion current profiles for commercial PCBs manufactured outside the U.S.

D. Nuclear Magnetic Resonance

Although the acknowledged sensitivity of nuclear magnetic resonance (NMR) is much less than that demonstrated for chromatographic and mass spectrometric methods, it can function as a powerful analytical tool in providing data on molecular structure, electronic charge distribution, and molecular interactions. The first report of an attempted identification of PCBs by proton NMR was conducted at 60 MHz.[80] This study concluded that in the absence of reference standards of the individual congeners, identification of certain specific isomers belonging to the PCB family was possible based on a priori arguments involving chemical shifts and coupling constants. In fact, the higher chlorinated congeners were the simplest to identify based on the small number of coupled protons expected in such structures. Welti and Sissons[81] then reported proton NMR data at 220 MHz for a large number of PCBs

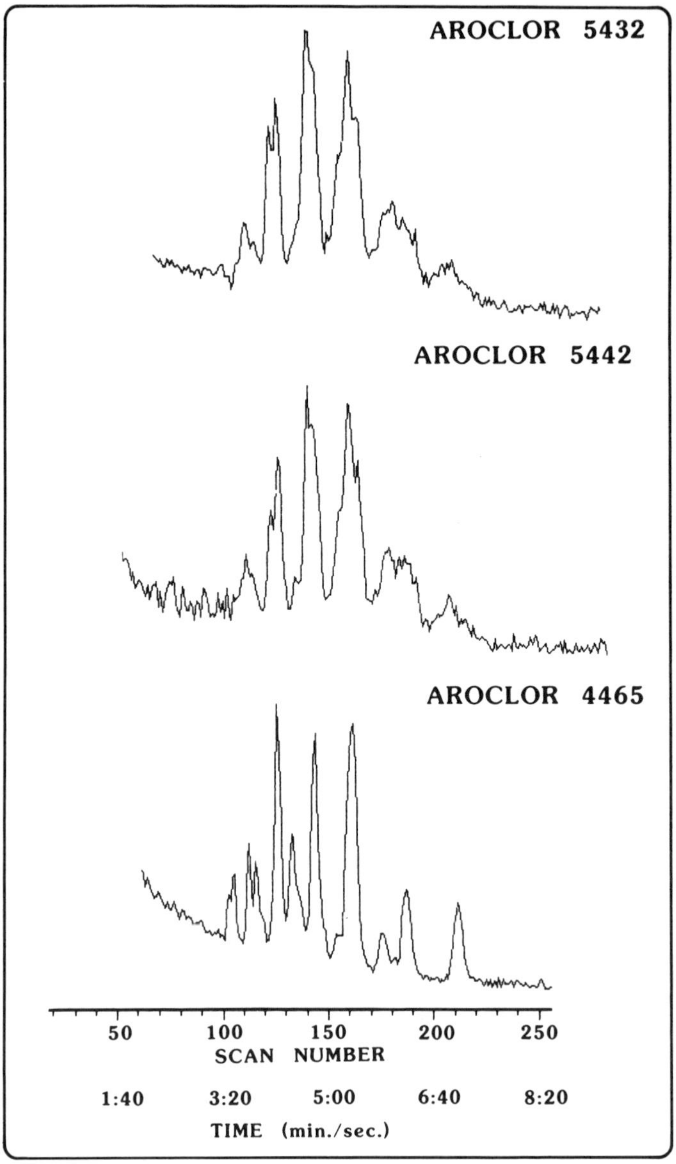

FIGURE 24. Total ion current profiles for polychlorinated terphenyls.

isolated from commercial standards such as the Aroclors®. This work fortified the preliminary studies by Bartle[80] and permitted less complicated spectral analysis. Symmetrically substituted dihalobiphenyls were extensively studied by Tarpley and Goldstein[82] who were able to account for chemical shifts in 2,2′-dichlorobiphenyl by application of an additivity scheme based on chlorobenzene and biphenyl. With technological advancements in instrumentation and the availability of Fourier transform (FT), studies were naturally then extended to the application of ^{13}C NMR to PCBs. Wilson and Anderson[83] reported the ^{13}C and ^{1}H NMR spectra for ten symmetrically substituted chlorinated biphenyls. These authors were able to predict the ^{13}C shieldings with reasonable precision from additive substituent parameters on benzene. Additional data on ^{13}C shieldings was then measured by Wilson[84] for 25 chlorinated biphenyls. Assignment of the protonated carbon atoms in 13 compounds was confirmed by selective proton coupling. Using additive procedures derived from data on chlorobenzene

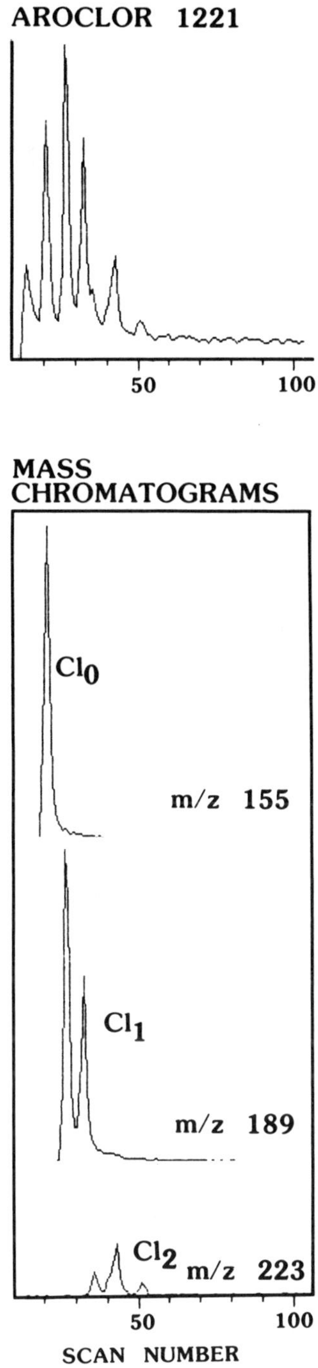

FIGURE 25. Total ion current with relevant mass chromatograms to illustrate the composition of Aroclor® 1221.

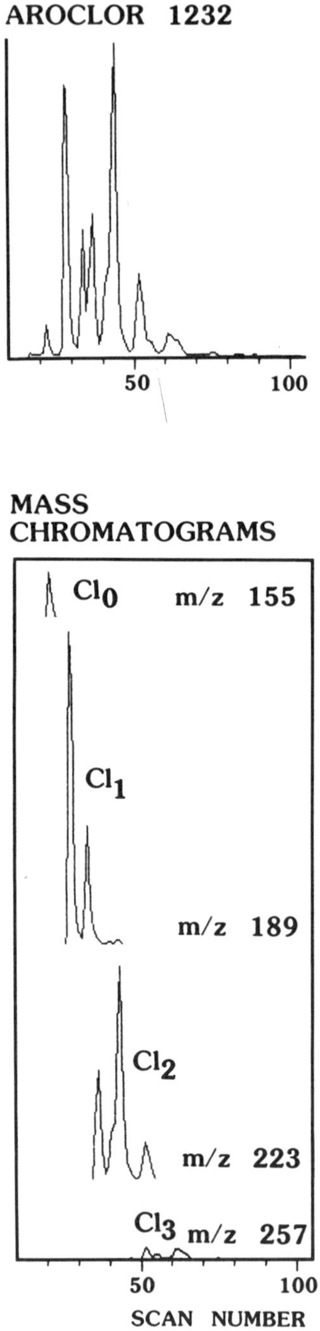

FIGURE 26. Total ion current with relevant mass chromatograms to illustrate the composition of Aroclor® 1232.

and 2-, 3-, and 4-chlorobiphenyl, the author was able to predict in an approximate fashion the effect of chlorine substitution on the ^{13}C shieldings. An interesting conclusion concerning the effects of a chlorine substituent was the discovery of their significant transmission through eight covalent bonds. The most significant finding, however, was the gross correlation of ^{13}C shieldings with total charge density which in turn permitted the division of the spectrum

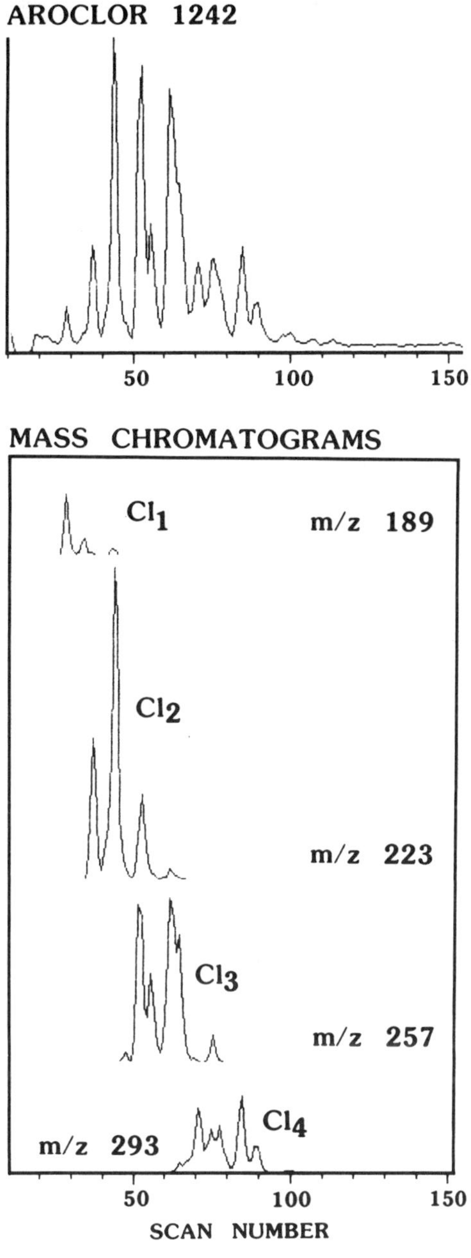

FIGURE 27. Total ion current with relevant mass chromatograms to illustrate the composition of Aroclor® 1242.

into three distinct sections: 135 to 144 ppm for C-1 carbons, 130 to 135 ppm for C-Cl carbons, and 120 to 130 ppm for C-H carbons. Levy and Hewitt[85] employed this analytical approach in the semiquantitative analysis by ^{13}C NMR (69.7 MHz) of Aroclor® 1221. The total chlorine content obtained by NMR was 1.4 atoms of chlorine per molecule. Based on GC/MS and EC/GC studies the quantitative value obtained was 1.2 atoms per molecule (i.e., 21% Cl by weight).

Rather than attempt to illustrate typical spectra obtained for certain individual congeners with tables of chemical shifts and coupling constants, four Aroclors® have been selected as

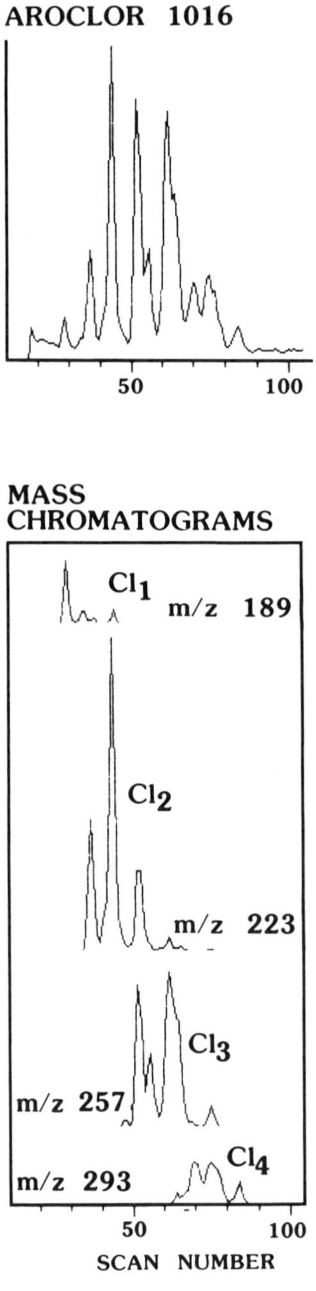

FIGURE 28. Total ion current with relevant mass chromatograms to illustrate the composition of Aroclor® 1016.

representative reference materials and their ^{1}H (270 MHz) and ^{13}C (69.7 MHz) NMR spectra are presented in Figures 40 and 41. These spectra were obtained on a Bruker WH270 spectrometer. Samples were dissolved in chloroform-d_1 (270 mg/mℓ) for the ^{13}C measurements and in methylene chloride-d_2 (280 mg/mℓ) for the ^{1}H measurements. ^{13}C spectra were recorded with proton broad band decoupling. In addition, the spin relaxation reagent chromium acetylacetonate (0.052 *M)* was present for all the ^{13}C measurements. The concentration of the relaxation reagent and the ^{13}C data acquisition parameters were selected so that

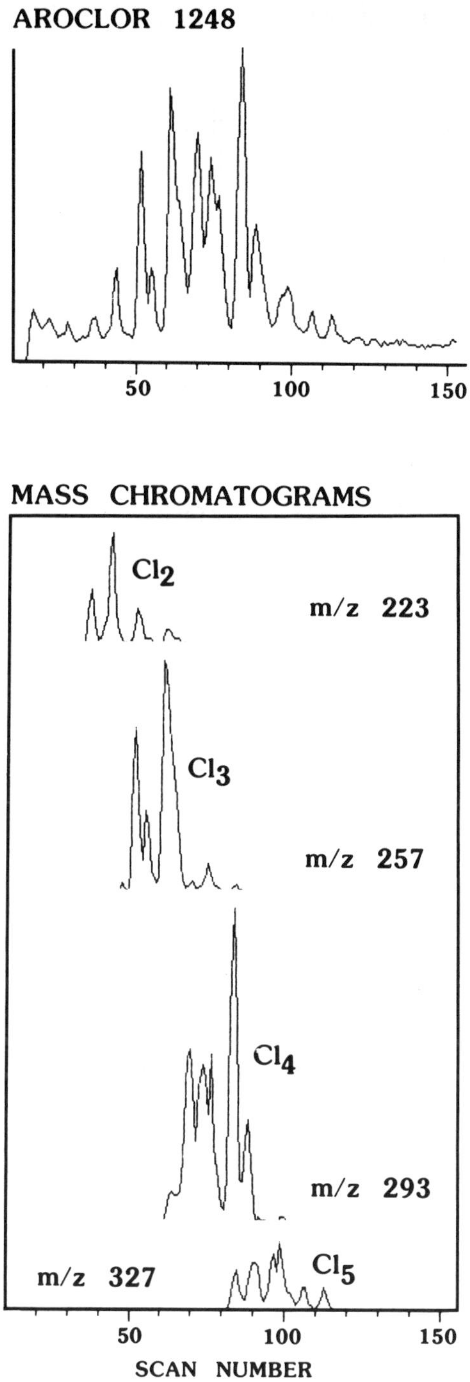

FIGURE 29. Total ion current with relevant mass chromatograms to illustrate the composition of Aroclor® 1248.

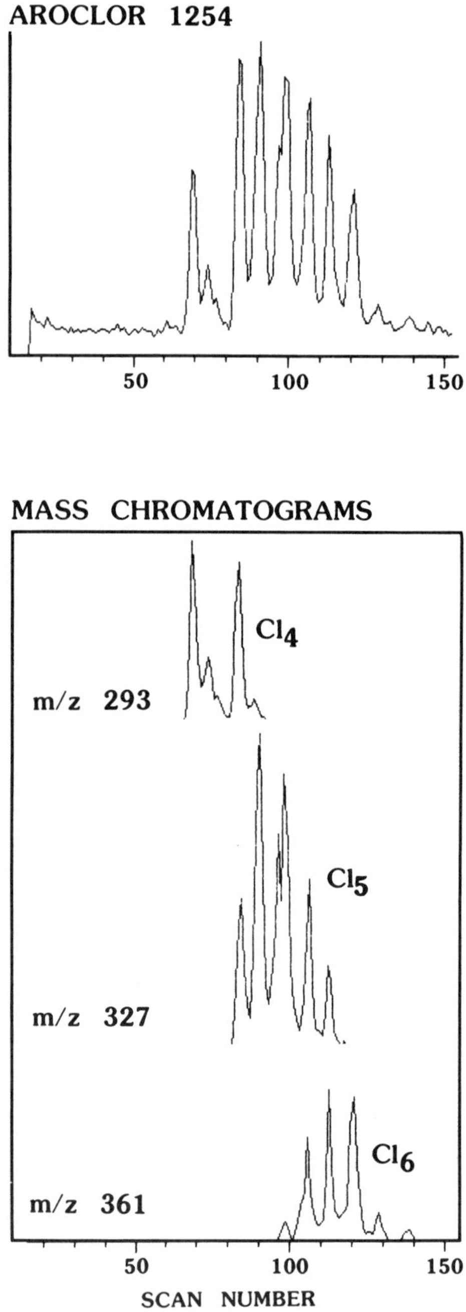

FIGURE 30. Total ion current with relevant mass chromatograms to illustrate the composition of Aroclor® 1254.

sensitivity was optimized under conditions where quantitative data was obtained. The ^{1}H data was recorded under quantitative conditions simply by using a long relaxation time. Chemical shifts were reported in ppm downfield from internal TMS for both ^{13}C and ^{1}H spectra.

These spectra of selected Aroclors® have been presented to characterize or fingerprint rather than provide a detailed structural analysis of the data. In the case of ^{1}H NMR spectra

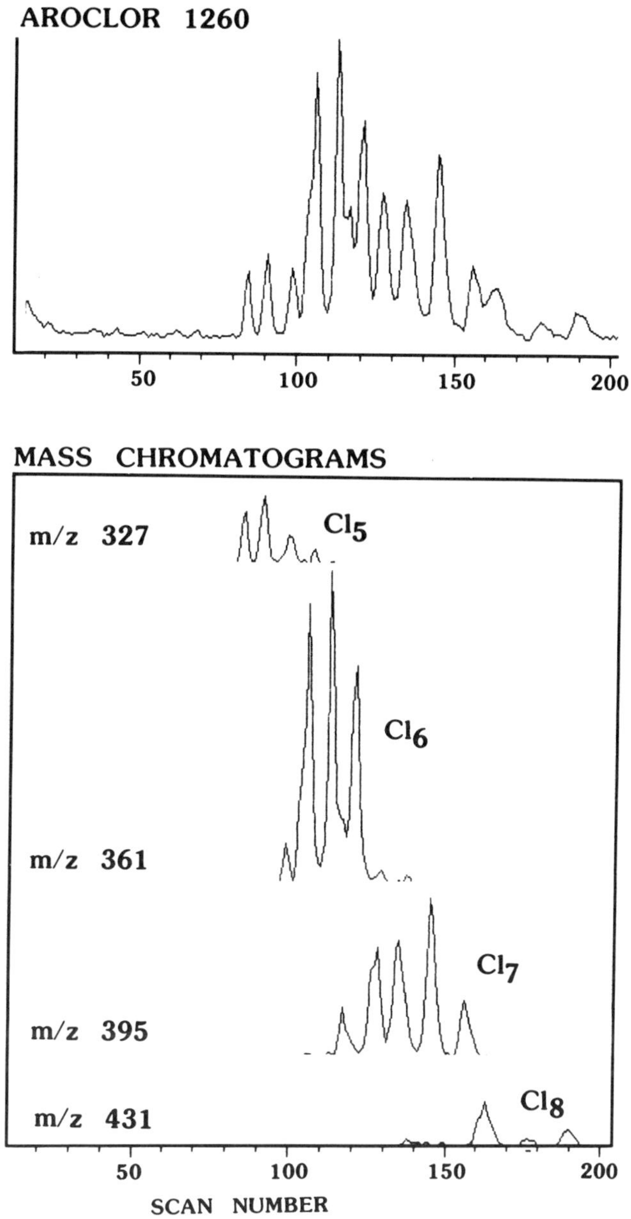

FIGURE 31. Total ion current with relevant mass chromatograms to illustrate the composition of Aroclor® 1260.

(Figure 40) the trend in chemical shift of the various protons *(ortho* to *para)* is downfield as the %Cl by weight increases. Chlorine substituents have the expected deshielding effect on the protons in adjacent positions. Application of these fingerprints might well serve to indicate the degree of chlorination in the sample and hence provide proof for the absence or presence of particular congeners.

In the case of the ^{13}C NMR spectra (Figure 41) the gross correlation of charge density and ^{13}C shielding established by Wilson has been applied to differentiate the different types of carbon atoms present within the complex mixtures. The approach to semiquantitative analysis used by Levy and Hewitt[85] has also been applied to these spectra. Results from this

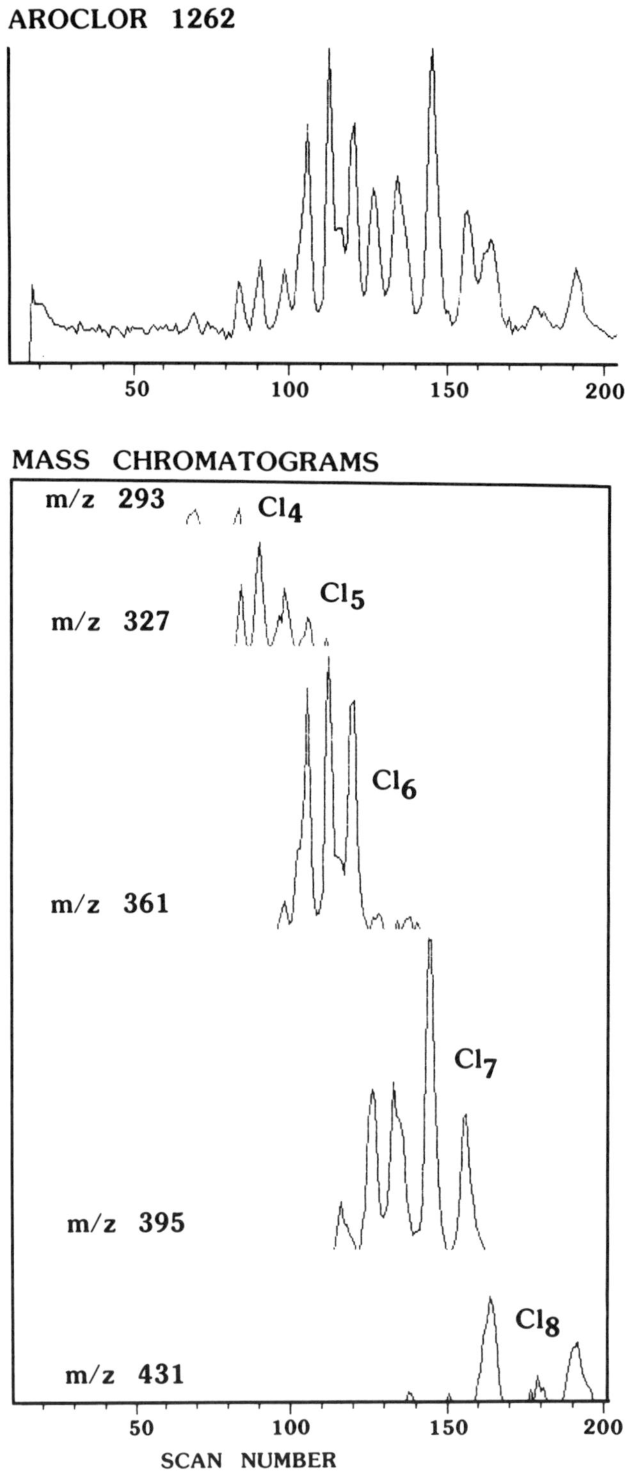

FIGURE 32. Total ion current with relevant mass chromatograms to illustrate the composition of Aroclor® 1262.

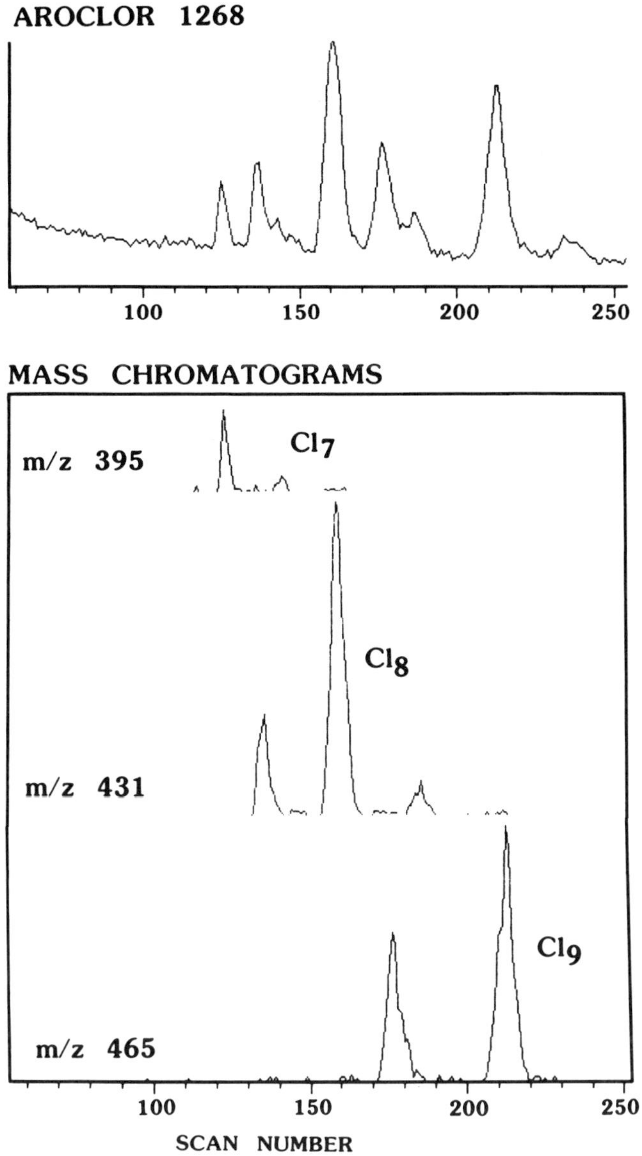

FIGURE 33. Total ion current with relevant mass chromatograms to illustrate the composition of Aroclor® 1268.

approach are listed in Table 4. The semiquantitative agreement is somewhat discouraging. When the range for the C-Cl was reduced to 131.5 to 135.5 ppm then the results all came within 1% of the theoretical values for %Cl by weight. While there are good reasons to support reducing the range, the uncertainty in deciding the true boundary still exists. Chemical shift values to substantiate both boundaries have been published and for this reason the quantitative analysis by ^{13}C NMR of PCBs is not recommended. Closer examination of the spectra, however, do give a vast amount of information concerning the nature of the carbon atoms in the samples studied. The reduction in the type of C-1 carbon atoms observed (135 to 140 ppm) as %Cl increased reflected the reduction in the number of congeners present and hence a resultant increase in the population of certain other congeners with common C-1 carbons. Conversely, the spectra demonstrated the clear trend in the disappearance of

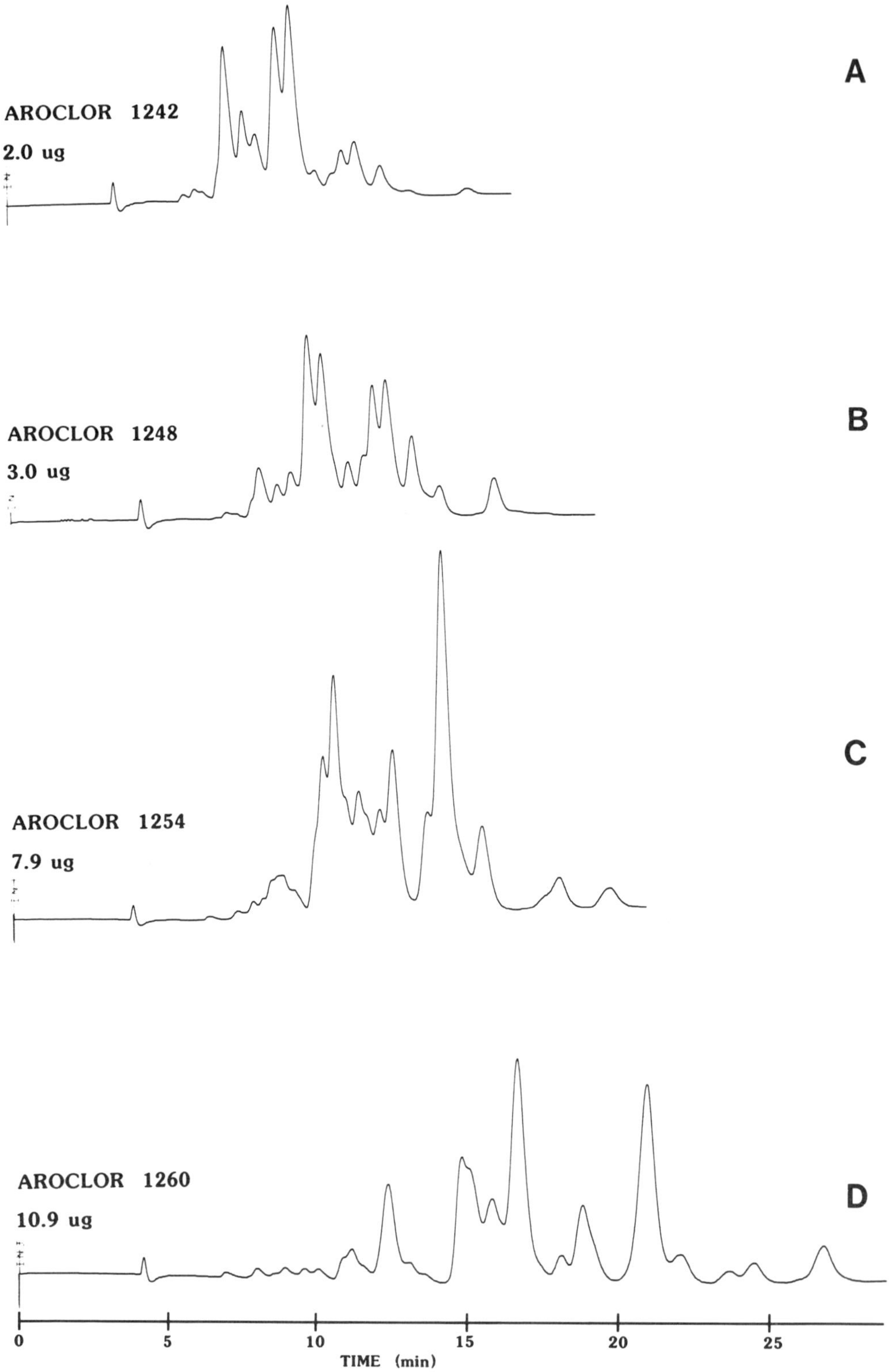

FIGURE 34. Reversed-phase LC elution patterns observed for various Aroclors®: mobile phase, 5% water in methanol, 0.5 mℓ/min; Altex ultrasphere ODS 5 μm; 254 nm fixed wavelength detector.

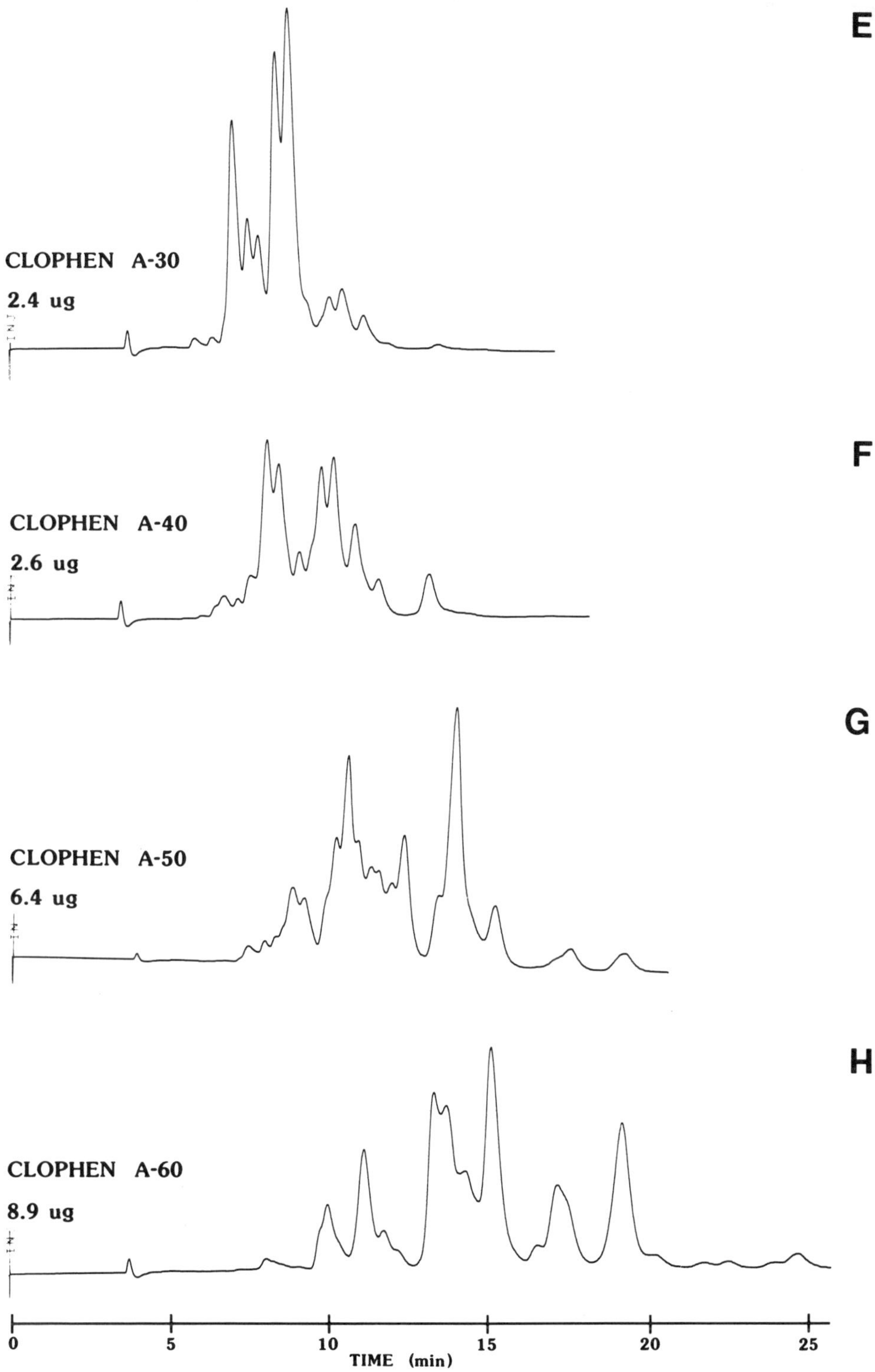

FIGURE 35. Reversed-phase LC elution patterns observed for various Clophens®: mobile phase, 5% water in methanol, 0.5 mℓ/min; Altex ultrasphere ODS 5 μm; 254 nm fixed wavelength detector.

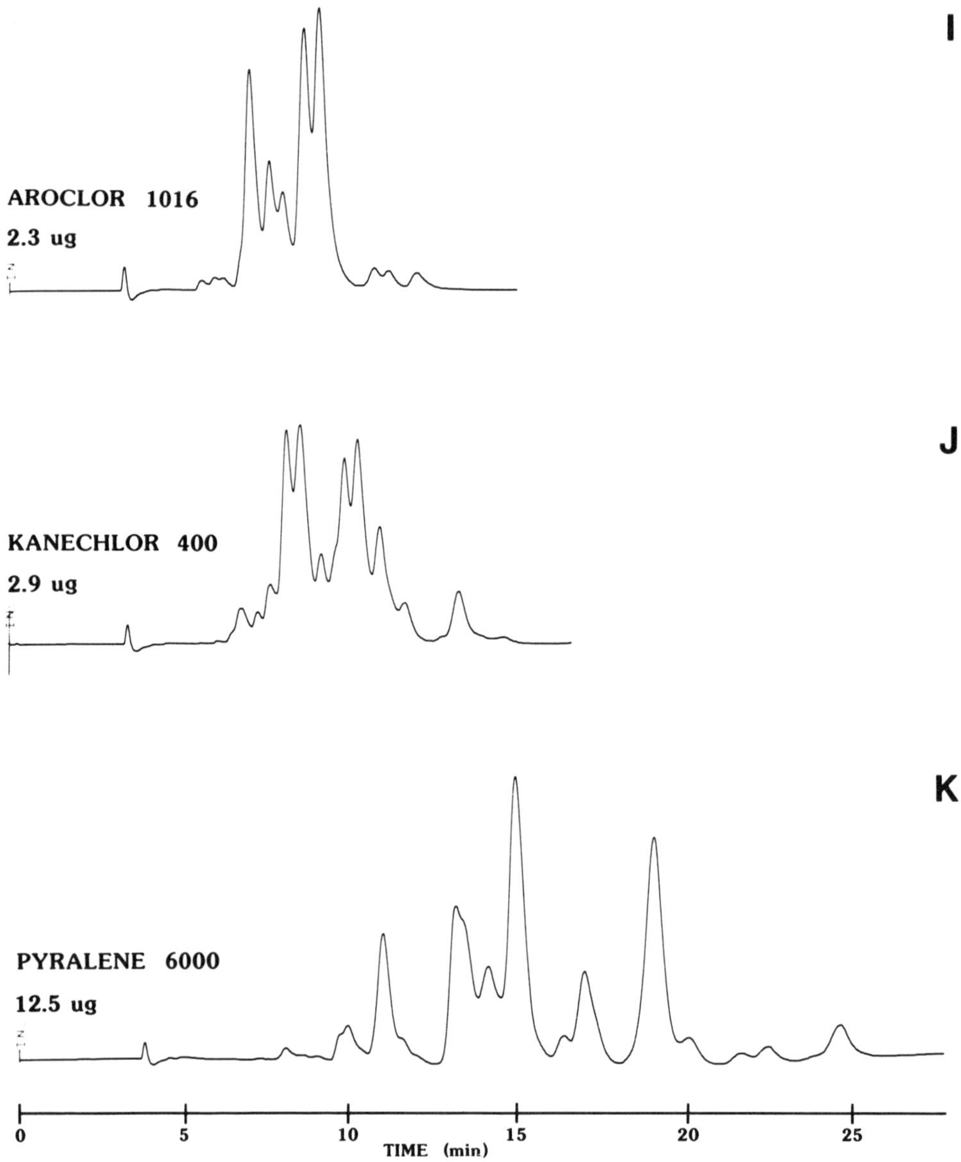

FIGURE 36. Reversed-phase LC elution patterns observed for various other commercial PCB mixtures: mobile phase, 5% water in methanol, 0.5 mℓ/min; Altex ultrasphere ODS 5 μm; 254 nm fixed wavelength detector.

carbons bearing a proton as %Cl increased. However, the chemical shift zone for carbon atoms bearing a chlorine (130 to 135 ppm) was so complex as to discourage interpretation. At best these spectra can be a useful addition to the library of spectral data on PCBs for use in qualitative identification only.

IV. APPROACHES TO QUANTITATION

With the widespread availability of EC as the most cost-effective detector for incurred PCB residues at the parts per million level, quantitation becomes a complex issue.[86] Unfortunately, the majority of PCBs found in environmental samples only resemble certain Aroclors® in chromatographic pattern but do not match them exactly (Section III.A). Because

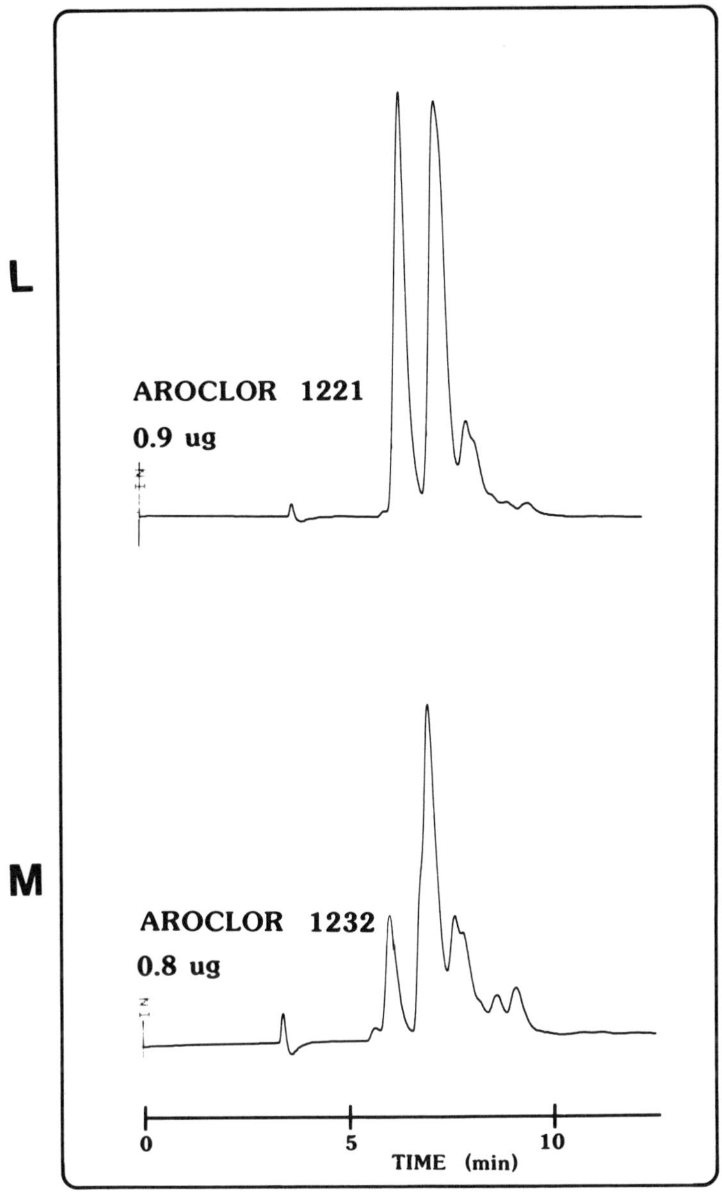

FIGURE 37. Reversed-phase LC elution patterns observed for various other Aroclors®: mobile phase, 5% water in methanol, 0.5 mℓ/min; Altex ultrasphere ODS 5 μm; 254 nm fixed wavelength detector.

of the disproportionality of detector response, the choice of a suitable reference standard can significantly affect the final analytical result. Having made the necessary judgmental decision as to the most suitable standard, quantitation by comparison of profile areas can be executed. Very often, however, the PCB residue may be a composite of two or more reference standards, and a mixture of reference materials may be required. Considerable improvement in the reliability of results obtained by EC was provided through an interlaboratory study[87,88] utilizing a peak-by-peak area comparison (retention times) procedure rather than a total pattern concept. This approach originally suggested by Webb and McCall[86] was highly dependent on properly characterized reference materials (i.e., weight percentage by peak) and illustrated the improved precision and accuracy over existing methods.

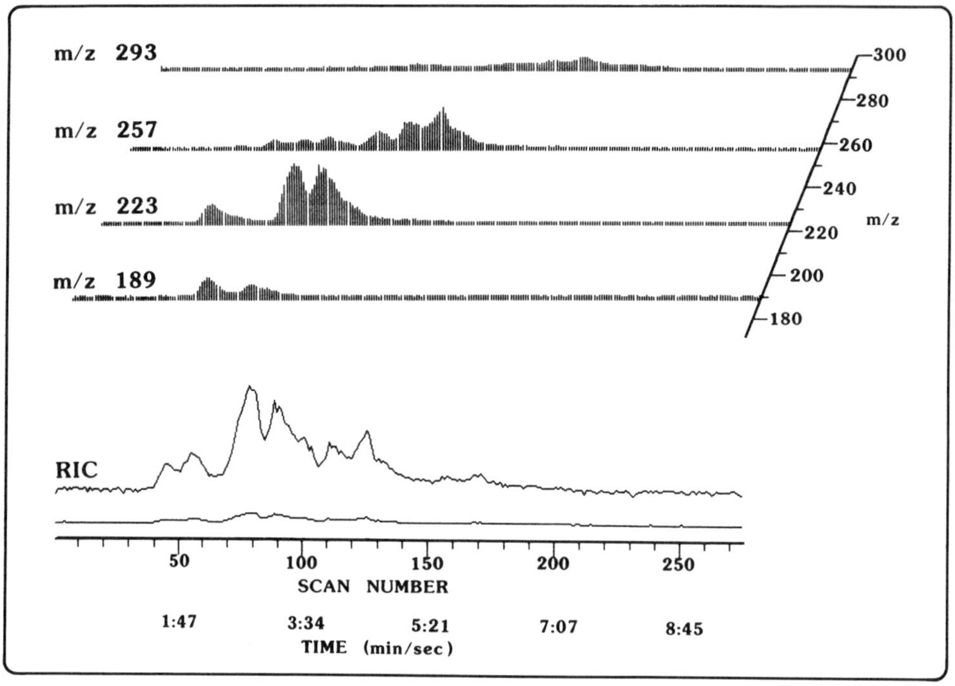

FIGURE 38. Total ion current profile and relevant mass chromatograms for Aroclor® 1242 by LC/MS.

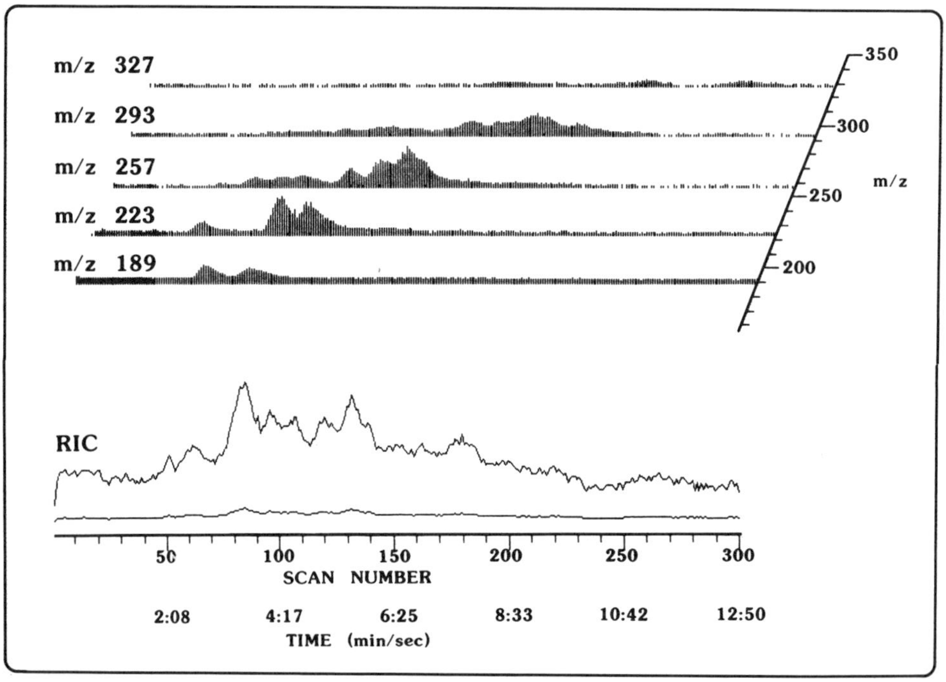

FIGURE 39. Total ion current profile and relevant mass chromatograms for Aroclor® 1248 by LC/MS.

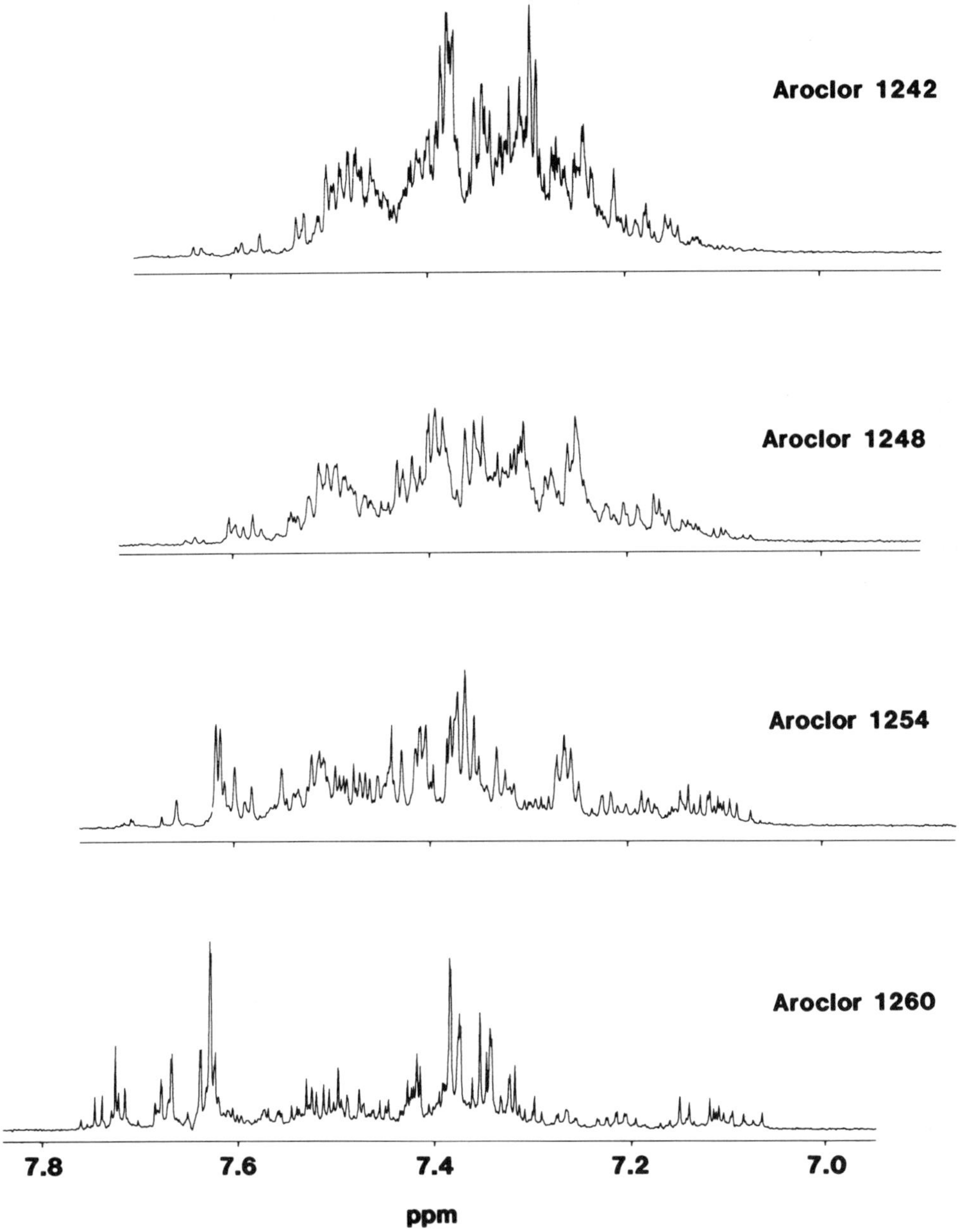

FIGURE 40. Proton magnetic resonance (270 MHz) spectra of various Aroclors®

Use of halogen-specific detectors such as HECD greatly reduces the potential error of disproportionality of response between reference standard and sample. However, their lack of availability in many laboratories has precluded their use in quantitative analysis. HECD has contributed to the vital characterization of Aroclors® fundamental to the analytical approach adopted for quantitation by EC via peak-by-peak comparisons. Additionally, the use of GC/MS-CI with representative congeners of PCBs has also assisted in providing quantitative analysis[63] of five Aroclor® standards (Figure 42) recommended for use in quantitation via EC.

To assess the relative merits of these techniques, the composition of Aroclor® 1248 was

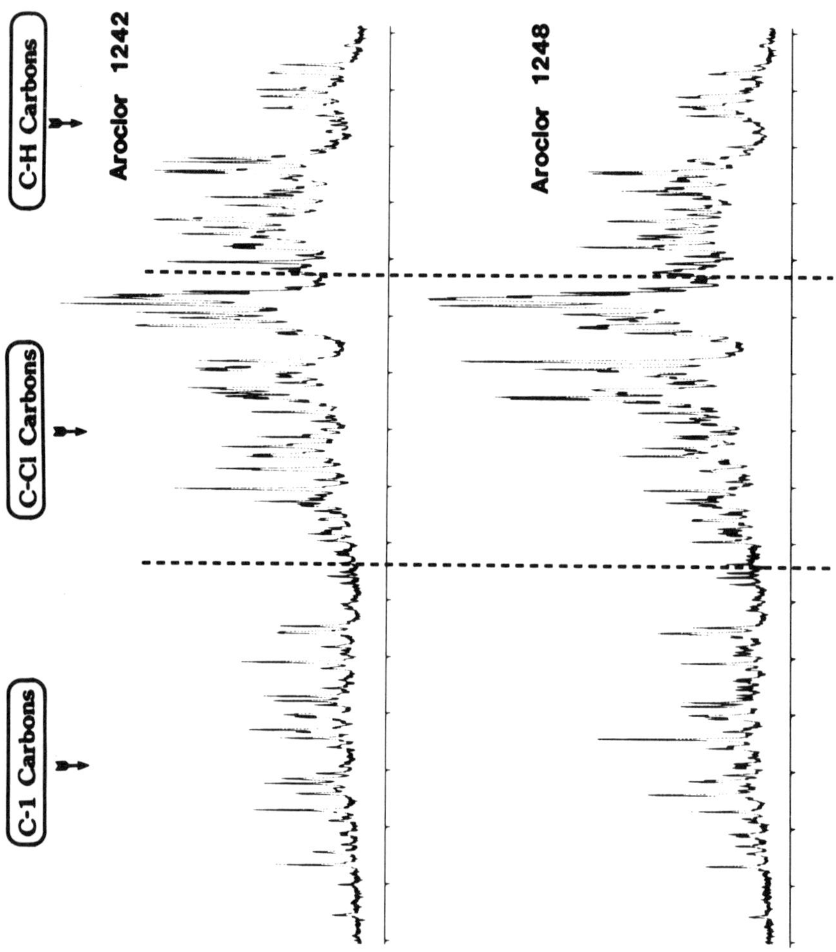
C-1 Carbons
C-Cl Carbons
C-H Carbons
Aroclor 1242
Aroclor 1248

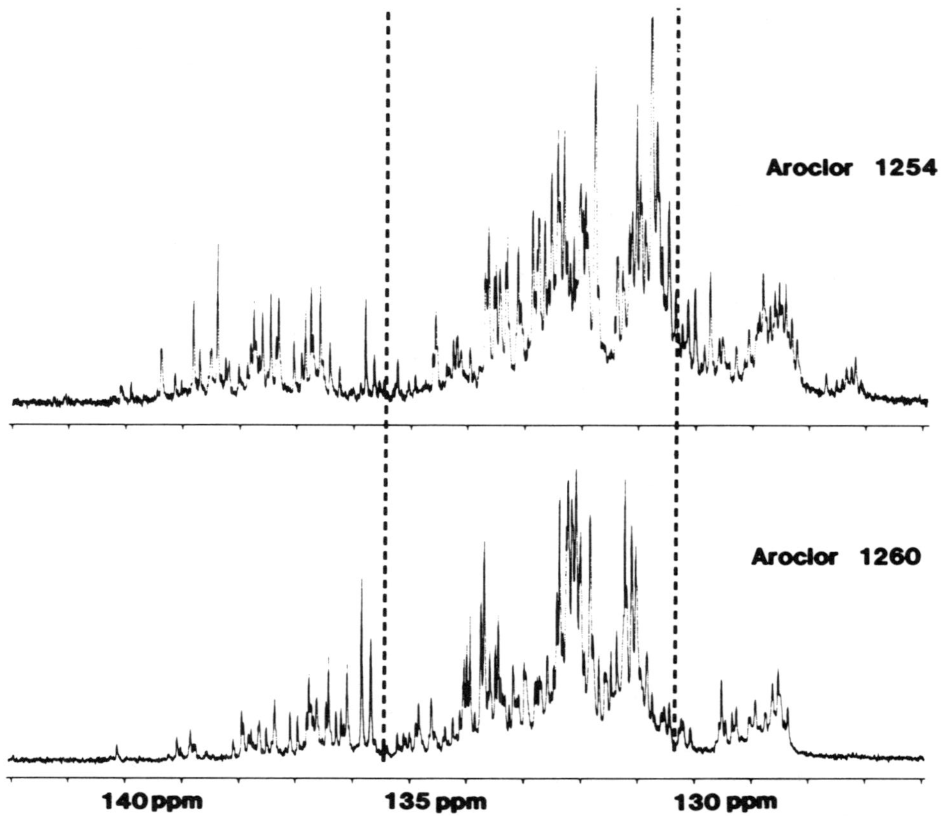

FIGURE 41. ^{13}C magnetic resonance (69.7 MHz) of various Aroclors®.

Table 4
SEMIQUANTITATIVE ANALYSIS OF AROCLORS® BY ^{13}C NMR

	Area measurements (integration range in ppm)				
				%Cl	
Aroclor®	C-1 (135.5—142)	C-Cl (135.5—130.25)	C-H (130.25—126)	Calc. theor.	Actual found
1242	2.1	5.6	4.3	57	42
1248	2.0	6.8	3.2	62	48
1254	1.9	7.8	2.3	65	54
1260	1.9	8.3	1.8	66	60

determined by EC, HECD, and GC/MS-CI (Table 5).[63] The peak identification system employed by Sawyer[87] has been used to provide a numerical value for each of the elution peaks in the profile (retention values for *p,p'*-DDE = 100, *p,p'*-DDT = 168, and aldrin = 54 constitute a reference system). It is recognized that both EC and HECD share the same basic principle of detection, i.e., increased sensitivity to the total number of chlorines in the molecule. Therefore, it is not surprising that a peak-by-peak quantitation of Aroclor® 1248 by both these methods provided good overall agreement. In several instances, however, the correlation must be considered as poor (peaks 21, 28, 37, and 40). These results are experimental evidence of disproportionality in spite of similar sensitivity to molar chlorine. In the case of results via GC/MS-CI, the inability to properly quantitate the last five peaks separately due to their low intensity elution profiles was evident. In general, however, the agreement between GC/MS-CI and/or EC and HECD was poor. In the case of peaks containing Cl_2 to Cl_4, the values by GC/MS were much larger. This was an indication of the greater sensitivity of MS to lower molecular weight compounds while for EC and HECD the reverse was true. Irrespective of the factors governing the mechanism of detection, all three methods did provide a total chlorine content extremely close to that injected on column. A conclusion to be gathered from this example is that the summation of individual area measurements usually results in an averaging of the disproportionality factors experienced by each method.

The variety of problems encountered in quantitation of PCB residues has focused attention on developing alternate approaches involving measurement of a single species rather than a mixture. Such brave attempts have included carbon-skeleton GC where PCBs are reduced on-column (5% Pt at 180°C with H_2 as carrier gas) to biphenyl,[89] perchlorination of PCBs by antimony pentachloride to give decachlorobiphenyl,[90] and microcoulometric determination of total chlorine as HCl.[91]

Perhaps the most innovative approach to quantitation has been provided by GC/MS where representative congeners from each molecular weight class are utilized as single standards to quantitate all members of their respective families.[63] In the analysis of a fish sample contaminated by PCBs and DDE (Figure 43), the ability to manipulate the recorded total ion current data base into various mass chromatograms (Figure 44) provided a clear separation of PCBs from DDE without further chemical treatment.[42] This ability to assign the observed PCB residue to Aroclor® 1254 was then easily established and the quantitation carried out with a high degree of confidence. This key determinative step performed by GC/MS seems to be free from previously encountered difficulties.

In conclusion, the routine use of EC for primary determination of PCBs and a whole host of other organochlorine-containing pesticides is still mandatory based on sensitivity. The gross differences in sample residue composition from that of a recognized standard can often contribute to the complexity of an analysis. Such incidences result in other novel approaches

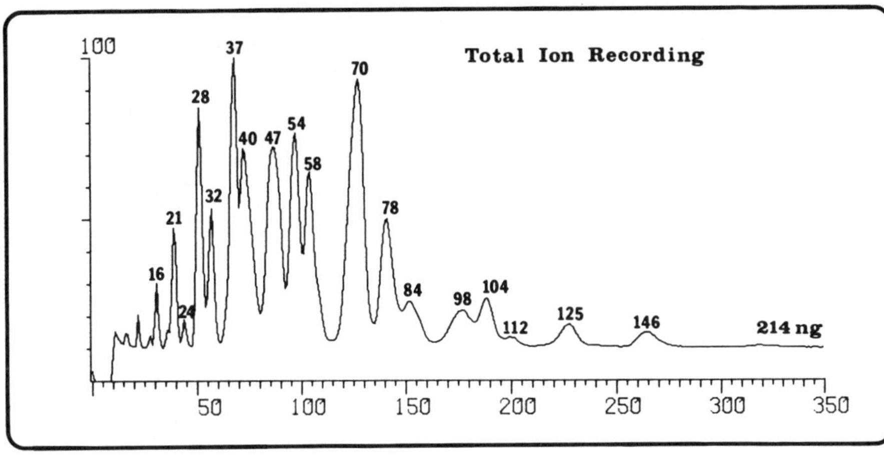

Peak Identification R_{DDE}(x 100)	Biphenyls (ng) Cl_1	Cl_2	Cl_3	Cl_4	Cl_5	chlorine (ng)
16	0·1	1·6				0·52
21		4·3				1·35
24			1·0			0·41
28		1·4	18·0			7·82
32		0·3	8·1			3·42
37			21·8			8·94
40			13·1	8·2		9·3
47			3·9	35·3		18·5
54			3·6	21·6		11·82
58			1·9	18·0		9·4
70				29·4	3·3	15·86
78				9·0		4·85
84					5·0	2·7
98 104					6·5	3·5
112 125 146					1·8	0·97
Totals =	0·1	7·6	71·4	121·5	16·6	99·36
	Polychlorinated Biphenyls			=	217·2 ng	

FIGURE 42. Total ion current by GC/MS of Aroclor® 1248 with quantitative results calculated as chlorinated biphenyls and chlorine.[63]

to quantitation which avoid these complexities. Whatever the analytical approach adopted, it should be clearly understood that good judgment on the part of the performing analyst is still required to ensure reliable reporting of PCBs in the literature.

ACKNOWLEDGMENTS

The authors wish to express their gratitude to Dr. Fred Evans of the National Center for Toxicological Research for providing the high field strength NMR data.

Table 5
QUANTITATIVE COMPARISON OF CHLORINE DETERMINATIONS FOR AROCLOR® 1248 REFERENCE STANDARD BY EC, HECD, AND GC/MS-CI[a] [63]

Peak ident. R_{DDE}(× 100)	No. Cl_x	%Cl via EC	%Cl via HECD	Quantitative analysis (ng) EC	HECD	GCMS
16	2,3	0.3	0.2	0.31	0.21	0.52
21	2	1.1	0.7	1.14	0.73	1.35
24	3	0.2	0.2	0.21	0.21	0.41
28	2,3	6.0	5.1	6.21	5.28	7.82
32	2,3	2.6	2.2	2.69	2.28	3.42
37	3	8.7	7.6	9.00	7.87	8.94
40	3,4	7.4	6.8	7.66	7.04	9.30
47	3,4	15.7	15.8	16.24	16.37	18.50
54	3,4	9.3	9.3	9.62	9.62	11.82
58	3,4	8.3	8.4	8.59	8.70	9.40
70	4,5	18.2	18.9	18.83	19.58	15.86
78	4	6.4	6.6	6.62	6.84	4.85
84	5	4.6	5.3	4.76	5.49	2.70
98	5	3.4	3.9	3.52	4.04	3.50
104	5	3.3	3.8	3.42	3.94	
112	5	1.0	1.1	1.04	1.14	
125	5	2.3	2.6	2.38	2.69	0.97
146	5	1.2	1.4	1.24	1.45	
Total (ng)				103.48	103.49	99.36

[a] 214 ng Aroclor® 1248 injected in each case; 48.4% Cl; 103.5 ng Cl injected.

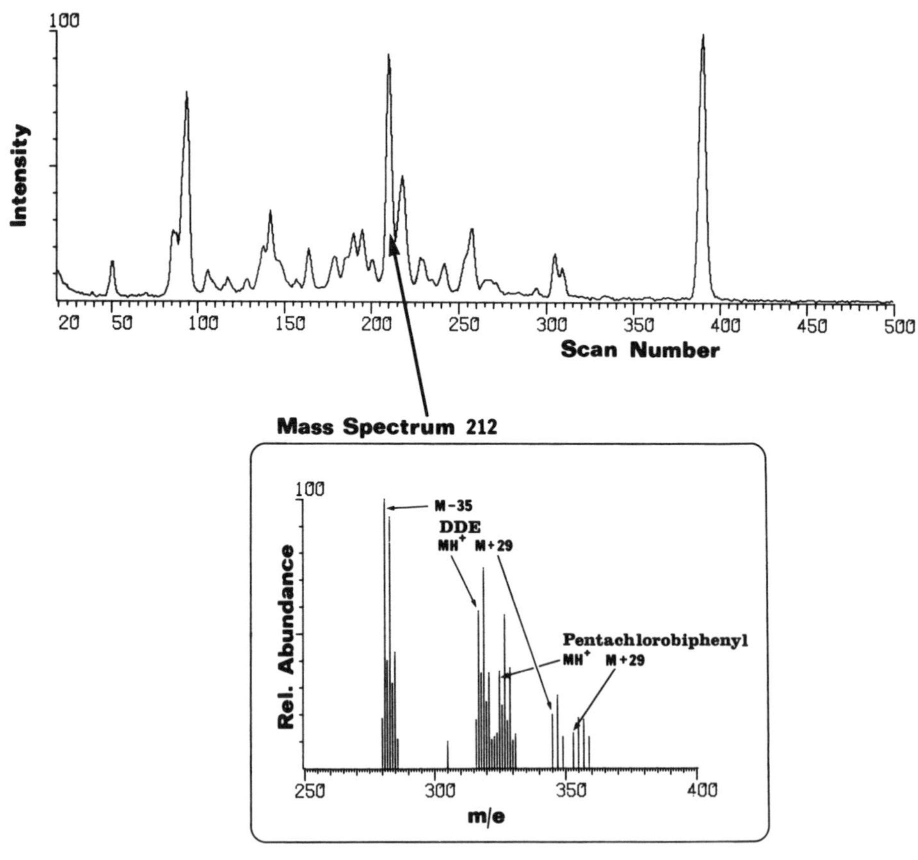

FIGURE 43. GC/MS-CI analysis of an extract of salmon illustrating the co-elution of DDE with pentachlorobiphenyl.[37]

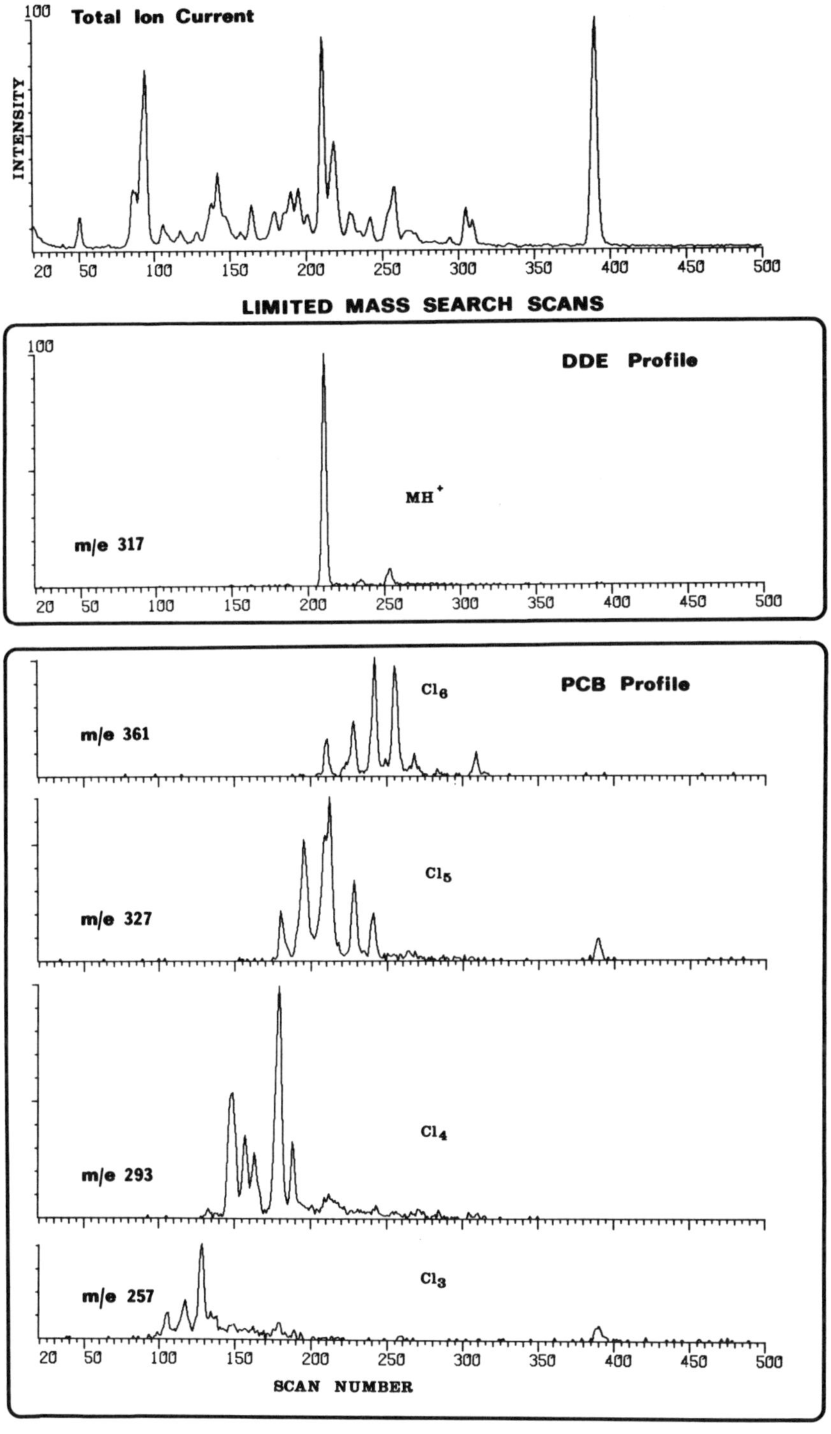

FIGURE 44. GC/MS-CI analysis of an extract of salmon and subsequent mass chromatograms for profile of the PCB content.[63]

REFERENCES

1. **Schmidt, H. and Schultz, G.,** *Ann. Chem.*, 207, 338, 1881.
2. Aroclor Plasticizers, Technical Bulletin O/PL-306A, Monsanto Chemical Co, St. Louis, 1981.
3. **Nisbet, I. C. T. and Satofin, A. E.,** *Environ. Health Perspect.*, 1, 21, 1972.
4. **Jensen, S.,** *New Sci.*, 32, 612, 1966.
5. **Holmes, D. C., Simmens, J. M., and Tatton, J. O. G.,** *Nature (London)*, 216, 227, 1967.
6. **Risebrough, R. W., Reichle, P., and Olcott, H. S.,** *Bull. Environ. Contam. Toxicol.*, 4, 192, 1969.
7. **Koeman, J. H., Debrauw, M. C., and DeVos, R. H.,** *Nature (London)*, 221, 1126, 1969.
8. **Richardson, B. J. and Waid, J. S.,** *Search*, 13, 17, 1982.
9. **Risebrough, R. W., Reichle, P., Herman, S. G., Peakall, D. B., and Kirven, M. N.,** *Nature (London)*, 220, 1098, 1968.
10. **Westoo, G., Noren, K., and Anderson, M.,** *Var Foda*, 2—3, 10, 1970.
11. **Biros, F. J., Walker, A. C., and Medbury, A.,** *Bull. Environ. Contam. Toxicol.*, 5, 317, 1970.
12. **Risebrough, R. W. and Brodine, V.,** *Environment*, 12, 958, 1970.
13. **Saeki, S., Tsutsui, A., Oguri, K., Yoshimura, H., and Hamana, M.,** *Fukuoka Acta Med.*, 62, 20, 1971.
14. *Fed. Reg.*, 37(No. 54), March 18, 1972.
15. **Skrentny, R. F., Hemken, R. W., and Dorough, H. W.,** *Bull. Environ. Contam. Toxicol.*, 6, 409, 1971.
16. *Fed. Reg.*, 38, 18096, July 6, 1973.
17. **Winston, S. G. and Gerstner, H. B.,** Polychlorinated Biphenyls, Polybrominated Biphenyls and Their Contaminants — A Literature Compilation, 1965—1977, ORNL/TIRC-78/2, Natural Library of Medicine, March 1978.
18. Polychlorinated Biphenyls and the Environment, Interdepartmental Task Force on PCBs, Washington, D.C., COM-72-10419, National Technical Information Services, Springfield, Va., 1972.
19. **Bird, F. H., Chawan, C. B., and Genty, R. W.,** *Poult. Sci.*, 57, 538, 1978.
20. **Fukano, S. and Doguchi, M.,** *Bull. Environ. Contam. Toxicol.*, 17, 613, 1977.
21. **Greenlee, W. F. and Poland, A.,** *J. Biol. Chem.*, 254, 9814, 1979.
22. **Ghiasuddin, A. M., Menzer, R. E., and Nelson, J. O.,** *Toxicol. Appl. Pharmacol.*, 36, 187, 1976.
23. **Dymerski, P., Kennedy, M., and Kaminsky, L.,** Trace origins in analysis: new frontier, *Anal. Chem.*, 519, 685, 1979.
24. **Norback, D. H., Britt, J., Hsia, M. T., and Allen, J. R.,** *Fed. Proc.*, 37, 299, 1978.
25. **Gardner, A. M., Righter, H. F., and Roach, J. A.,** *J. Assoc. Off. Anal. Chem.*, 59, 273, 1976.
26. **Jensen, S. and Jansson, B.,** *Ambio*, 5, 257, 1976.
27. National Conference on PCBs, EPA-560/6-75-004, U.S. Environmental Protection Agency, Washington, D.C., March 1976.
28. *Fed. Reg.*, 42, 17487, April 1, 1977.
29. *Fed. Reg.*, 44, 38330, June 29, 1979.
30. PCBs: An Alert for Food and Feed Facilities, U.S. Environmental Protection Agency, Washington, D.C., December 1979.
31. *Fed. Reg.*, 45, 30980, May 9, 1980.
32. **Cairns, T. and Siegmund, E. G.,** *Anal. Chem.*, 53, 1183A, 1981.
33. **Sissons, D. and Welti, D.,** *J. Chromatogr.*, 60, 15, 1971.
34. **Fishbein, L.,** *J. Chromatogr.*, 68, 345, 1972.
35. **Cook, J. W.,** *Environ. Health Perspect.*, 1, 3, 1972.
36. **Fishbein, L.,** *J. Chromatogr.*, 68, 345, 1972.
37. **Cairns, T.,** *Bull. Assoc. Food Drug Off.*, 42, 3, 1978.
38. **Wilson, N. K.,** *J. Am. Chem. Soc.*, 97(13), 3573, 1975.
39. **Brinkman, U. A. Th., Seetz, J. W. F. L., and Reymer, H. G. M.,** *J. Chromatogr.*, 116, 353, 1976.
40. **Zitko, V.,** *Int. J. Environ. Anal. Chem.*, 1, 221, 1972.
41. **Sawyer, L. D.,** *J. Assoc. Off. Anal. Chem.*, 61, 282, 1978.
42. **Amour, J. and Burke, J.,** *J. Assoc. Off. Anal. Chem.*, 53, 761, 1970.
43. **Masumoto, H. T.,** *J. Assoc. Off. Anal. Chem.*, 55, 1092, 1972.
44. **Erney, D. R.,** *J. Assoc. Off. Anal. Chem.*, 57, 5576, 1974.
45. **Bidleman, T. F., Matthews, J. R., Olney, C. E., and Rice, C. P.,** *J. Assoc. Off. Anal. Chem.*, 61, 820, 1978.
46. **Park, C. K., Lee, C. Y., and Park, R. D.,** *Han'guk Sikmul Poho Hakhoe Chi*, 16, 221, 1977.
47. **Young, S. J. V., Finsterwalder, C., and Burke, J. A.,** *J. Assoc. Off. Anal. Chem.*, 56, 957, 1973.
48. **Mulhern, B. M., Cromartie, E., Reichel, W. L., and Belisle, A. A.,** *J. Assoc. Off. Anal. Chem.*, 54, 548, 1970.
49. **Trotter, W. J.,** *J. Assoc. Off. Anal. Chem.*, 58, 461, 1975.

50. **Miles, J. R. W.,** *J. Assoc. Off. Anal. Chem.,* 55, 1039, 1972.
51. **Mitzutani, T., Matsumoto, M., and Kaguka, E.,** *J. Hyg. Chem.,* 19, 287, 1973.
52. **Gothe, R.,** *Bull. Environ. Contam. Toxicol.,* 11, 451, 1974.
53. **Westoo, G. and Noren, K.,** *Acta Chem. Scand.,* 24, 1639, 1970.
54. **Jensen, S. and Sundstrom, G.,** *Ambio,* 3, 70, 1974.
55. **Fehringer, N. V. and Westfall, J. E.,** *J. Chromatogr.,* 57, 397, 1971.
56. **deVos, R. H. and Peet, E. W.,** *Bull. Environ. Contam. Toxicol.,* 6, 1164, 1971.
57. **Natusch, D. F. S. and Thrope, T. M.,** *Anal. Chem.,* 45, 1184A, 1973.
58. **Rote, J. W. and Murphy, P. G.,** *Bull. Environ. Contam. Toxicol.,* 6, 377, 1971.
59. **Cairns, T. and Siegmund, E. G.,** *Biomed. Mass Spectrom.,* 8, 569, 1981.
60. **Casida, J. E. and Saleh, M. A.,** Toxaphene Composition and Toxicology, EPA-600/1-78-060, U.S. Environmental Protection Agency, Washington, D.C., September 1978.
61. **Cochrane, W. P., Lawrence, J. F., Lee, T. W., Maybury, R. B., and Wilson, B. P.,** *J. Assoc. Off. Anal. Chem.,* 58, 1051, 1975.
62. **William, D. T. and Blanchfield, B. J.,** *J. Assoc. Off. Anal. Chem.,* 55, 93, 1972.
63. **Cairns, T. and Siegmund, E. G.,** *Anal. Chem.,* 53, 1599, 1981.
64. **Hutzinger, O., Jamieson, J., and Zitko, V.,** *Nature (London),* 226, 664, 1970.
65. **Webb, R. S. and McCall, A.,** *J. Assoc. Off. Anal. Chem.,* 55, 746, 1972.
66. **Stalling, D. and Huckins, J.,** *J. Assoc. Off. Anal. Chem.,* 54, 801, 1971.
67. **Safe, S. and Hutzinger, O.,** *Nature (London),* 232, 641, 1971.
68. **Safe, S. and Hutzinger, O.,** *Chem. Commun.,* 446, 1971.
69. **Safe, S. and Hutzinger, O.,** *J. Chem. Soc. Perkin Trans. 1:,* 5, 686, 1972.
70. **Rote, J. W. and Morris, W. J.,** *J. Assoc. Off. Anal. Chem.,* 56, 188, 1973.
71. **Byrne, S. H., Schmidt, J. A., and Johnson, P. E.,** *J. Chromatogr. Sci.,* 9, 592, 1971.
72. **Eisenbeiss, F. and Sieper, H.,** *J. Chromatogr.,* 83, 439, 1973.
73. **Seiber, J. N.,** *J. Chromatogr.,* 94, 151, 1974.
74. **Willmott, F. W. and Dolphin, R. J.,** *J. Chromatogr. Sci.,* 12, 695, 1974.
75. **Aitzetmuller, K.,** *J. Chromatogr.,* 107, 411, 1975.
76. **Frei, R. W., Lawrence, J. F., Hope, J., and Cassidy, R. M.,** *J. Chromatogr. Sci.,* 12, 40, 1974.
77. **Chiosi, S., Evidente, A., Randazzo, G., and Scalorbi, O.,** *J. Liquid Chromatogr.,* 5, 1653, 1982.
78. **Kirkland, J. J.,** *Modern Practice of Liquid Chromatography,* Wiley Interscience, New York, 1971, chap. 5.
79. **Cairns, T., Siegmund, E. G., and Doose, G. M.,** *Anal. Chem.,* 54, 953, 1982.
80. **Bartle, K. D.,** *J. Assoc. Off. Anal. Chem.,* 55, 1101, 1972.
81. **Welti, D. and Sissons, D.,** *Org. Magnetic Reson.,* 4, 309, 1972.
82. **Tarpley, A. R. and Goldstein, J. H.,** *J. Phys. Chem.,* 75, 421, 1971.
83. **Wilson, N. K. and Anderson, M.,** *Mass Spectrometry and NMR Spectroscopy in Pesticide Chemistry,* Haque, R. and Biros, F. J., Eds., Plenum Press, New York, 1974, 197.
84. **Wilson, N. K.,** *J. Am. Chem. Soc.,* 3573, 1975.
85. **Levy, G. C. and Hewitt, J. M.,** *J. Assoc. Off. Anal. Chem.,* 60, 241, 1977.
86. **Webb, R. G. and McCall, A. C.,** *J. Chromatogr. Sci.,* 11, 366, 1973.
87. **Sawyer, L. D.,** *J. Assoc. Off. Anal. Chem.,* 61, 282, 1978.
88. **Sawyer, L. D.,** *J. Assoc. Off. Anal. Chem.,* 61, 272, 1978.
89. **Cooke, M., Nickless, G., Prescott, A. M., and Roberts, J. D.,** *J. Chromatogr.,* 156, 293, 1978.
90. **Robbins, A. L. and Willhite, C. R.,** *Bull. Environ. Contam. Toxicol.,* 21, 428, 1979.
91. **Donkin, P., Mann, S. V., and Hamilton, E. I.,** *Anal. Chem. Acta,* 88, 289, 1977.

Chapter 2

CHEMISTRY AND PROPERTIES OF PCBs IN RELATION TO ENVIRONMENTAL EFFECTS

B. L. Sawhney

TABLE OF CONTENTS

I. STRUCTURE AND PREPARATION OF PCBs

A. Nomenclature

PCBs are a class of synthetic chlorinated organic compounds with biphenyl as the basic structural unit. Chlorination of the group can produce 209 possible chlorobiphenyls (congeners) substituted with 1 to 10 chlorine atoms. Systematic numbering and structures of these chlorobiphenyls from Ballschmiter and Zell[1] are given in Table 1. According to the IUPAC definitive rules for nomenclature of organic chemistry, one ring system in the biphenyl ring assembly is assigned unprimed numbers and the other primed numbers as illustrated in Figure 1.

The order for assigning priorities to the substituents in the ring assembly, as Cl in PCBs is (1) unprimed number is assigned lower order than the corresponding primed number, as 2 vs. 2′; (2) lower number is assigned to a point of attachment in equivalent position, as 2 vs. 6, for a substituent in one of the *ortho* positions; (3) when the number of substituents in the two ring systems is the same, unprimed numbers are assigned to the ring system with smaller numbered substituents. These rules are illustrated in Figure 2 for pentachlorobiphenyl. Nomenclatures of chlorinated biphenyls containing additional substituents as hydroxyl, amino, and carboxyl groups are further described in a comprehensive account of the chemistry of PCBs by Hutzinger et al.[2] These principal functions are cited as the suffix; chlorobiphenylol, chlorobiphenyl amine, chlorobiphenyl carboxylic acid and so on.

B. Synthesis of Individual Chlorobiphenyls

Synthesis of most of the 209 chlorobiphenyls has been accomplished and many of these are now available. Methods used for the preparation of individual chlorobiphenyls include (1) phenylation or arylation of aromatic compounds, (2) condensation reactions, and (3) chlorination of biphenyl.

Arylation of a number of aromatic compounds as aroyl peroxides, carboxylic acids, diazonium salts, nitrosoamines, aryltriazenes, phenylhydrazines, and others has been attained. The reactions between the selected aromatic compounds and the substrate, benzene, usually involve free radicals provided either by the aromatic compound or the free radical precursors used in the reaction. Examples of arylation in the presence of benzene and chlorinated benzene are given in Figures 3A and B.

A number of condensation reactions involving aromatic halides, phenylsulfinates, aryl peroxides and acid anhydrides, chlorobenzenes, and other aromatic compounds have been described by Hutzinger et al.[2] Condensation of two aromatic halides in the presence of finely divided Cu is probably the most widely used procedure for the preparation of symmetrical chlorobiphenyls. Figure 4A illustrates formation of hexachlorobiphenyl from trichloroiodobenzene. Copper is usually mixed with the halide and heated for several hours. Mixtures of symmetrical and asymmetrical chlorobiphenyls are produced by this reaction when two different halides are used (Figure 4B).

Chlorination of biphenyl can occur either by substitution of Cl for H (H-Cl) to form chlorobiphenyls or by addition

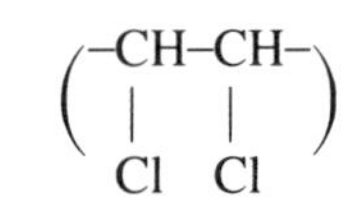

to form hexachlorocyclohexyl benzenes ($C_{12}H_{10}Cl_6$). The products formed by the addition of Cl are, however, thermally unstable and decompose to produce chlorobiphenyls during the chlorination process. Direct chlorination produces mixtures of chlorobiphenyls, depending upon the chlorine to biphenyl ratio, the catalyst used, and the temperature of the experiment. For example, in the presence of iron filings, equivalent proportions of biphenyl and chlorine produced about 32% 2-chlorobiphenyl, 26% 4-chlorobiphenyl, and 2% dichlorobiphenyl.[3] In the presence of bentonite clay as catalyst, 70% of the product was *o*-chlorobiphenyl.[4] Chlorination of biphenyl with liquid chlorine in the presence of $FeCl_3$ produced a mixture of nona- and decachlorobiphenyls,[5] while at high temperatures, in the presence of $SbCl_5$[6] or $TiCl_4$[7], decachlorobiphenyl was produced. Replacements of amino, hydroxyl, nitro, sulfonic acid, and others by chlorine also produce chlorinated biphenyls.

C. Commercial PCBs

Chlorination of biphenyl in the presence of a catalyst, such as iron filings or iron chloride is used for industrial preparations of PCBs. The chlorination process produces mixtures of chlorobiphenyls which are influenced by the ratio of chlorine to biphenyl. The crude product resulting from chlorination of biphenyls is purified to remove color, catalyst, and traces of HCl by alkali treatment and distillation. The purified PCB preparations are generally viscous liquids. In the U.S., PCBs have been manufactured by Monsanto under the trade name Aroclors®. The most common Aroclor® preparations include 1242, 1248, 1254, and 1260. The first two digits are the number of carbon atoms in the biphenyl group and the last two digits give approximate %Cl content in the PCB preparation. GC analysis of the Aroclors® (and other commercial preparations sold under different trade names around the world (see Chapter 1), show that they are complex mixtures of different chlorobiphenyls. Gas chromatograms of each preparation show a number of peaks. Use of glass capillary columns in GC has permitted better resolution of the peaks, resulting in over 60 peaks from Aroclor® 1248 and over 90 peaks from Aroclors® 1254 and 1260.[8] Using high-resolution thin-film glass capillary GC, Ballschmiter and Zell[1] have identified about 200 chlorobiphenyls in a commercial PCB preparation, Clophen®. Determinations of the composition of different peaks using a conductivity detector show that Aroclors® with lower chlorine content contain less chlorinated biphenyls with peaks having smaller retention times.[9,10] Aroclor® 1242 gave peaks corresponding to compounds containing 1 to 5 chlorine atoms, while peaks from Aroclor® 1254 contained 3 to 6 chlorine atoms and those from Aroclor® 1260 contained 5, 6, and 7 chlorine atoms.

Analyses of chlorobiphenyls in commercial preparations using capillary GC/MS,[11] based on retention times and molecular ions permit definitive identification and quantitative estimation of different chlorobiphenyls in the sample. Relative percentages of about 100 different PCBs in the Aroclors®, thus determined, can be used as secondary standards for environmental samples.

II. PROPERTIES OF PCBs

A. General Physical Properties

PCBs are among the most stable organic compounds known. They have a low dielectric constant and high heat capacity which render them ideal for use in electrical capacitors and transformers. While most individual chlorobiphenyls are solids at room temperature, commercial preparations are generally resins or viscous liquids of density greater than water. Table 2 shows some selected properties of four common Aroclors® and Table 3 gives amounts of chlorobiphenyls of different chlorine contents in the Aroclors®. Other properties of PCBs, solubility, vaporization, and sorption, which are important in controlling their transport and distribution in the environment are discussed in detail below.

B. Solubility

Solubility of PCBs in water and in organic solvents, as lipids, greatly influences their transport and persistence in the environment. Solubilities of a number of individual chlo-

Table 1
SYSTEMATIC NUMBERING OF PCB COMPOUNDS

No.	Structure	No.	Structure	No.	Structure	No.	Structure
	Monochlorobiphenyls		Tetrachlorobiphenyls		Pentachlorobiphenyls		Hexachlorobiphenyls
1	2	52	2,2′,5,5′	105	2,3,3′,4,4′	161	2,3,3′,4,5′,6
2	3	53	2,2′,5,6′	106	2,3,3′,4,5	162	2,3,3′,4′,5,5′
3	4	54	2,2′,6,6′	107	2,3,3′,4′,5	163	2,3,3′,4′,5,6
		55	2,3,3′,4	108	2,3,3′,4,5′	164	2,3,′,4′,5′,6
	Dichlorobiphenyls	56	2,3,3′,4′	109	2,3,3′,4,6	165	2,3,3′,5,5′,6
4	2,2′	57	2,3,3′,5	110	2,3,3′,4′,6	166	2,3,4,4′,5,6
5	2,3	58	2,3,3′,5′	111	2,3,3′,5,5′	167	2,3′,4,4′,5,5′
6	2,3′	59	2,3,3′,6	112	2,3,3′,5,6	168	2,3′,4,4′,5′,6
7	2,4	60	2,3,4,4′	113	2,3,3′,5′,6	169	3,3′,4,4′,5,5′
8	2,4′	61	2,3,4,5	114	2,3,4,4′,5		
9	2,5	62	2,3,4,6	115	2,3,4,4′,6		Heptachlorobiphenyls
10	2,6	63	2,3,4′,5	116	2,3,4,5,6	170	2,2′,3,3′,4,4′,5
11	3,3′	64	2,3,4′,6	117	2,3,4′,5,6	171	2,2′,3,3′,4,4′,6
12	3,4	65	2,3,5,6	118	2,3′,4,4′,5	172	2,2′,3,3′,4,5,5′
13	3,4′	66	2,3′,4,4′	119	2,3′,4,4′,6	173	2,2′,3,3′,4,5,6
14	3,5	67	2,3′,4,5	120	2,3′,4,5,5′	174	2,2′,3,3′,4,5,6′
15	4,4′	68	2,3′,4,5′	121	2,3′,4,5′,6	175	2,2′,3,3′,4,5′,6
		69	2,3′,4,6	122	2′,3,3′,4,5	176	2,2′,3,3′,4,6,6′
	Trichlorobiphenyls	70	2,3′,4′,5	123	2′,3,4,4′,5	177	2,2′,3,3′,4′,5,6
16	2,2′,3	71	2,3′,4′,6	124	2′,3,4,5,5′	178	2,2′,3,3′,5,5′,6
17	2,2′,4	72	2,3′,5,5′	125	2′,3,4,5,6′	179	2,2′,3,3′,5,6,6′
18	2,2′,5	73	2,3′,5′,6	126	3,3′,4,4′,5	180	2,2′,3,4,4′,5,5′
19	2,2′,6	74	2,4,4′,5	127	3,3′,4,5,5′	181	2,2′,3,4,4′,5,6
20	2,3,3′	75	2,4,4′,6			182	2,2′,3,4,4′,5,6′
21	2,3,4	76	2′,3,4,5		Hexachlorobiphenyls	183	2,2′,3,4,4′,5′,6
22	2,3,4′	77	3,3′,4,4′	128	2,2′,3,3′,4,4′	184	2,2′,3,4,4′,6,6′
23	2,3,5	78	3,3′,4,5	129	2,2′,3,3′,4,5	185	2,2′,3,4,5,5′,6
24	2,3,6	79	3,3′,4,5′	130	2,2′,3,3′,4,5′	186	2,2′,3,4,5,6,6′
25	2,3′,4	80	3,3′,5,5′	131	2,2′,3,3′,4,6	187	2,2′,3,4′,5,5′,6
26	2,3′,5	81	3,4,4′,5	132	2,2′,3,3′,4,6′	188	2,2′,3,4′,5,6,6′

27	2,3′,6			133	2,2′,3,3′,5,5′	189	2,3,3′,4,4′,5,5′
28	2,4,4′		Pentachlorobiphenyls	134	2,2′,3,3′,5,6	190	2,3,3′,4,4′,5,6
29	2,4,5	82	2,2′,3,3′,4	135	2,2′,3,3′,5,6′	191	2,3,3′,4,4′,5′,6
30	2,4,6	83	2,2′,3,3′,5	136	2,2′,3,3′,6,6′	192	2,3,3′,4,5,5′,6
31	2,4′,5	84	2,2′,3,3′,6	137	2,2′,3,4,4′,5	193	2,3,3′,4′,5,5′,6
32	2,4′,6	85	2,2′,3,4,4′	138	2,2′,3,4,4′,5′		
33	2′,3,4	86	2,2′,3,4,5	139	2,2′,3,4,4′,6		Octachlorobiphenyls
34	2′,3,5	87	2,2′,3,4,5′	140	2,2′,3,4,4′,6′	194	2,2′,3,3′,4,4′,5,5′
35	3,3′,4	88	2,2′,3,4,6	141	2,2′,3,4,5,5′	195	2,2′,3,3′,4,4′,5,6
36	3,3′,5	89	2,2′,3,4,6	142	2,2′,3,4,5,6	196	2,2′,3,3′,4,4′,5′,6
37	3,4,4′	90	2,2′,3,4′,5	143	2,2′,3,4,5,6′	197	2,2′,3,3′,4,4′,6,6′
38	3,4,5	91	2,2′,3,4′,6	144	2,2′,3,4,5′,6	198	2,2′,3,3′,4,5,5′,6
39	3,4′,5	92	2,2′,3,5,5′	145	2,2′,3,4,6,6′	199	2,2′,3,3′,4,5,6,6′
		93	2,2′,3,5,6	146	2,2′,3,4′,5,5′	200	2,2′,3,3′,4,5′,6,6′
	Tetrachlorobiphenyls	94	2,2′,3,5,6′	147	2,2′,3,4′,5,6	201	2,2′,3,3′,4′,5,5′,6
40	2,2′,3,3′	95	2,2′,3,5′,6	148	2,2′,3,4′,5,6′	202	2,2′,3,3′,5,5′,6,6′
41	2,2′,3,4	96	2,2′,3,6,6′	149	2,2′,3,4′,5′,6	203	2,2′,3,4,4′,5,5′,6
42	2,2′,3,4′	97	2,2′,3′,4,5	150	2,2′,3,4′,6,6′	204	2,2′,3,4,4′,5,6,6′
43	2,2′,3,5	98	2,2′,3′,4,6	151	2,2′,3,5,5′,6	205	2,3,3′,4,4′,5,5′,6
44	2,2′,3,5′	99	2,2′,4,4′,5	152	2,2′,3,5,6,6′		
45	2,2′,3,6	100	2,2′,4,4′,6	153	2,2′,4,4′,5,5′		Nonachlorobiphenyls
46	2,2′,3,6′	101	2,2′,4,5,5′	154	2,2′,4,4′,5,6	206	2,2′,3,3′,4,4′,5,5′,6
47	2,2′,4,4′	102	2,2′,4,5,6′	155	2,2′,4,4′,6,6′	207	2,2′,3,3′,4,4′,5,6,6′
48	2,2′,4,5	103	2,2′,4,5′,6	156	2,3,3′,4,4′,5	208	2,2′,3,3′,4,5,5′,6,6′
49	2,2′,4,5′	104	2,2′,4,6,6′	157	2,3,3′,4,4′,5		
50	2,2′,4,6			158	2,3,3′,4,4′,6		Decachlorobiphenyl
51	2,2′,4,6′			159	2,3,3′,4,5,5′	209	2,2′,3,3′,4,4′,5,5′,6,6′
				160	2,3,3′,4,5,6		

From Ballschmiter, K. and Zell, M., *Fresenius' Z. Anal. Chem.*, 302, 20, 1980. With permission.

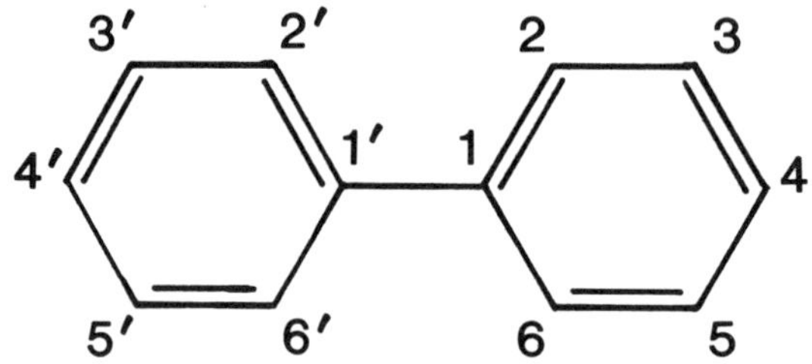

FIGURE 1. Numbering in the biphenyl ring system.

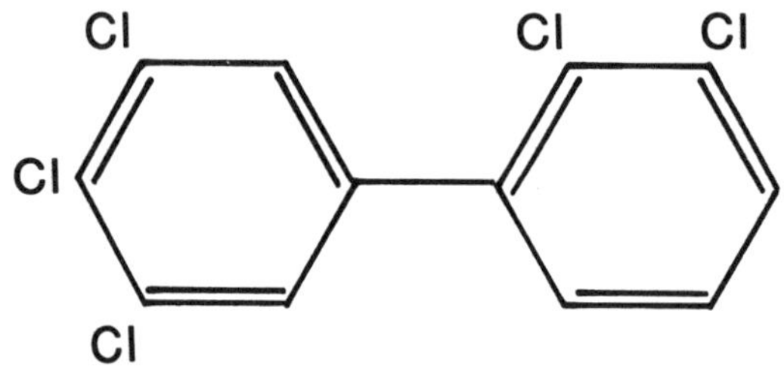

FIGURE 2. Structure of 2,3,3′,4′,5′-pentachlorobiphenyl.

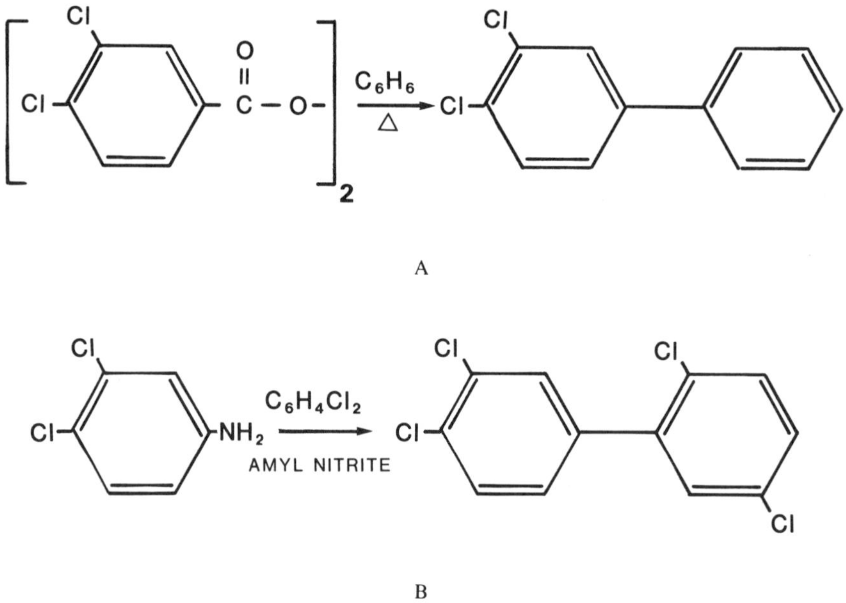

FIGURE 3. (A) Arylation of 3,3′,4,4′-tetrachlorobenzoyl peroxide to 3,4-dichlorobiphenyl.[2] (B) Arylation of 3,4-dichloroaniline to 2,3′,4′,5-tetrachlorobiphenyl.[2]

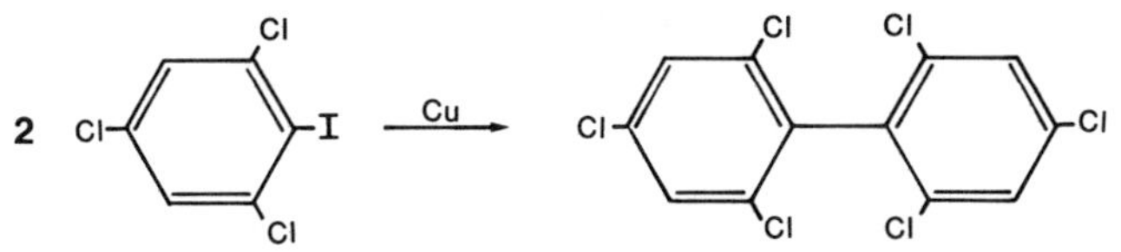

A

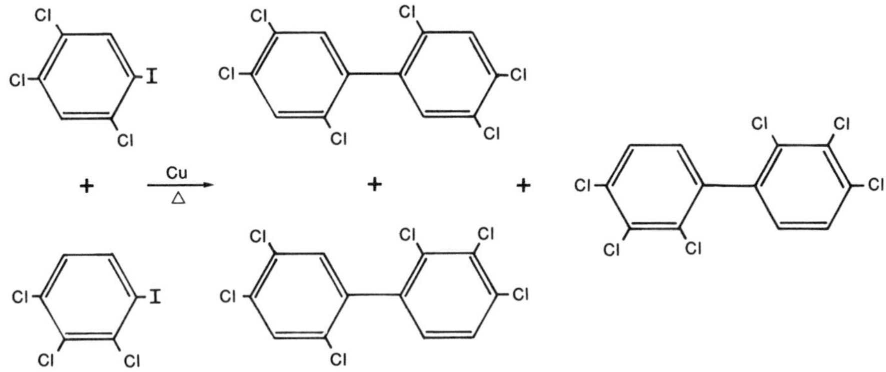

B

FIGURE 4. (A) Condensation of 2,4,6-trichloroiodobenzene to 2,2′,4,4′,6,6′-hexachlorobiphenyl.[2] (B) Condensation of trichloroiodobenzenes to hexachlorobiphenyls.[2]

Table 2
PHYSICAL CHARACTERISTICS OF FOUR COMMON AROCLORS®[2]

Aroclor®	Dielectric constant (at 1000 c) 25°C	100°C	Distillation range (°C)	Density
1242	5.8	4.9	325—366	1.38
1248	5.6	4.6	340—375	1.44
1254	5.0	4.3	365—390	1.53
1260	4.3	3.7	385—420	1.62

robiphenyls as well as of commercial preparations have been tabulated by Hutzinger et al.[2] Because of very low solubility of PCBs in water, a common method used for solubility determination involves placing a quantity of the chlorobiphenyl in a large quantity of water which is continuously stirred. Aliquots are then withdrawn periodically, centrifuged, and chlorobiphenyl in the supernatant estimated. This procedure is continued until an equilibrium is attained, which is a slow process.[12]

Solubility of PCBs in water generally decreases with increase in the degree of chlorination. Individual chlorobiphenyls vary in their solubility from about 6 ppm for monochlorobiphenyl to as low as 0.007 ppm for octachlorobiphenyl.[2] Decachlorobiphenyl, despite its higher chlorine content, is about twice as soluble as the octachlorobiphenyl. The solubilities also vary among chlorobiphenyls with the same number of chlorines; for example, solubilities

Table 3
MOLECULAR WEIGHT, Cl CONTENT, AND AMOUNTS OF CHLOROBIPHENYLS IN FOUR COMMON AROCLORS®[2]

			% in Aroclors®			
Chlorobiphenyl	**Mol wt**	**%Cl**	**1242**	**1248**	**1254**	**1260**
Monochlorobiphenyl	188.65	18.79	3	—	—	—
Dichlorobiphenyl	223.10	31.77	13	2	—	—
Trichlorobiphenyl	257.54	41.30	28	18	—	—
Tetrachlorobiphenyl	291.99	48.56	30	40	11	—
Pentachlorobiphenyl	326.43	54.30	22	36	49	12
Hexachlorobiphenyl	360.88	58.93	4	4	34	38
Heptachlorobiphenyl	395.32	62.77	—	—	6	41
Octachlorobiphenyl	429.77	65.98	—	—	—	8
Nonachlorobiphenyl	464.21	68.73	—	—	—	1

of 2,4-, 2,2′-, and 2,4′-dichlorobiphenyls are 1.40, 1.50, and 1.88 ppm, respectively, while that of 4,4′-dichlorobiphenyl is only 0.08 ppm.[2] Chlorine substitutions in different positions of the biphenyl ring system are apparently responsible for the differences in solubilities of chlorobiphenyls with same chlorine content.

Solubilities of chlorobiphenyls reported by different workers are similar in most instances. For example, solubility of 2,4,5,2′,5′-pentachlorobiphenyl is 10,[13] 20,[14] and 31 ppb.[2] Solubility of 2,4,5,2′,4′,5′-hexachlorobiphenyl, on the other hand, differed by an order of magnitude in two reports — Hutzinger et al.[2] reported a solubility of 8.8 ppb while Haque and Schmedding[15] reported 0.9 ppb.

Solubilities of various Aroclors® reported by Hutzinger et al.[2] are Aroclors® 1242 = 200 ppb, 1248 = 100 ppb, 1254 = 40 ppb, and 1260 = 25 ppb. Solubility of Aroclor® 1254 determined by Haque et al.[12] was about 56 ppb, similar to 40 ppb above. Because less chlorinated biphenyls are generally more soluble than more chlorinated biphenyls, aqueous solutions of Aroclors® show greater proportions of the lower chlorinated biphenyls than the standard Aroclors® from which the solutions are made.[2,12] However, some anomalies are observed because of differential solubilities of certain PCB congeners, as noted above. Ratios of GC peak heights from saturated aqueous solution of Aroclor® 1254 at 26°C to the corresponding peaks from the standard Aroclor® 1254 varied from about 5 for the first two peaks to less than 1 for peaks 8 through 13. Peak number 4 had a value of 14 for this ratio, presumably because of high solubility of the isomer corresponding to the peak.[2]

Solubilities of PCBs are greatly influenced by the environment. For example, aqueous phases in the environment generally contain dissolved organic substances which probably enhance the concentration of PCBs in solution. Zitko[16] obtained solubilities of Aroclor® 1254 in "fresh water" in the range of 0.3 to 3 ppm, which are much higher than 0.012 ppm reported by MacKay and Wolkoff,[17] and 0.056 ppm by Haque et al.[12] and were attributed by these authors to the presence of organic material in water. Conversely, sorption of PCBs on soil or sediment surfaces in the aquatic environment would decrease their solution concentration. A recent study shows that the movement of PCBs from dialysis membrane bags suspended in water was much less from bags with particles than from those without particles.[18]

C. Vapor Pressure and Vaporization

PCBs have very low vapor pressures which, like their solubility in water, decrease with increased chlorination. Vaporization rates for six common Aroclors® (Table 4) show that the vaporization rates decrease about 200-fold from Aroclor® 1221 which consists primarily of mono- and dichlorobiphenyls, to 1260 which consists essentially of penta-, hexa-,

Table 4
VAPORIZATION RATES OF SIX AROCLORS® MEASURED AT 100°C[2]

Aroclor® (12.28 cm^2 surface)	Vaporization rate ($g/cm^2/hr$)
1221	0.00174
1232	0.000874
1242	0.000338
1248	0.000152
1254	0.000053
1260	0.000009

hepta-, and octachlorobiphenyls. Haque et al.[12] observed a vaporization rate of 0.0000036 $g/cm^2/hr$ for Aroclor® 1254 at 60°C as compared to 0.000053 $g/cm^2/hr$ at 100°C (Table 4).

In environmental samples where PCBs are sorbed on soil or sediment surfaces, the rate of vaporization of PCBs is greatly reduced. The vaporization rate depends upon the sorption surface. Haque et al.[12] observed that about 60% of Aroclor® 1254 sorbed by Ottawa sand was lost by vaporization in a 4-week period, while no significant loss occurred from Woodburn soil in the same period. Also, the less chlorinated biphenyls in the Aroclor® showed the most loss and the more chlorinated biphenyls the least loss in accordance with their vapor pressures. The role of the various components of soils or sediments in affecting the vaporization of PCBs from their surfaces is not well understood. However, characteristics such as surface area, organic matter content, type of clay, and pH of the medium which are important in considerations of environmental samples, would likely influence the vaporization of sorbed PCBs.

While vaporization of PCBs sorbed on soil and sediment surfaces is reduced, their vaporization from aqueous solutions is anomalously high, given their low vapor pressure and high molecular weight. Mechanism of this high vaporization rate is not known. However, the losses from aqueous solutions of PCBs and other chlorinated hydrocarbons of very low solubility have been described in terms of fugacity, the escaping tendency of the solute, by MacKay.[19] Fugacity appears to be a useful concept in considering losses of organic materials from aqueous solutions. Based on this approach, the half-life of PCBs in a well-mixed 1-m deep body of water such as a fast-flowing river was calculated to be about 10 hr.[20] Evaporation of PCBs from contaminated rivers and lakes may, thus, represent a major means of their transport into the environment.

D. Sorption Reactions

Transport and fate of PCBs in aquatic systems and their partitioning in different compartments of the environment depend to a large degree on the sorption reactions. These reactions have received considerable attention and sorption-desorption of a number of individual chlorobiphenyls as well as of commercial PCB preparations with sorbents of different characteristics have been studied. Generally, sorption increases with increase in chlorine content of the chlorobiphenyl, and with surface area and organic carbon of the sorbent.

Haque and Schmedding,[15] using individual chlorobiphenyls, found that sorption by four different sorbents, a sand, a soil, a clay, and a humic acid sample occurred in the order: hexachlorobiphenyl > tetrachlorobiphenyl > dichlorobiphenyl. Not only are the chlorobiphenyls with higher chlorine content sorbed in larger quantity but they are also held more tightly on sorbent surfaces. In experiments on desorption of Aroclor® 1254, Haque et al.[12] observed that vapor loss of individual chlorobiphenyls in Aroclor® 1254 decreased with increased chlorination in accordance with their vapor pressures.

Hiraizumi et al.,[21] using a number of adsorbents, observed that the sorption of PCBs was linearly related to the specific surface areas of the adsorbents. The relationships between the PCBs sorbed and their concentrations in solutions have generally been described in terms of equilibrium models such as the Freundlich model:

$$C_{ad} = K \cdot Co^{1/n}$$

where C_{ad} is the amount adsorbed (microgram per gram), Co is the equilibrium solution concentration (microgram per liter), and K and l/n are constants. For the adsorption and desorption of PCBs, however, the observed isotherms are nonsingular, with desorption showing marked hysteresis. Horzempa and Toro[22] found values of K for desorption of hexachlorobiphenyl to be two to three times the value for adsorption by a lake sediment and two clay minerals, montmorillonite and kaolinite. The sorption appeared to be correlated both with the surface area and the organic matter content. The nonsingular behavior which may result from lack of attainment of equilibrium,[23] besides other poorly understood factors, clearly demonstrates that PCBs are strongly held on the surfaces of the sorbents. Indeed, experiments show that PCBs sorbed by soils remain immobile against leaching with water[24,25] or landfill leachate.[25] Similarly, PCBs are strongly sorbed on sediment surfaces and are thus transported downstream in river systems.[26,27]

A number of investigations show that sorption of PCBs by sediments and soils is related to their total organic carbon.[28-31] These correlations have led to the suggestion that sorption of PCBs and other nonionic organic compounds of low aqueous solubility is due to solute partitioning in the organic matter.[31-33] Consequently, partitioning of an organic compound, as PCBs, between organic matter of soils and sediments and water should correlate well with its partition coefficient between water and an immiscible solvent, as *n*-octanol. Octanol/water partition coefficients and water solubilities of a number of organic compounds, including PCBs, covering more than eight orders of magnitude in solubility (10^{-3} to 10^4) and six orders of magnitude in partition coefficient (10 to 10^7) have been demonstrated.[32] Thus, estimates of distribution coefficient or sorption of these organic compounds by environmental samples as soils and sediments can be obtained either from their aqueous solubilities or *n*-octanol/water partition coefficients. Although these relationships apply to sorbents with appreciable organic surface coatings, in sorbents with little or no organic matter present, sorption must proceed through other mechanisms.

III. PROCESSES AND PRODUCTS OF PCBs DEGRADATION

A. Photochemical Degradation

One possible route of environmental breakdown of PCBs is photochemical degradation. Early laboratory experiments on photolysis of PCBs were conducted with mercury lamps as UV sources, emitting radiations at a wavelength of about 254 nm. These photodegradations resulted in dechlorination of PCBs. PCBs do absorb in the 280 to 300 nm range,[34] which is the high energy end of the solar spectrum reaching the earth's surface. Many experiments have been performed with the high energy UV radiation and UV fluorescent lamps simulating sunlight, producing less chlorinated PCBs than the starting material. Safe and Hutzinger[35] observed that 2,2′,4,4′,6,6′-hexachlorobiphenyl photolyses readily at 310 nm, with stepwise loss of chlorine, producing molecular ions corresponding to di-, tri-, tetra-, and pentachlorobiphenyls. Photolysis of 3,3′,4,4′-tetrachlorobiphenyl at 300 nm produced 3,4,3′-trichlorobiphenyl and 4,4′-dichlorobiphenyl, while that of 4,4′-dichlorobiphenyl produced 4-chlorobiphenyl. No further dechlorination of 4-chlorobiphenyl occurred because of its lack of absorption of the wavelength used.[36] Although photolysis results mainly in dechlorination of PCBs, examples of photo-induced isomerization and of condensation of the individual chlorobiphenyls have been reported.

Photochemical degradation of PCBs is influenced by the degree of chlorination, position of chlorine substitution in the ring, and the solvent used for PCB dissolution. Photolysis of PCBs has been studied in the gaseous, solution, and solid phases. In the solution phase, most photolysis experiments have been done with PCBs dissolved in organic solvents or in aqueous organic mixtures because of very low solubility of PCBs in water. In solid phase, PCBs have been photolyzed as thin films on glass plates or as layers adsorbed on silica.

Chlorobiphenyls with higher chlorine content undergo photolysis faster than those with lower chlorine content.[2,37] For example, irradiation at 310 nm wavelength of 0.1% solutions of tetra-, hexa-, and octachlorobiphenyls in hexane for 24 hr under N_2 degraded about 70% of tetra-, 96% of hexa-, and more than 99% of octachlorobiphenyl.[2] Bunce et al.[38] observed increased photodecomposition with increased chlorine content and with increased irradiation time. Ruzo et al.[39] studied photodecomposition of six tetrachlorobiphenyls at 300 nm in cyclohexane and found di- and trichlorinated biphenyls as the major products. They further observed that in 20 hr, all *o*-chlorines produced dechlorinated products. In the absence of *o*-chlorines, *m*-chlorines were cleaved; but when an *o*-chlorine was present, less than 1% of the altered product arose from loss of *m*-chlorine. No cleavage of the *p*-chlorine occurred during this time period.

The primary process in photoreaction is reductive dechlorination, which appears to occur by C–Cl bond cleavage, producing a biphenyl free radical species which abstracts H from the solvent as:

$$\mathrm{ArCl} \rightarrow \mathrm{Ar}^{\cdot}\ \mathrm{Cl}^{\cdot} \rightarrow \mathrm{ArH}$$

Consequently, photolysis occurs readily in solvents as hexane and isopropyl alcohol whose hydrogens are lost easily by free radical attack.

Ruzo et al.[39] suggested the following scheme to illustrate the photochemical mechanism where excited triplet is considered responsible for the photoreaction and abstraction of hydrogen from the solvent.

$^{0}[\mathrm{P\text{-}Cl}]$	$\rightarrow$	$^{1}[\mathrm{P\text{-}Cl}^{*}]$	$\rightarrow$	$^{3}[\mathrm{P\text{-}Cl}]^{*}$
PCB in ground state		Excited singlet		Excited reactive triplet

In accordance with this mechanism, photolysis of 2,2′,4,4′-tetrachlorobiphenyl in methanol and cyclohexane solutions is illustrated in Figure 5. As expected, formation of HCl was detected during photolysis in both solvents. Although dechlorination was the major reaction in both solvents, small quantities (<3%) of methoxylated products were observed in methanol solution.

The greater photolysis of the *o*-chlorine[38,39] is probably due to steric hindrance to the preferred excited state geometry. Wagner[40] inferred a planar structure for the excited triplet from its inefficiency in quenching ketones. The "crowding" by the *o*-substituents causes greater twisting of the inter-ring bond that decreases its double bond character. Thus, *o*-chlorines are easily lost, relieving the strain on the bond and producing greater quantum yields of the products.

As with individual chlorobiphenyls, the more highly chlorinated commercial PCBs also photolyze preferentially in the environmental samples.[38] These authors estimated that in natural waters, the loss of PCBs may be of the order of 10 to 1000 g/Km^{-2}/year. Their estimates also show that in shallow waters, at least one chlorine from chlorobiphenyls of high chlorine content is photolyzed annually; apparently photolysis plays a significant role in PCB chemistry in the environment. These predictions are based upon concentrations of PCBs dissolved in water. It is generally contended, however, that nonpolar compounds, as PCBs, form surface films where concentrations of the compound are much greater and can

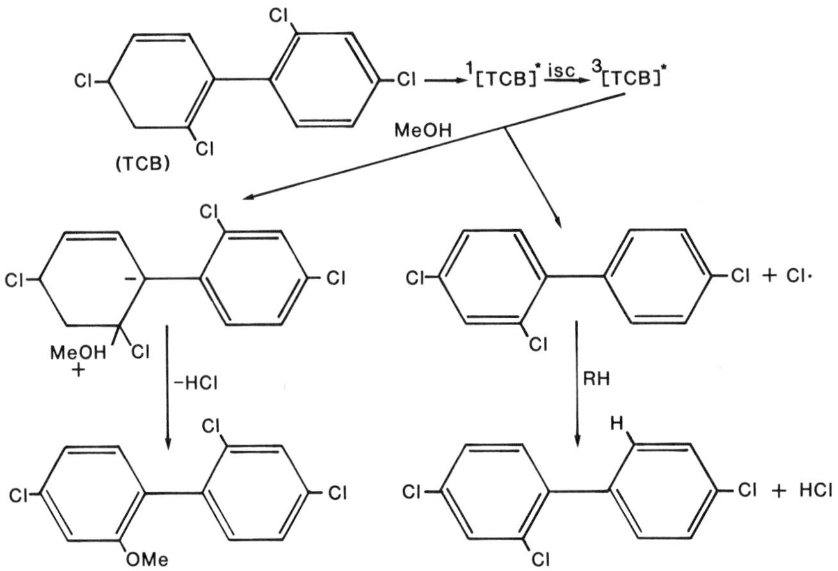

FIGURE 5. Photolysis of 2,2′,4,4′-tetrachlorobiphenyl in methanol and cyclohexane solutions.[39]

be as high as 10^6 times the concentration in the bulk solution.[38] Consequently, much larger amounts of PCBs can be photolyzed from water bodies.

Environmental photodegradation of PCBs can also increase in the presence of organic compounds that sensitize the photoreaction. Zepp et al.[41] reported that natural substances as humic or other suspended materials can induce or accelerate photolysis of various types of synthetic chemicals, including PCB. Occhiucci and Patachiola[42] have recently observed that photodegradation of PCBs adsorbed on silica gel or montmorillonite clay was considerably enhanced by the addition of triethylamine to the system. The sensitization has been attributed to the formation of an excited charge-transfer complex. Sensitization of the photoreaction in PCBs has also been observed in the presence of aliphatic and aromatic amines,[43,44] and titanium oxide.[45]

B. Biodegradation

Investigations of biodegradation of PCBs in soils, sediments, and lakes and rivers show that both aerobic and anaerobic microorganisms decompose and metabolize PCBs. Pal et al.[46] reviewed biodegradation of PCBs in soil-plant systems and compiled microorganism sources and products of biodegradation of a number of chlorinated biphenyls and commercial PCB preparations. Experiments by Wong and Kaiser[47] show stimulation of bacterial growth in moderate concentration (500 ppm) of Aroclors® 1221, 1242, and 1254 as carbon and energy sources, demonstrating PCBs metabolization by the bacteria. Similarly, growth of *Escherichia coli* was stimulated in the presence of Aroclor® 1242 at the low concentrations of 0.01 and 0.1 ppm. At high concentration, 10,000 ppm Aroclor® 1254, soil microbial activity as measured by CO_2 evolution was inhibited and no degradation of the Aroclor® was observed during a 60-day period.[48] Likewise, Moein et al.[49] detected no degradation of Aroclor® 1254 in a soil during a 2-year period following a spill of transformer fluid. Clearly, while small concentration of PCBs can stimulate bacterial growth by providing carbon, large concentrations are toxic to microorganisms.

A number of reports show that microbial degradation of the lower chlorinated biphenyls occurs at a faster rate than the higher chlorinated biphenyls. Furukawa and Matsumura,[50] using a microbial strain *Alkaligenes* sp. from a lake sediment, found that 2-, 3-, and 4-

monochlorobiphenyls, 2,3′-, 2,4′-, 3,3′-, 2,4-, 2,6-, and 3,4-dichlorobiphenyls, and 2,4,5- and 2,4,4′-trichlorobiphenyls degraded completely within 20 hr. In contrast, tetra- and pentachlorobiphenyls degraded very slowly. Using different and mixed microbial strains, Metcalf et al.,[51] Baxter et al.,[52] Clark et al.,[53] Liu,[54] and Hankin and Sawhney,[55] also observed increased degradation in lower chlorinated biphenyls.

It has also been observed that some chlorobiphenyls that do not degrade easily when present alone in a culture medium do so when present in a mixture or when biphenyl is added to the substrate.[52] The enhanced degradation is likely due to co-metabolism, which is commonly observed in microbial processes.[56]

Not only does the degree of chlorination influence biodegradation, but the environment also affects biodegradation. For example, Iwata et al.,[57] studied PCB degradation in six different soils and found no degradation of Aroclor® 1254 in two soils, little degradation in two, and significant degradation in the remaining two soils. In a recent study of PCBs degradation in eight soil cultures, Hankin and Sawhney[55] observed that all soils but one degraded at least 70% of Aroclor® 1248 in 14 days, while only three soils degraded 1254. In a 112-day incubation period, over 90% of Aroclor® 1248 and about 40% of Aroclor® 1254 were degraded in some soils, while no Aroclor® 1260 degraded in any soil. Growth of aerobic organisms increased with time as did degradation of the Aroclors®.

Although hydroxylation and ring cleavage are the main mechanisms in biodegradation of PCBs, producing less chlorinated chlorobiphenyls than in the starting materials, other products of PCB biodegradation have also been identified. Degradation of mono- and dichlorobiphenyls by two *Achromobacter* cultures produced *p*-chlorobenzoic acid.[58] Degradation of 4-chlorobiphenyl by a Gram negative bacterium in the presence of nitrate produced four compounds, 2- and 4-hydroxy-4′-chlorobiphenyl and 2- and 4-hydroxymononitro-4′-chlorobiphenyl.[59] The mechanism for the formation of these compounds is illustrated in Figure 6. The proposed mechanism involves the formation of arene oxide intermediates which have high reactivity with biological materials.

Incubation of commercial PCB preparation Aroclor® 1242 with a bacterial culture from a harbor produced a number of alkenes and alkyl benzenes.[60] Under anaerobic conditions, bacterial dehalogenation of halobenzoates, the degradation products of PCBs, proceeded to give CH_4 + CO_2.[61] Clearly, biological degradation of PCBs proceeds via different, as yet not well understood mechanisms depending upon the environmental conditions.

As PCBs are soluble in lipids, they are accumulated by a number of organisms, following lipid/water partition coefficients. Chlorobiphenyls show selective bioaccumulation and degradation, which are affected by both the chlorobiphenyl and the animal species. Hansen[62] has reviewed various factors, including animal species, size of adipose compartments, enzyme activity etc., that affect bioaccumulation and biodegradation by animals. Accumulation ratios of chlorobiphenyls vary among different animals as well as in lipids from different parts of an animal.[63]

While the biodegradation of PCBs has been studied extensively, no information on the fate of PCBs sorbed or taken up by plants is available. Indeed, controversy persists as to whether PCBs are translocated through the plants or are merely sorbed on plant surfaces from PCB vapors or PCB-bearing dust particles. Although long-term alteration in PCBs sorbed by perennial plants or trees is not known, the PCBs in plants growing on contaminated soils increases as the chlorine content of the chlorobiphenyl decreases[64-66] (Figure 7).

C. Thermal Degradation

Because of the stability and potential toxicity of PCBs, numerous laboratory experiments have been conducted for the combustion and complete destruction of these compounds for safe disposal of industrial wastes and used products. While pyrolysis at below 700°C produces various toxic materials, higher temperatures decompose PCBs completely. Buser and Rappe[67]

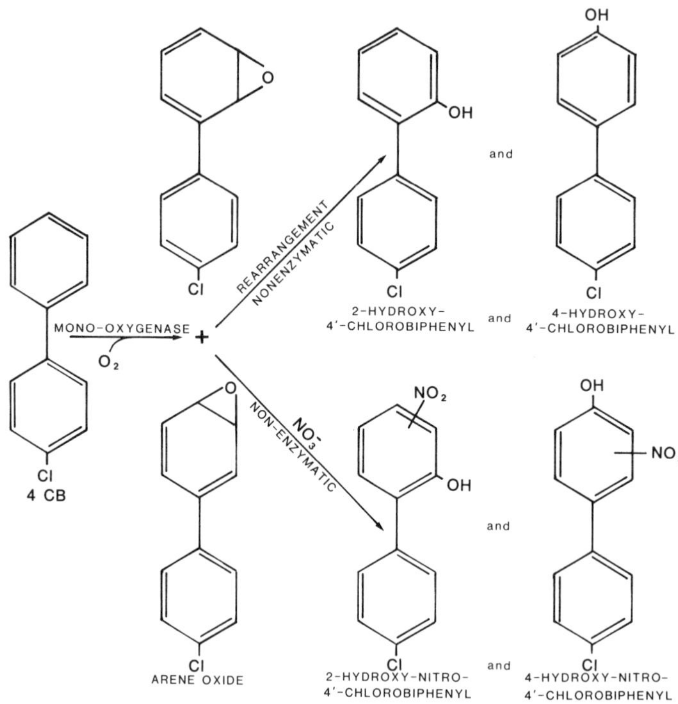

FIGURE 6. Biodegradation of 4-chlorobiphenyl by a Gram negative bacterium in presence of nitrate.[59]

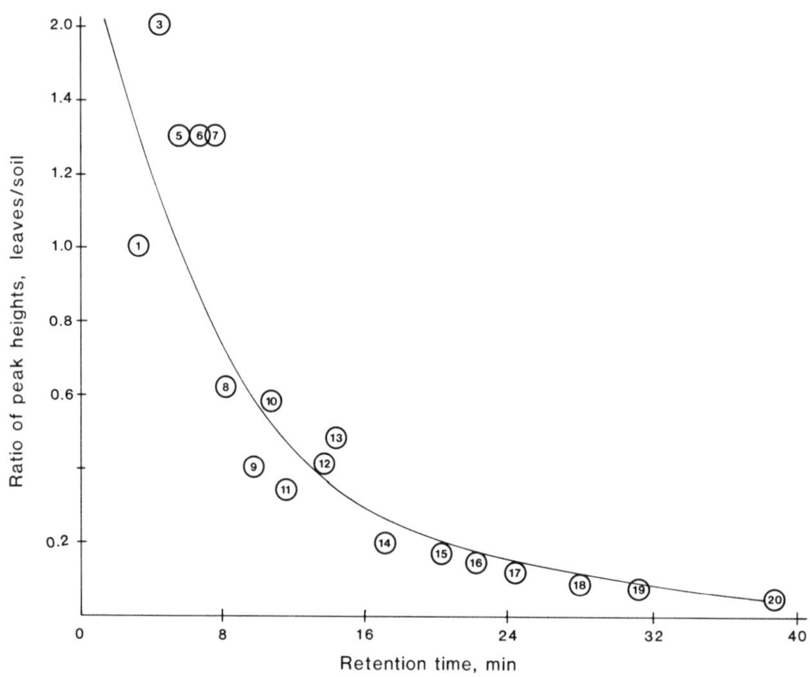

FIGURE 7. Ratios of GC peak heights from extracts of beet leaves grown on a soil contaminated with PCBs to the corresponding peaks from soil extracts. Peak numbers are given in circles. (From Sawhney, B. L. and Hankin, L., *J. Food Protect.*, 47, 232, 1984. With permission.)

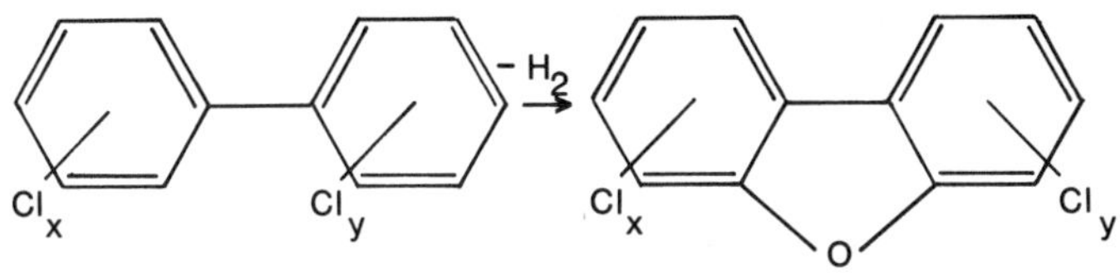

FIGURE 8. Thermal conversion of PCBs to polychlorinated dibenzofurans.

conducted pyrolysis of 18 individual chlorobiphenyls including tetra-, penta-, hexa-, hepta-, octa-, nona-, and decachlorobiphenyls at 600°C. Using high resolution GC/MS for their separation and identification, they identified several highly toxic polychlorinated dibenzofurans (PCDFs) in the products and discussed different schemes for their formation. The yield of PCDFs ranged from 0.1 to several percent of the starting PCBs. The conversion of PCBs to PCDFs involves intramolecular cyclization with loss of H_2 (as shown in Figure 8) or HCl and loss or rearrangement of chlorine.

Pyrolysis of commercial PCBs, Aroclor® 1248 by Morita et al.[68] at 300°C and Aroclor® 1254 by Buser et al.[69] at 550 to 650°C formed PCDFs, while above 700°C, the PCBs were completely destroyed. PCDFs as well as chlorobenzenes and *p*-dioxins in municipal incinerator wastes[70-72] have been attributed to chlorophenols and thermal synthesis from phenols and chlorine, and they are not necessarily formed from pyrolysis of PCBs. Mechanisms for their formation are discussed by Choudhry et al.[73] Many chemical procedures of complete dechlorination of PCBs using catalysts have also been developed. Examples are 5% platinum or paladium,[74] nickel boride in alcohol with excess sodium borohydride,[75] and $LiAlH_4$.[76]

Yamasaki et al.[77] reported complete destruction of PCBs in the presence of a methanol and sodium hydroxide solution at 300 to 320°C and 180 kg/cm^2 pressure. They suggested that the method can be adopted on an industrial scale, using continuous pipeline capillary system. Based upon PCBs incineration in accordance with the EPA test protocols in high efficiency boilers with a thermal destruction efficiency of >99%, Hunt et al.[78] have recently concluded that the method provides a satisfactory means of disposing waste oil contaminated with 50 to 500 ppm PCBs. Thermal destruction, as well as other physicochemical and biological methods, for low and high level PCB contamination have recently been discussed by Ackerman et al.[79]

REFERENCES

1. **Ballschmiter, K. and Zell, M.,** Analysis of polychlorinated biphenyls (PCB) by glass capillary gas chromatography, *Fresenius' Z. Anal. Chem.*, 302, 20, 1980.
2. **Hutzinger, O., Safe, S., and Zitko, V.,** *The Chemistry of PCBs*, CRC Press, Cleveland, 1974.
3. **Jenkins, R. L., McCallough, R., and Booth, C. F.,** Synthesis in the biphenyl series, *Ind. Eng. Chem.*, 22, 31, 1931.
4. **Scheidt, F. M.,** Selective monochlorination of biphenyl with clay catalysts, German Patent 2,056,441, 1971, *Chem. Abstr.*, 75, 35379, 1971.
5. **Lawlor, F. E.,** Substitution chlorination of aromatic compounds with liquid chlorine, U.S. Patent 2,608,532, 1952, *Chem. Abstr.*, 48, 3390, 1954.
6. **Webb, R. G. and McCall, A. C.,** Identities of polychlorinated biphenyl isomers in Aroclors, *J. Assoc. Off. Anal. Chem.*, 55, 746, 1972.
7. Sandoz Ltd., Hallogenated vat dyes, Belgian Patent 665,732, 1965, *Chem. Abstr.*, 64, 19847, 1966.
8. **Albro, P. W., Corbet, J. T., and Schroeder, J. L.,** Quantitative characterization of polychlorinated biphenyl mixtures (Aroclors 1248, 1254 and 1260) by gas chromatography using capillary columns, *J. Chromatogr.*, 205, 103, 1981.
9. **Webb, R. G. and McCall, A. C.,** Quantitative PCB standards for electron capture gas chromatography, *J. Chromatogr. Sci.*, 11, 366, 1973.

10. **Sawyer, L. D.,** Quantitation of polychlorinated biphenyl residues by electron capture gas-liquid chromatography: reference material characterization and preliminary study, *J. Assoc. Off. Anal. Chem.*, 61, 272, 1978.
11. **Tuinstra, L. G. M. Th. and Traag, W. A.,** Capillary gas chromatographic-mass spectrometric determination of individual chlorobiphenyls in technical Aroclors, *J. Assoc. Off. Anal. Chem.*, 66, 1708, 1983.
12. **Haque, R., Schmedding, D. W., and Freed, V. H.,** Aqueous solubility, adsorption, and vapor behavior of polychlorinated biphenyl Aroclor 1254, *Environ. Sci. Technol.*, 8, 139, 1974.
13. **Chiou, C. T., Freed, V. H., Schmedding, D. W., and Kohnert, R. L.,** Partition coefficient and bioaccumulation of selected organic chemicals, *Environ. Sci. Technol.*, 11, 475, 1977.
14. **Metcalf, R. L., Sanborn, J. R., Lu, P., and Nye, D.,** Laboratory model ecosystem studies of the degradation and fate of radiolabeled tri-, tetra-, and pentachlorobiphenyl compared with DDE, *Arch. Environ. Contam. Toxicol.*, 3, 151, 1975.
15. **Haque, R. and Schmedding, D.,** Studies on the adsorption of selected polychlorinated biphenyl isomers on several surfaces, *J. Environ. Sci. Health,* B11, 129, 1979.
16. **Zitko, V.,** Polychlorinated biphenyls (PCB) solubilized in water by nonionic surfactants for studies of toxicity to aquatic animals, *Bull. Environ. Contam. Toxicol.*, 5, 279, 1970.
17. **Mackay, D. and Wolkoff, A. W.,** Rate of evaporation of low solubility contaminants from water bodies to atmosphere, *Curr. Res.*, 7, 611, 1973.
18. **Nau-Ritter, G. M. and Wurster, C. F.,** Sorption of polychlorinated biphenyls (PCB) to clay particulates and effects of desorption on phytoplankton, *Water Res.*, 17, 383, 1983.
19. **Mackay, D.,** Finding fugacity feasible, *Environ. Sci. Technol.*, 13, 1218, 1979.
20. **Mackay, D. and Leinonew, P. J.,** Rate of evaporation of low solubility contaminants from water bodies to atmosphere, *Environ. Sci. Technol.*, 9, 1178, 1975.
21. **Hiraizumi, Y., Takahashi, M., and Nishimura, H.,** Adsorption of polychlorinated biphenyl onto sea bed sediment, marine plankton, and other adsorbing agents, *Environ. Sci. Technol.*, 13, 580, 1979.
22. **Horzempa, L. M. and Di Tora, D. M.,** The extent of reversibility of polychlorinated biphenyl adsorption, *Water Res.*, 17, 851, 1983.
23. **Rao, P. S. C. and Davidson, J. M.,** Estimation of pesticide retention and transformation parameters required in nonpoint source pollutant models, in *Environmental Impact of Nonpoint Source Pollution,* Sect. 1, Overcash, M. R. and Davidson, J. M., Eds., Ann Arbor Scientific Publications, Ann Arbor, Mich., 1980, chap. 1.
24. **Filonow, A. B., Jacobs, L. W., and Mortland, M. M.,** Fate of polybrominated biphenyls (PBB's) in soils. Retention of hexabromobiphenyl in four Michigan soils, *J. Agric. Food Chem.*, 24, 1201, 1976.
25. **Griffin, R. A. and Chou, S. F. J.,** Movement of PCB's and other persistent compounds through soil, *Water Sci. Technol.*, 13, 1153, 1981.
26. **Sawhney, B. L., Frink, C. R., and Glowa, W.,** PCBs in the Housatonic River: determination and distribution, *J. Environ. Qual.*, 10, 444, 1981.
27. **Frink, C. R., Sawhney, B. L., Kulp, K. R., and Fredette, C. G.,** Polychlorinated biphenyls in Housatonic River sediments in Massachusetts and Connecticut: determination, distribution and transport, *Conn. Agric. Exp. Stn. Bull.*, 800, 20, 1982.
28. **Karickhoff, S. W., Brown, D. S., and Scott, T. A.,** Sorption of hydrophobic pollutants on natural sediments, *Water Res.*, 13, 241, 1979.
29. **Sawhney, B. L.,** Sorption of PCBs by soils, *Abstr. 12th Int. Soil Sci. Congr.*, New Delhi, 1982, 33.
30. **Weber, W. J., Voice, T. C., Pirbazari, M., Hunt, G. E., and Ulanoff, D. M.,** Sorption of hydrophobic compounds by sediments, soils and suspended solids. II. Sorbent evaluation studies, *Water Res.*, 17, 1443, 1983.
31. **Lambert, S.M.,** Omega (Ω) a useful index of soil sorption equilibria, *J. Agric. Food Chem.*, 16, 340, 1968.
32. **Chiou, C. T., Peters, L. J., and Freed, V. H.,** A physical concept of soil-water equilibria for non ionic organic compounds, *Science,* 206, 831, 1979.
33. **Pavlou, S. P. and Dexter, R. N.,** Distribution of polychlorinated biphenyls (PCB) in estuarine ecosystems. Testing the concept of equilibrium partitioning in the marine environment, *Environ. Sci. Technol.*, 13(1), 65, 1979.
34. **MacNeil, J. D., Safe, S., and Hutzinger, O.,** The ultraviolet absorption spectra of some chlorinated biphenyls, *Bull. Environ. Contam. Toxicol.*, 15, 66, 1976.
35. **Safe, S. and Hutzinger, O.,** Polychlorinated biphenyls: photolysis of 2,4,6,2′,4′,6′-hexachlorobiphenyl, *Nature (London),* 232, 641, 1971.
36. **Ruzo, L. O., Zavik, M. J., and Schuetz, R. D.,** Polychlorinated biphenyls: photolysis of 3,4,3′,4′-tetrachlorobiphenyl and 4,4′-dichlorobiphenyl in solution, *Bull. Environ. Contam. Toxicol.*, 8, 217, 1972.
37. **Hannan, E. J., Bills, D. D., and Herring, J. L.,** Analysis of polychlorinated biphenyls by gas chromatography and ultraviolet irradiation, *J. Agric. Food Chem.*, 21, 87, 1973.

38. **Bunce, N. J., Kumar, Y., and Brownlee, B. G.,** An assessment of the impact of solar degradation of polychlorinated biphenyls in the aquatic environment, *Chemosphere,* 7, 155, 1978.
39. **Ruzo, L. O., Zabik, M. J., and Schuetz, R. D.,** Photochemistry of bioactive compounds. Photochemical processes of polychlorinated biphenyls, *J. Am. Chem. Soc.,* 96, 3809, 1974.
40. **Wagner, P. J.,** Conformational changes involved in the singlet-triplet transitions of biphenyl, *J. Am. Chem. Soc.,* 89, 2820, 1967.
41. **Zepp, R. G., Baughman, G. L., and Schlotzhauer, P. F.,** Comparison of photochemical behavior of various humic substances in water: sunlight induced reactions of aquatic pollutants photosensitized by humic substances, *Chemosphere,* 10, 109, 1981.
42. **Occhiucci, G. and Patacchiola, A.,** Sensitized photodegradation of adsorbed polychlorobiphenyls (PCB's), *Chemosphere,* 11, 255, 1982.
43. **Ohashi, M., Tsujimoto, K., and Seki, K.,** Photoreduction of 4-chlorobiphenyl by aliphatic amines, *J. Chem. Soc. Chem. Commun.,* 384, 1973.
44. **Nordblom, G. D. and Miller, L. L.,** Photoreduction of 4,4′-dichlorobiphenyl, *J. Agric. Food Chem.,* 22, 57, 1974.
45. **Carey, J. H., Lawrence, J., and Tosine, H. M.,** Photodechlorination of PCB's in the presence of titanium dioxide in aqueous suspensions, *Bull. Environ. Contam. Toxicol.,* 16, 697, 1976.
46. **Pal, D., Weber, J. B., and Overcash, M. R.,** Fate of polychlorinated biphenyls (PCBs) in soil-plant systems, *Residue Rev.,* 74, 45, 1980.
47. **Wong, P. T. S. and Kaiser, K. L. E.,** Bacterial degradation of polychlorinated biphenyls. II. Rate studies, *Bull. Environ. Contam. Toxicol.,* 13, 249, 1975.
48. **Nissen, T. V.,** Stability of PCB in the soil, *Tidsskr. Planteavl,* 77, 533, 1973.
49. **Moein, G. J., Smith, A. J., Jr., and Stewart, P. L.,** Follow-up study of the distribution and fate of PCBs and benzenes in soil and ground water samples after an accidental spill of transformer fluid, *Proc. Natl. Conf. on Control of Hazardous Material Spills,* Information Transfer, Rockville, Md., 1976, 368.
50. **Furukawa, K. and Matsumura, F.,** Microbial metabolism of polychlorinated biphenyls. Studies on the relative degradability of polychlorinated biphenyl components by *Alkaligenes* sp., *J. Agric. Food. Chem.,* 24, 251, 1976.
51. **Metcalf, R. L., Sanborn, J. R., Lu, P.-Y., and Nye, D.,** Laboratory model ecosystem studies of the degradation and fate of radiolabeled tri-, tetra-, and pentachlorobiphenyl compared with DDE, *Arch. Environ. Contam. Toxicol.,* 3, 151, 1975.
52. **Baxter, R. A., Gilbert, P. E., Lidgett, R. A., Mainprize, J. H., and Vodden, H. A.,** The degradation of polychlorinated biphenyls by micro-organisms, *Sci. Total Environ.,* 4, 53, 1975.
53. **Clark, R. R., Chian, E. S. K., and Griffin, R. A.,** Degradation of polychlorinated biphenyls by mixed microbial cultures, *Appl. Environ. Microbiol.,* 37, 680, 1979.
54. **Liu, D.,** Assessment of continuous biodegradation of commercial PCB formulations, *Bull. Environ. Contam. Toxicol.,* 29, 200, 1982.
55. **Hankin, L. and Sawhney, B. L.,** Microbial degradation of PCBs in soil, *Soil Sci.,* 137, 401, 1984.
56. **Jacobson, S. N. and Alexander, M.,** Enhancement of the microbial dehalogenation of a model chlorinated compound, *Appl. Environ. Microbiol.,* 42, 1062, 1981.
57. **Iwata, Y., Westlake, W. E., and Gunther, F. A.,** Varying persistence of polychlorinated biphenyls in six California soils under laboratory conditions, *Bull. Environ. Contam. Toxicol.,* 9, 204, 1973.
58. **Ahmed, M. and Focht, D. D.,** Degradation of polychlorinated biphenyls by two species of *Achromobacter, Can. J. Microbiol.,* 19, 47, 1973.
59. **Sylvestre, M., Masse, R., Messier, F., Fauteux, J., Bisaillon, J.-G., and Beaudet, R.,** Bacterial nitration of 4-chlorobiphenyl, *Appl. Environ. Microbiol.,* 44, 871, 1982.
60. **Kaiser, K. L. E. and Wong, P. T. S.,** Bacterial degradation of polychlorinated biphenyls. I. Identification of some metabolic products from Aroclor 1242, *Bull. Environ. Contam. Toxicol.,* 11, 291, 1974.
61. **Suflita, J. M., Horowitz, A., Shelton, D. R., and Tiedje, J. M.,** Dehalogenation: a novel pathway for the anaerobic biodegradation of haloaromatic compounds, *Science,* 218, 1115, 1982.
62. **Hansen, L. G.,** Selective accumulation and depletion of polychlorinated biphenyl components: food animal implications, *Ann. N.Y. Acad. Sci.,* 320, 238, 1979.
63. **Hansen, L. G., Tuinstra, L. G. M. Th., Kan, C. A., Strik, J. J. T. W. A., and Koeman, J. H.,** Accumulation of chlorobiphenyls in chicken fat and liver after feeding Aroclor 1254 directly or fat from swine fed Aroclor 1254, *J. Agric. Food. Chem.,* 31, 254, 1983.
64. **Iwata, Y. and Gunther, F. A.,** Translocation of the polychlorinated biphenyl Aroclor 1254 from soil into carrots under field conditions, *Arch. Environ. Contam. Toxicol.,* 4, 44, 1976.
65. **Suzuki, M., Aizawa, N., Okano, G., and Takahashi, T.,** Translocation of polychlorobiphenyls in soil into plants: a study by a method of culture of soybean sprouts, *Arch. Environ. Contam. Toxicol.,* 5, 343, 1977.

66. **Sawhney, B. L. and Hankin, L.,** Plant contamination by PCBs from amended soils, *J. Food Protect.,* 47, 232, 1984.
67. **Buser, H. R. and Rappe, C.,** Formation of polychlorinated dibenzofurans (PCDFs) from the pyrolysis of individual PCB isomers, *Chemosphere,* 3, 157, 1979.
68. **Morita, M., Nakagawa, J., and Rappe, C.,** Polychlorinated dibenzofuran (PCDF) formation from PCB mixture by heat and oxygen, *Bull. Environ. Contam. Toxicol.,* 19, 665, 1978.
69. **Buser, H. R., Bosshardt, H.-P., and Rappe, C.,** Formation of polychlorinated dibenzofurans (PCDFs) from the pyrolysis of PCBs, *Chemosphere,* 1, 109, 1978.
70. **Olie, K., Lustenhouwer, J. W. A., and Hutzinger, O.,** Polychlorinated dibenzo-p-dioxins and related compounds in incinerator effluents, in *Chlorinated Dioxins and Related Compounds,* Hutzinger, O., Frei, R. W., Merian, E., and Pocchiari, F., Eds., Pergamon Press, New York, 1980, 227.
71. **Liberti, A. and Brocco, D.,** Formation of polychlorodibenzodioxins and polychlorodibenzofurans in urban incinerators emissions, in *Chlorinated Dioxins and Related Compounds,* Hutzinger, O., Frei, R. W., Merian, E., and Pocchiari, F., Eds., Pergamon Press, New York, 1980, 245.
72. **Cavallaro, A., Luciani, L., Ceroni, G., Rocchi, I., Invernizzi, G., and Gorni, A.,** Summary of results of PCDDs analyses from incinerator effluents, *Chemosphere,* 11, 859, 1982.
73. **Choudhry, G. G., Olie, K., and Hutzinger, O.,** Mechanisms in the thermal formation of chlorinated compounds including polychlorinated dibenzo-p-dioxins, in *Chlorinated Dioxins and Related Compounds,* Hutzinger, O., Frei, R. W., Merian, E., and Pocchiari, F., Eds., Pergamon Press, New York, 1980, 275.
74. **Berg, O. W., Diosady, P. L., and Rees, G. A. V.,** Column chromatographic separation of polychlorinated biphenyls from chlorinated hydrocarbon pesticides, and their subsequent gas chromatographic quantitation in terms of derivatives, *Bull. Environ. Contam. Toxicol.,* 7, 338, 1972.
75. **Dennis, W. H., Jr., Chang, Y. H., and Cooper, W. J.,** Catalytic dechlorination of organochlorine compounds. V. Polychlorinated biphenyls — Aroclors 1254, *Bull. Environ. Contam. Toxicol.,* 22, 750, 1979.
76. **De Kok, A., Geerdink, R. B., Frei, R. W., and Brinkman, U. A. Th.,** The use of dechlorination in the analysis of polychlorinated biphenyls and related classes of compounds, *Int. J. Environ. Anal. Chem.,* 9, 301, 1981.
77. **Yamasaki, N., Yasui, T., and Matsuoka, K.,** Hydrothermal decomposition of polychlorinated biphenyls, *Environ. Sci. Technol.,* 14, 550, 1980.
78. **Hunt, G. T., Wolf, P., and Fennelly, P. F.,** Incineration of polychlorinated biphenyls in high-efficiency boilers: a viable disposal option, *Environ. Sci. Technol.,* 18, 171, 1984.
79. **Ackerman, D. G., Scinto, L. L., Bakshi, P. S., Delumyea, R. G., Johnson, R. J., Richard, G., Takata, A. M., and Sworzyn, E. M.,** *Destruction and Disposal of PCBs by Thermal and Non-Thermal Methods,* Noyes Data Corp., Park Ridge, Ill., 1983.

Chapter 3

THE RELIABILITY OF PCB ANALYSIS

A. V. Holden

TABLE OF CONTENTS

I. INTRODUCTION

Chemical analysis for polychlorinated biphenyls (PCBs), since their identification in environmental samples by Jensen[1] in 1966, has been almost exclusively performed by gas liquid chromatography (GLC), with the aid of the electron capture (EC) detector which is particularly sensitive to organochlorine compounds. This detector, developed by Lovelock,[2] proved very effective in detecting and quantifying organochlorine pesticides such as hexachlorocyclohexane (HCH), aldrin, dieldrin, and DDT and its breakdown products. Each individual compound, as it emerges from the GC column, produces a signal from the EC detector, the magnitude of which is proportional to the amount of the compound present, and hence is related to the concentration in the solution injected. The time elapsing between injection and elution, for a particular temperature of operation, is characteristic of the compound present.

PCBs however, normally occur in mixtures of many individual isomers and homologs, each mixture in GLC analysis producing a pattern of serially eluted peaks emerging from the electron capture detector. Although analyses for PCBs are commonly reported as single values, such values are only obtained by comparing the peak pattern with that from a standard PCB mixture (normally a commercial product), and no information is given of the relative proportions of the different components in the mixture. For various reasons the single value can only be regarded as a compromise, and in many situations it may be both inaccurate and misleading. This chapter discusses the problems of accurate PCB analysis (hardly attainable at the present time) and indicates the developments likely to lead to both improved accuracy and more detailed information regarding individual PCB compounds.

II. COMPOSITION OF PCB RESIDUES

The mixtures of PCB residues which may be subjected to chemical analysis can vary widely, depending on several factors. The commercially produced mixtures intended for a wide variety of uses, according to the physical and chemical properties of these mixtures, vary with the degree of chlorination to which the parent biphenyl has been subjected. Some change in the composition of a commercial product may occur during use, e.g., in heat exchangers or as hydraulic fluids. On discharge to the environment, the mixtures, whether used or not, can undergo further modification in respect of the proportions of individual compounds present, as a result of photodegradation or hydrolysis, preferential solubility, absorption or evaporation, or mixing with other PCB products manufactured at a different level of chlorination. Finally, the group of PCB residues extracted from an environmental sample will usually be mixed with coextracted organochlorine compounds such as pesticides which can interfere in GLC analysis unless steps are taken to avoid this.

A total of 209 chlorinated biphenyl isomers and homologs is theoretically possible. This total consists of

3 Monochloro-
12 Dichloro-
24 Trichloro-
42 Tetrachloro-
46 Pentachloro-
42 Hexachloro-
24 Heptachloro-
12 Octachloro-
3 Nonachloro-
1 Decachloro-

Commercial producers have manufactured mixtures which have generally resulted from the chlorination of biphenyl to give products containing approximately 20, 30, 40, 50, 60, and 70% of chlorine by weight. The actual percentage of chlorine varies to some extent between manufacturers, and to a smaller degree between batches from an individual manufacturer. Monsanto Ltd. (U.S.), who marketed a series of products under the trade name Aroclor®, listed nine separate products based on PCBs with 21, 32, 42, 48, 54, 60, 62, and 68% by weight of chlorine. (Other products were wholly or partly based on polychlorinated terphenyls). Other manufacturers produced similar mixtures, for example, a 50% chlorinated product which bore a fairly close resemblance to the 48 or 54% chlorinated Aroclor®, but a detailed chemical analysis would almost certainly have revealed some differences in the proportions of different individual PCBs.

Several studies have been made to determine the identities of the individual PCBs in commercial products. Webb and McCall[3] examined the Aroclor® series 1221, 1232, 1242, 1248, and 1254, in which the first two digits denote biphenyls and the last two the degree of chlorination. Aroclor® 1221 contained mainly mono- and dichlorophenyls, 1232 mostly trichlorophenyls with some mono- and dichlorophenyls; 1242 contained most of the compounds present in 1232 but also several tetra- and pentachlorobiphenyls. Aroclor® 1248 extended the mixture to include more pentachloro compounds and a few hexachloro compounds, while Aroclor® 1254 consisted primarily of tetra-, penta-, and hexachlorobiphenyls. A more recent study by Ballschmiter and Zell,[4] using the Clophen® series of products (Bayer, Federal Republic of Germany), demonstrated the presence of the following compounds in the 30, 50, and 60% chlorinated products.

	Clophen® A-30	**Clophen® A-50**	**Clophen® A-60**
1 Chloro-	1	0	0
2 Chloro-	9	1	4
3 Chloro-	16	14	7
4 Chloro-	18	13	4
5 Chloro-	10	19	14
6 Chloro-	1	22	22
7 Chloro-	0	0	10
8 Chloro-	0	0	1

As the individual compounds listed were those separated on a particular GLC stationary phase, and a few could not be identified from their specific retention indexes, the true numbers of compounds present will be somewhat greater. It is clear, however, that the 50 to 60% chlorinated preparations will contain at least 60 to 70 different compounds although some will be present only in very small proportions. Improvements in the ability to separate the individual PCBs over a decade are demonstrated by the fact that Webb and McCall[3] in 1971 recorded 33 compounds in Aroclor® 1242, whereas Ballschmiter and Zell[4] in 1980 presented a chromatogram of the same product in which at least 74 peaks could be distinguished.

It is clear that, even in the analysis of a mixture of PCBs in an unused commercial product, the complete separation, identification, and quantification of the individual compounds will be difficult, but the comparison of a chromatogram from a sample with that from a standard solution of the identical commercial product will normally provide a sufficient basis for quantification in terms of total PCB, expressed on the basis of the reference product. When, however, the product has undergone some form of degradation, selective removal of compounds, or mixing with other organochlorines, the analysis by GLC becomes more difficult. Comparison of a chromatogram from a sample with that of a standard commercial product can only provide a crude assessment of the total quantity of PCBs present, information which is often of doubtful value, and may be misleading. In particular, it must be emphasized that quantification in terms of a particular product does not imply that contamination by that product has necessarily occurred.

The analysis of organochlorine residues by packed column GLC, which has been used by the majority of analysts since the electron capture detector became available over 20 years ago is, in the hands of most analysts, capable of separating at best only some 15 to 20 peaks on the chromatogram resulting from the injection of a PCB sample. As stated above the PCB products most commonly manufactured are likely to contain at least 60 to 70 individual compounds and packed column chromatography is thus inadequate for accurate analysis. However, at the present time the number of individual PCB compounds available as standards for quantification is much smaller than the number which can be separated by capillary column chromatography, so that the latter technique, while a considerable improvement, cannot yet be fully exploited.

Although chromatograms from the most efficient packed columns may show the presence of 15 to 20 separate peaks, many of these are very small and not measurable with sufficient accuracy. Consequently, quantification of the PCB complex using such chromatograms is based usually on the measurement of the heights or areas of at most five or six peaks (some of which may be produced by more than one individual PCB compound). Care must be taken, with environmental samples, to avoid the use of any PCB peak which coincides or overlaps with the *p,p'*-DDT degradation product *p,p'*-DDE, usually coextracted with PCBs. If the analytical technique used fails to separate or eliminate DDT and TDE (another degradation product of DDT) these too may interfere with the PCB chromatogram.

III. ANALYTICAL PROCEDURES FOR ENVIRONMENTAL SAMPLES

Although there are circumstances when the analysis of commercial PCB products is required, either in their original or used form (as in heat exchangers), most analysts are likely to be confronted with the task of determining PCBs in environmental samples following contamination resulting from some form of discharge. The organochlorine residues extracted will usually be a mixture of PCBs and various pesticidal compounds, many of which have chemical properties similar to those of PCBs. This section is therefore directed primarily toward the problems arising from the analysis of environmental samples, or of agricultural or marine products for human consumption.

A. Levels of Determination

Concentrations of total PCB in environmental materials can range from several thousand milligrams per kilogram in lipid samples to the order of about 1 ng/kg in sea water and perhaps 0.1 ng/m^3 in clean air. All materials, whatever their origin, must be extracted, and the final extracts concentrated or diluted to provide concentrations sufficient for analysis at levels easily determinable by GLC techniques. The minimum acceptable response produced by most electron capture detectors occurs with the injection of about 100 pg of total PCB (e.g., 5 μℓ of a solution containing 0.02 μg/mℓ), but only four or five of the largest peaks will be measurable. If individual isomers are to be determined, the EC detector may require up to 10 pg of each for an adequate response, and the quantity of a mixture of PCBs required will thus be a minimum of several nanograms. With the smaller volumes (1 to 2 μℓ) injected for splitless capillary column GLC analysis the total PCB concentration required in the extract for injection may be at least 5 μg/mℓ, if a large number of individual PCBs are to be quantified.

Some environmental samples, e.g., the organs of birds or fish, will limit the quantity of PCBs available for analysis, and hence the concentration in the final extract. With other materials, e.g., air and water, the quantity of samples may be unlimited, but difficulties can arise in handling large volumes for extraction. (Both air and water samples can be extracted by passage through suitable adsorbents in specially designed apparatus.)

Irrespective of the nature of the sample material, when low concentrations of PCBs are to be determined adequate precautions must be taken to ensure that no extraneous contamination occurs either in the sampling or subsequent handling. Plastic materials (except PTFE), paints and lubricants, or hydraulic fluids must be avoided, and all solvents and adsorbents used in the analyses must be free of detectable contaminants likely to interfere in the determination of PCBs. Aliquots of all solvents used should give no measurable response on GLC analysis after 100-fold concentration. Dummy runs through the entire extraction, separation, and concentration procedures should provide residue-free chromatograms from the analysis of the final concentrate.

Extraction of biological sample material is usually achieved using Soxhlet extractors and a boiling solvent such as *n*-hexane. It is essential that the solvent penetrates the sample on each cycle, and to ensure this the biological material is often ground to a powder with pure dry sodium sulfate beforehand. However, even with such precautions, some samples such as animal fats, may require periods of extraction up to 16 hr to achieve complete removal of organochlorines and lipids.

Some sample materials containing appreciable quantities of water, if not dried beforehand, will require extraction with a polar solvent such as acetone or diethyl ether, perhaps mixed with hexane. After extraction the polar solvent is removed by evaporation, or washing with residue-free distilled water, until a final extract in hexane is obtained.

B. Separation from Interfering Residues

The initial extract will normally contain organochlorines together with pigments, fats, and other co-extractives which must be eliminated prior to GLC analysis. Three procedures have been developed and widely used over the past 25 years — solvent partition, sulfuric acid or alkaline treatment, and adsorption chromatography.

Solvent partition involves dispersion of the organochlorine residues and fats in a mixture of hexane and either dimethylformamide[5] or acetonitrile,[6] the organochlorines remaining largely in the hexane fraction. The method was developed for pesticide analysis, and it is doubtful whether it is now sufficiently efficient to ensure complete extraction of PCBs and some other organochlorines.

Sulfuric acid treatment[7] of the hexane extract results in the destruction of fats and pigments, usually forming water soluble compounds which can be removed by shaking with distilled water, although many analysts remove aliquots of the hexane layer without such washing. Hydrolysis of fats with alcoholic potassium hydroxide is sometimes used, the alkali and hydrolysis products being removed by an aqueous wash. Both procedures will destroy certain pesticides, but PCBs are unaffected.

Adsorption chromatography employs columns of either Florisil[8] or alumina,[9] through which a volume of the hexane extract is passed to remove the lipids and pigments. Separation of the various organochlorines into several fractions can also be achieved on columns of Florisil[8] or silica[9,10] by the use of solvents of different polarity, the PCB group usually being separated from most interfering organochlorine pesticides except *p,p'*-DDE. However, in the presence of chlorinated naphthalenes, chlordane, and toxaphene more elaborate techniques are required.

Adsorption chromatography with fractionation is perhaps the most widely used technique, but a somewhat similar procedure using gel permeation chromatography[11,12] deserves more attention, as it can be automated. It can remove lipids and separate organochlorines in the same process.

C. Gas Liquid Chromatography

In this section only an outline can be given of the main procedures used for GLC analysis. Most analysts over the past 20 years have employed packed glass columns up to 2 m in length and 2- to 6-mm bore. As indicated earlier, injections of PCB extracts using such columns produce chromatograms showing a maximum of about 20 peaks but often only 6 or 7 peaks. The resolution of individual peaks can be expressed in terms of the number of equivalent theoretical plates calculated from the width and elution time of a specified peak, the higher number denoting greater separation efficiency. Uthe and Musial[13] compared the results submitted by 23 analysts in an international comparative study, and found that the GLC columns used produced plate counts ranging from 320 to 3600. (Capillary columns achieve separations equivalent to plate counts exceeding 10,000 and often up to 100,000.)

The stationary phases used in packed GLC columns, all stable at the high operating temperatures, have different compound separation characteristics, but with the large number of organochlorine compounds likely to be present in any environmental sample it is inevitable that several compounds will elute at the same times as others. However, for the determination of total PCB little difference in the final measurement is evident from the use of the various stationary phases available. The degree of resolution achieved depends primarily on the care taken in preparing and coating the solid support used for packing the columns, rather than on the selection of the stationary phase.

In recent years capillary column chromatography has been developed, using glass or silica columns 25 to 50 m in length with a bore of 0.2 to 0.3 mm. The column temperature is usually programmed to improve the separation of the various organochlorine compounds, particularly those which are more volatile. With such columns it is possible to separate over 100 different peaks from an injection of a mixture of commercial PCB products.[13] Approximately half of the peaks are quantifiable by peak height, but electronic area integration would probably enable most peaks to be measured.

Although the separation of individual PCB compounds is much more effective with capillary columns than with packed columns, some overlapping or coincidence of peaks will still occur, irrespective of the particular stationary phase chosen. If other organochlorine compounds are present, and a few pesticide residues may be eluted from adsorption columns with PCBs despite the use of different solvents, the possibility of confusion between peaks emerging from the capillary column at the same time is increased.[14]

The identification of individual PCB compounds on capillary GC chromatograms has been studied by several workers,[4,14-16] using mass spectrometric and NMR spectroscopic techniques to establish the molecular structure of individual PCB components, and confirming the identity of some by comparison with the injection of known synthetically prepared compounds.

Ballschmiter and Zell[4] state that retention indexes, which they measured using a series of *n*-alkyltrichloroacetates as reference homologs, depend strongly on the separation characteristics of the individual glass capillary column used, involving both the type of stationary phase and the nature of the glass surface on which it is coated. No standard PCB retention indexes can be defined at present, but the matching of peaks with those from reference compounds enables other unknown peaks to be identified through retention indexes on a particular column. Examples of capillary chromatograms are shown in Figures 1 and 2.

The 209 possible PCB compounds are systematically numbered,[4] and their positions calculated in the form of retention indexes. The precision of these calculations, which serve to identify individual compounds not available as reference compounds, is high. Ballschmiter and Zell[4] measured the retention indexes of 43 available compounds, which agreed well with those measured as PCB components of three commercial (Clophen®) products. A further 66 retention indexes were calculated, and were in good agreement with those measured from other components of the Clophen® mixtures.

Chromatograms obtained by Ballschmiter and Zell[4] on capillary columns coated with Apiezon L show a number of examples of coincidence (or possible coincidence) of individual PCB components despite the generally good separation of peaks (see also examples in Figures 1 and 2). For Aroclor® 1254, ten peaks combining two different components and one

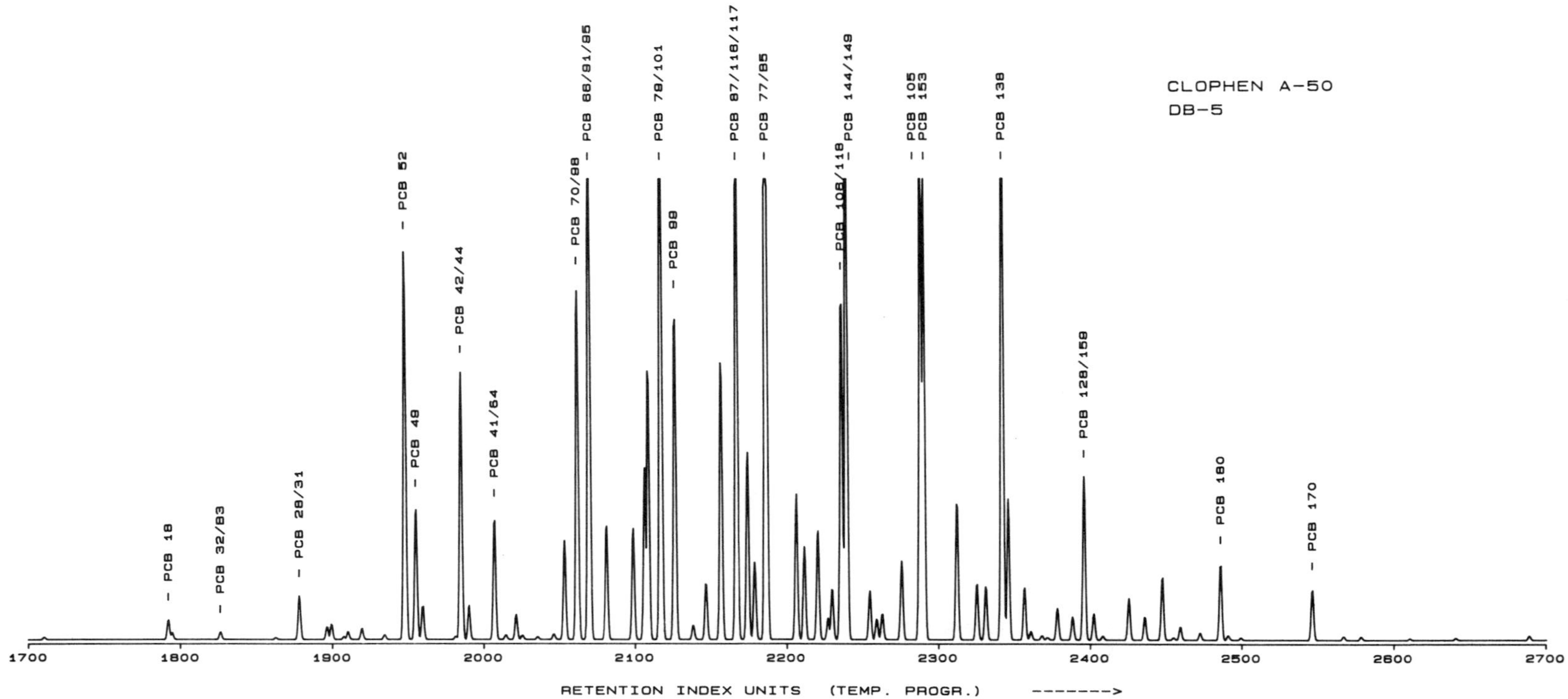

FIGURE 1. Capillary column chromatogram of Clophen® A-50. Instrument: Hewlett-Packard 5880A. Column: 60 m × 0.33 mm bore, fused silica. Stationary phase: DB-5, polymethyl-5% phenyl-siloxane; temperature programmed. (Courtesy of Prof. K. Ballschmiter.)

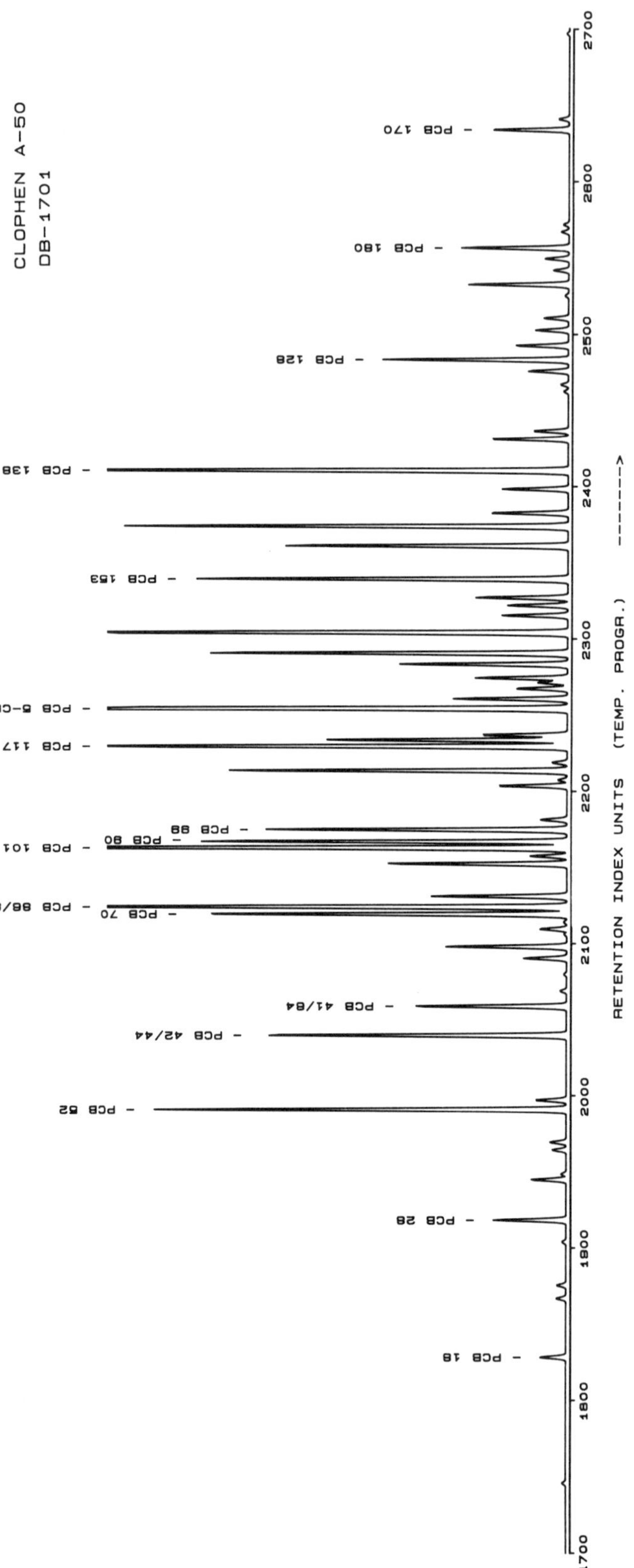

FIGURE 2. Capillary column chromatogram of Clophen® A-50. Instrument: Hewlett-Packard 5880A. Column: 30 m × 0.25 mm bore, fused silica. Stationary phase: DB-1701, 7% phenyl, 7% cyanopropyl, 86% methyl silicones; temperature programmed. (Courtesy of Prof. K. Ballschmiter.)

combining four components occur. In the latter case two are pentachloro- and two hexachlorobiphenyls, the grouping also appearing in the chromatographs of Aroclor® 1242 and 1260. Employing a different stationary phase (SE 30), other examples of coincidence of different components in the Clophen® series of products were found by the same authors. It is thus evident that even the improved techniques of capillary chromatography are not yet able to achieve complete separation of all PCB components, and thus full quantification of PCB mixtures.

There is ample evidence that the pattern of PCB peaks isolated from environmental samples usually differs in detail from those of commercial mixtures, although a rough matching with the 54 to 60% chlorinated types is sometimes possible. For approximate quantification of total PCB this level of agreement may be sufficient, but if the question of toxicity is important it will become necessary to know both the concentrations of individual PCB components and their relative toxicities. At the present time too few have been synthesized, and the quantities available are too small for sufficient research on toxicity to be undertaken. Yet, with the recognition that the proportions of PCB components in environmental samples differ in detail from those found in commercial mixtures, it is unwise to assume that the toxicity of any mixture of residues found in the environment would be similar to that of a concentration of a commercial product calculated on a total PCB basis. More information is required on the mechanism of toxic action of PCBs, perhaps leading to an indication of the molecular structures most likely to be toxic. This in turn would enable analysts to confine their measurements to the compounds of greatest interest. (The toxicity of impurities in commercial products, particularly the chlorinated dibenzofurans, must also be considered.[17,18])

IV. ANALYTICAL INTERCOMPARISON STUDIES

In the foregoing section the ultimate need for the determination of individual PCB compounds has been stressed. At present, however, too few laboratories are equipped with the essential instrumentation, and our knowledge of the techniques of separation and identification is inadequate. For the past 15 years laboratories in different countries have been developing their expertise in the general field of organochlorine analysis, primarily for the ubiquitous types of organochlorine pesticide residues, and total PCB determinations have been attempted as part of such analyses. The degree of variability experienced among analysts in respect of total PCB determinations is, not surprisingly, rather greater than can be achieved for certain easily separated and determinable pesticides using the same analytical procedure.

Holden[19] reviewed the results of a number of international organochlorine intercomparison exercises conducted on behalf of OECD (Organisation for Economic Co-operation and Development) up to 1972. In these studies the analysts were asked to use their customary analytical procedures, and their own reference standards, no attempt being made to specify any common procedure. In 1969 a solution of a commercial PCB mixture in hexane, sealed in glass ampoules, was distributed and 14 laboratories reported their analytical results. A 60% chlorinated mixture (similar to that in the sample) was used by 12 laboratories as reference, obtaining a mean value of 10.17 mg total PCB per liter, as compared with the true value of 9.8 mg/ℓ. Two laboratories using a reference identical to that of the sample obtained values of 10.0 mg/ℓ, and the coefficient of variation for the 16 laboratories was ±10.2%. Two other laboratories employing a 50% chlorinated reference mixture reported a concentration of 8.0 mg/ℓ, an underestimate of only about 20%.

When an environmental sample of cormorant tissue heavily contaminated by PCBs was analyzed by the laboratories, the reference solutions chosen included 50, 54, and 60% chlorinated mixtures. Fourteen laboratories reported concentrations of PCBs between 240 and 525 mg/kg, but eight basing their calculations on a 60% chlorinated reference found 279 to 462 mg/kg. At this early period of PCB analysis, and for such high concentrations, the agreement was reasonably good.

In 1972, a sample of fish oil spiked with organochlorines including PCBs was distributed, and on this occasion 14 laboratories reported concentrations averaging 9.23 mg/kg, with a coefficient of variation of ±12.0%. Unlike the solution in hexane, both the fish oil and earlier cormorant tissue required a full clean-up and separation procedure. The methods of measuring PCB concentrations were varied, using peak heights or peak areas of from one to nine peaks.

One criticism made of the exercise using spiked fish oil was that the concentrations present were unrealistically high by comparison with those the analysts were accustomed to expect from environmental samples. (This situation arose because the spike levels were intentionally an order of magnitude greater than the residue levels in the matrix oil, to ensure that the difference from the matrix could be estimated with reasonable accuracy). In 1974 a vegetable oil virtually free of organochlorine contamination was used as a matrix and spiked with lower concentrations of organochlorines. This sample was distributed both among OECD laboratories and to several laboratories representing countries belonging to the International Council for the Exploration of the Sea (ICES).

The results of the PCB determinations by 24 analysts in the OECD exercise[20] gave a mean concentration of 1.100 mg/kg, with a mean percentage recovery of 97.3% and a coefficient of variation of 15.4%. (Individual pesticide residues were also determined with coefficients of variation in the range 10.3 to 28.1%.) The ICES analysts,[21] reporting from eight laboratories, found a mean concentration of 1.100 mg/kg with a mean percentage recovery of 96.3% and a coefficient of variation of 9.0%. (The range of coefficients of variation for other organochlorine residues in this smaller group of analysts was 6.5 to 40.6%.)

The agreement between analysts was considered to be lower than desirable, when organochlorine residue determinations are made in the interests of acceptability of the source material for human consumption. A coefficient of variation of ±40% implies that the results from 19 out of 20 analysts may span a range covering an order of magnitude. Even a coefficient of variation of ±20% could lead to substantial disagreement as to whether a commodity was unfit for consumption. Further intercomparison exercises have therefore been considered essential to examine whether improved analytical techniques and greater experience would increase the level of agreement among analysts.

In 1978, samples of a fish oil, considered to contain a relatively low level of organochlorine contamination and not spiked with additional residues, were distributed to 43 laboratories in 18 ICES countries. Many analysts had difficulty in conducting the analysis of this sample, and over a period of 10 months 30 laboratories reported,[22] 2 of them attempting the analysis by each of 2 methods. A total of 28 values were reported for the total PCB content, but it was found that those analysts using a sulfuric acid clean-up technique obtained results significantly different from those using other methods. The 13 values from sulfuric acid treatment averaged 863.15 μg/kg, with a coefficient of variation of ±50.0%, while the 15 values obtained by other methods averaged 451.40 μg/kg, with a coefficient of variation of ±45.3%. The coefficients of variation for seven pesticidal residues reported ranged from 32.7 to 71.1%. This exercise demonstrated the difficulties presented by the analysis of an environmental sample containing a relatively low level of contamination.

Six laboratories had used capillary GC columns in the 1978 exercise, but one could not quantify the PCBs, and the other five reported results averaging 472.75 μg/kg with a coefficient of variation of ±42.6%. Nevertheless, many analysts considered that progress would only be achieved in the direction of capillary GC analysis. In 1980, a further ICES exercise was conducted, again using a marine fish oil, in both spiked and unspiked form.[13] For the unspiked oil, 23 results (excluding one outlier) averaged 1.07 mg/kg of PCB, with a coefficient of variation of ±31%. The spiked oil results averaged 1.93 mg/kg with a coefficient of variation of ±21%. Uthe and Musial,[13] reporting on this exercise, were not able to confirm that sulfuric acid treatment gave higher values for PCB as only two analysts used the method, but they identified a difference in the means of the PCBs reported by laboratories using Florisil for sample clean-up as compared with alumina.

This exercise also established that neither the use of a common PCB standard solution nor a common method of calculation of the PCB concentration gave any significant improvement in agreement among analysts. The determination of PCB concentrations was influenced by the stationary phase of the GLC column used, and the use of peak heights rather than peak areas on the same chromatogram also gave somewhat different results.

One further intercomparison exercise conducted among ICES analysts was confined to those using capillary GC.[23] Five laboratories analyzed the spiked and unspiked herring oil used in the 1980 ICES exercise, and reported the individual PCB isomers, identified by their IUPAC numbers, found in both samples and in Aroclor® 1254. A total of 35 different PCB compounds was reported, but individual laboratories reported different compounds, between 15 and 22 compounds being listed by each analyst. Of the 35 isomers, 32 were found in the Aroclor® sample, 15 to 20 again being reported by any one laboratory. The total amount of PCB isomers reported in the Aroclor® sample represented 52.2 to 88.1% by weight of the total Aroclor® present.

Only six isomers were reported by all five analysts, three of them coeluting with other isomers. Four analysts reported on a further four isomers. Uthe et al.[23] concluded that some analysts (or suppliers of reference isomers) wrongly identified some isomers, and that coelution of isomers on capillary columns was a source of error. It is possible that some isomers available commercially may not be pure, and the quantities available are often insufficient for the preparation of accurate reference standards.

Other intercomparison exercises conducted by the International Atomic Energy Agency (IAEA) on oyster and sea plant tissue,[24] and by the Intergovernmental Oceanographic Commission[25] on sea water have confirmed the current difficulty in approaching a satisfactory level of agreement among analysts undertaking organochlorine analysis including PCBs in environmental materials. For the reasons stated earlier, it will be necessary to move in the direction of more accurate techniques using capillary GLC, but it is essential that sufficient quantities of several of the more common isomers in pure form are made available at a reasonable cost for both standardization and toxicity testing.

V. FUTURE DEVELOPMENTS

The accuracy of the determination of individual organochlorine compounds, as in the case of any trace contaminant, decreases with the level of contamination. Horwitz et al.[26] showed that the interlaboratory coefficient of variation increases exponentially with decreasing concentration, and suggested that at the microgram per kilogram level of contamination it might be at least ±50%. Before analysts are asked to determine very low concentrations of PCBs in any samples, and attempt to achieve levels of precision and accuracy which may well be unattainable, those requiring the information should consider seriously whether less stringent demands would be sufficient. If toxicological data indicate that concentrations of total PCB below 5 mg/kg in material for human consumption are acceptable, screening at a 1 mg/kg limit of detection should be adequate, and the added difficulties presented by attempts to determine concentrations down to 0.1 mg/kg can be avoided.

However, if future research indicates that certain individual PCBs, perhaps minor components in commercial mixtures, are particularly toxic to life in some form, these compounds will require determination at significantly lower concentrations than are calculated for total PCB. Both the separation and measurement techniques necessary to achieve this will demand greater care in operation, with an associated increase in the cost of analysis.

It will also be necessary, for the better understanding of the processes which determine the movement of PCBs within our environment and the ultimate fate of individual components, to measure very low concentrations of some isomers (or possibly their degradation products) in air, water, soil, sediments, and biological material. It is not always essential that a high level of interlaboratory agreement is achieved, but within the individual laboratory undertaking such research, regular monitoring of the analytical precision and accuracy is necessary to ensure that the data produced are always comparable, particularly when the analysts change or the techniques are modified. Suggestions for the improvement of routine analytical techniques for organochlorines including PCBs have been made by Holden et al.[27] and are repeated in the following paragraphs.

Regular testing by intercalibration or intercomparison is highly desirable to ensure that the quality of analytical data is maintained. This should involve the analysis of appropriate matrixes containing PCBs, e.g., soil or biological samples, or aliquots of suitable absorbents such as resins which have been used to extract contaminants from air or water. For biological materials vegetable or fish oils provide a suitable matrix, particularly if spiking with known amounts of selected compounds is required, as effective mixing can be achieved using oils.

A number of sources of potential error in organochlorine analysis are possible including impure solvents, inadequate clean-up of sample extracts to remove lipids pigments and other interfering material, lack of separation of PCB residues from other organochlorine compounds, poor resolution of peaks or sensitivity in GLC analysis, failure to concentrate extracts accurately or sufficiently, and variability in the volume of extract injected into the chromatograph.

Improvements in techniques used for routine analysis which will help to eliminate or at least reduce the degree of error include:

1. The use of solvents of the highest quality. No significant GLC peaks which could interfere with the measurement or identification of PCB compounds should be present after 100-fold concentration of the solvent.
2. Extraction of biological tissues (suitably ground and in dried form) should be carried out in Soxhlet extractors for a minimum of 40 c.
3. Extracts should be cleaned up with absorbent columns, using Florisil or alumina.
4. Organochlorine residues in extracts must be subdivided into at least two fractions, PCB residues being separated from other compounds as completely as possible.
5. The PCB fraction (or fractions) must be concentrated to not less than 1 mℓ accurately measured (unless techniques are developed for the reduction to smaller volumes precisely measured).
6. Internal standards (e.g., the series of *n*-alkyltrichloroacetates) should be used to correct for errors at the GLC injection stage.
7. For the determination of individual PCB compounds, glass capillary columns with a resolving power equivalent to at least 50,000 theoretical plates should be used.
8. The sensitivity of the GLC detector should be sufficient to give a peak height response of five times the recorder baseline width for 10 pg of dieldrin or *p,p'*-DDE, or the equivalent area response for automatic integration.
9. The detector response should be within the linear range with frequent calibration.
10. Standard reference solutions should be renewed regularly, and solvent evaporation from these solutions minimized by refrigeration.

It is essential that each individual laboratory carries out regular calibration checks on the analytical procedure used, including the clean-up and pre-GLC separation stages. The international standardization of techniques is an unlikely prospect, but differences between laboratories in respect of performance, which currently give rise to coefficients of variation in PCB intercomparison studies of 30 to 50%, should be reduced significantly by better control in individual laboratories. The wider use of capillary columns, and the increased availability of individual PCB compounds as reference material, together with automated integration and data recording, are essential for the complete analysis of the spectrum of PCB residues in a sample. The conditions under which the analytical procedures are conducted, whether at the clean-up and separation stages before GLC analysis, at the GLC injection stage, or in temperature-programmed GLC analysis, must all be clearly specified and rigidly adhered to, if quantification at acceptable levels of accuracy and precision is to be achieved.

REFERENCES

1. **Jensen, S.,** The PCB story, *Ambio,* 1, 123, 1972.
2. **Lovelock, J. E. and Lipsky, S. R.,** Electron affinity spectroscopy — a new method for the identification of functional groups in chemical compounds separated by gas chromatography, *J. Am. Chem. Soc.,* 82, 431, 1960.
3. **Webb, R. G. and McCall, A. C.,** Quantitative PCB standards for electron capture gas chromatography, *J. Chromatogr. Sci.,* 11, 366, 1973.
4. **Ballschmiter, K. and Zell, M.,** Analysis of polychlorinated biphenyls (PCB) by glass capillary gas chromatography, *Fresenius' Z. Anal. Chem.,* 302, 20, 1980.
5. **de Faubert Maunder, M. J., Egan, H., Godley, E. W., Hammond, E. W., Roburn, J., and Thomson, J.,** Clean-up of animal fats and dairy products for the analysis of chlorinated pesticide residues, *Analyst,* 89, 168, 1964.
6. **Jones, L. R. and Riddick, J. A.,** Separation of organic insecticides from plant and animal tissues, *Anal. Chem.,* 24, 569, 1952.
7. **Jensen, S., Johnels, A. G., Olsson, M., and Otterlind, G.,** DDT and PCB in herring and cod from the Baltic, the Kattegat and the Skagerrak, *Ambio Special Report,* No. 1, 71, 1972.
8. Pesticide Analytical Manual, Vol. 2, U.S. Department of Health, Education and Welfare, Food and Drug Administration, Washington, D.C., 1975.
9. **Holden, A. V. and Marsden, K.,** Single-stage clean-up of animal tissue extracts for organochlorine residue analysis, *J. Chromatogr.,* 44, 481, 1969.
10. **Wells, D. E. and Johnstone, S. J.,** Method for the separation of organochlorine residues before gas-liquid chromatographic analysis, *J. Chromatogr.,* 140, 17, 1977.
11. **Tindle, R. C. and Stalling, D. L.,** Apparatus for automated gel permeation clean-up for pesticide residue analysis, *Anal. Chem.,* 44, 1768, 1972.
12. **Kuehl, D. W. and Leonard, E. N.,** Isolation of xenobiotic chemicals from tissue samples by gel permeation chromatography, *Anal. Chem.,* 50, 182, 1978.
13. **Uthe, J. F. and Musial, C. J.,** Report on the fourth ICES intercalibration study of PCBs in biological material, Cooperative Research Report No. 115, International Council for the Exploration of the Sea, Copenhagen, 1982.
14. **Zell, M. and Ballschmiter, K.,** Baseline studies of the global pollution. III. Trace analysis of polychlorinated biphenyls (PCB) by ECD glass capillary gas chromatography in environmental samples of different trophic levels, *Fresenius' Z. Anal. Chem.,* 304, 337, 1980.
15. **Sissons, D. and Welti, D.,** Structural identification of polychlorinated biphenyls in commercial mixtures by gas-liquid chromatography, nuclear magnetic resonance and mass spectrometry, *J. Chromatogr.,* 60, 15, 1971.
16. **Zell, M., Neu, H. J., and Ballschmiter, K.,** Single component analysis of polychlorinated biphenyl (PCB) — and chlorinated pesticide residues in marine fish samples, *Fresenius' Z. Anal. Chem.,* 292, 97, 1978.
17. **Vos, J. G. and Beems, R. B.,** Dermal toxicity studies of technical polychlorinated biphenyls and fractions thereof in rabbits, *Toxicol. Appl. Pharmacol.,* 19, 617, 1971.
18. **Bowes, G. W., Mulvihill, M. J., Simoneit, B. R. T., Burlingame, A. L., and Risebrough, R. W.,** Identification of chlorinated dibenzofurans in American polychlorinated biphenyls, *Nature (London),* 256, 305, 1975.
19. **Holden, A. V.,** The OECD international co-operative studies of organochlorine residues in wildlife, in *Environmental Quality and Safety,* Vol. 3, Coulston, F. and Korte, F., Eds., Georg Thieme, Stuttgart, 1975, 40.
20. Chemical Trends in Wildlife, An International Cooperative Study, Organisation for Economic Co-operation and Development, Paris, 1980, 26.

21. **Topping, G. and Holden, A. V.,** Report on intercalibration analyses in ICES North Sea and North Atlantic baseline studies. Co-operative Research Report No. 80, International Council for the Exploration of the Sea, Copenhagen, 1978.
22. **Holden, A. V. and Topping, G.,** Report on further intercalibration analysis in ICES pollution monitoring and baseline studies. Co-operative Research Report No. 108, International Council for the Exploration of the Sea, Copenhagen, 1981.
23. **Uthe, J. F., Musial, C. J., and Palmork, K. H.,** A report on an intercomparison study of the determination of chlorinated biphenyl (PCB) isomers in fish oil, Paper C.M. 1981/E:49, International Council for the Exploration of the Sea, Copenhagen, 1981.
24. **Anon.,** Activities of the International Laboratory of Marine Radioactivity, Report No. IAEA-187, International Atomic Energy Agency, Vienna, 1976.
25. **Andersen, N., Bewers, J. M., Duinker, J., Knap, A., Palmork, J., and Villeneuve, J.-P.,** Scientific report of the intercalibration exercise. The IOC/WMO/UNEP pilot project on monitoring background levels of selected pollutants in open ocean waters, Intergovernmental Oceanographic Commission Tech. Ser. 22, UNESCO, Paris, 1982.
26. **Horwitz, W., Kamps, L. R., and Bayer, K. W.,** Quality assurance in the analysis of foods for trace constituents, *J. Assoc. Off. Anal. Chem.*, 63, 1344, 1980.
27. **Holden, A. V., Topping, G., and Uthe, J. F.,** The use and relevance of analytical intercomparison exercises in monitoring the marine environment, *Can. J. Fish. Aquat. Sci.*, 1985, in press.

Chapter 4

ATMOSPHERIC TRANSPORT OF PCBs TO THE OCEANS

E. Atlas, T. Bidleman, and C. S. Giam

TABLE OF CONTENTS

I. INTRODUCTION

The atmosphere is increasingly being recognized as a significant pathway for the movement of chemicals in the environment.[1,2] Once airborne, stable chemicals such as polychlorinated biphenyls (PCBs) can be transported thousands of kilometers from their original sources, contaminating even remote regions of the earth. For example, PCB and other organochlorines have been detected in air from Enewetak Atoll[3] and Antarctica,[4] marine mammals from Antarctica,[5] and surface water from the major oceans and seas.[6-9] This chapter will discuss several aspects of aerial PCB transport to the oceans. First we will review sampling procedures for PCB in ambient air and atmospheric deposition, and will summarize some recent measurements of airborne PCB around the globe. This will provide data to estimate the atmospheric reservoir of PCB. Then we will examine rates and mechanisms of air-to-sea PCB transfer.

II. PCBs IN THE ATMOSPHERE

A. Collection Methods for PCBs in the Air

PCBs are found in the atmosphere as vapors or associated with particulate matter. As will be discussed later, the gaseous form predominates, typically comprising >90% of the total PCB concentration. PCB levels in ambient air are on the order of a few picograms per cubic meter to nanograms per cubic meter, and hundreds or even thousands of cubic meters of air must be sampled to obtain enough material for analysis. With a high volume (hi-vol) sampling system, particulate PCBs are collected on a precombusted glass- or quartz-fiber filter and an adsorbent column behind the filter is used to retain gaseous PCBs.

Within the last decade solid adsorbents have become popular for collecting organic vapors in ambient and workplace air. References to some of these applications for sampling PCB vapors in a variety of situations are given in Table 1. The list emphasizes the literature since 1977, and is not intended to be comprehensive but illustrative of the widespread usage of these sampling systems. Adsorbents are also useful for collecting many other types of high molecular weight organic vapors, including pesticides, phthalate esters, and polycyclic aromatic hydrocarbons (PAH). References to these applications and to earlier work can be found in the Table 1 citations and in several reviews on organic vapor sampling.[37-40]

An important consideration for an adsorbent is its blank value. Polymeric adsorbents are usually precleaned by Soxhlet extraction with chromatographic quality solvents followed by drying under vacuum or in organic-free air. Florisil has the advantage of being cleaned up easily by baking in a muffle furnace at 320°C.[32,35] With care it is possible to reduce electron-capturing impurities to a few nanograms, quantified as PCB. Some reported PCB blanks for different adsorbents are listed in Table 2. These generally range from <10 to 30 ng total PCBs for the polymeric materials and less for Florisil. A blank of 20 ng total PCBs corresponds to a detection limit of 0.03 to 0.04 ng/m for a 500 to 700 m^3 sample, a typical volume for a 24-hr air monitoring period. This is about two orders of magnitude below urban air PCB concentrations. In remote and oceanic regions, air volumes can be increased to raise the sample/blank ratio if collection efficiency can be maintained.

The ability of an adsorbent bed to quantitatively collect vapors can be determined in several ways. One approach to measuring collection efficiency is to vaporize a known quantity of chemical into an airstream which is passed through an adsorbent bed. The fraction collected is calculated by mass balance from the residues found in the sampling train. Sometimes retention efficiencies are measured instead by spiking the adsorbent with a known quantity of chemical and determining the percent retained after clean air has been pulled through the trap. Collection efficiencies in field work are often estimated by separately analyzing front and backup adsorbent traps. If the second trap contains much less material than the first

Table 1
SOLID ADSORBENTS USED FOR PCB VAPOR SAMPLING

Adsorbent	Air volume range (m^3)	Application	Ref.
Polyurethane foam	100—1600	Collection studies	10—14
	300—1600	Urban air	15—17
	400—4000	Ocean air	3, 4, 18—21
	8—16	Collection study	22
	1	Collection study	23, 24
	1	Home and workplace air	24, 25
XAD-2	70—3000	Collection studies	22, 26
	500—1800	Urban air	16, 27
	400—1100	Lake Michigan air	27
	200—400	Lake Superior air	28—30
	6—18	Collection study	22
	0.005—0.01	Collection study	31
Tenax-GC	300—1600	Urban air	15, 16
Florisil	8—20	Collection studies	22, 32
		Incinerator/capacitor mfg.	33
	10—15	Urban air	34
	75—150	Ocean air	3
	400—1500	Ocean air	35
Chromasorb® 102		Urban air, stack gases	36
Mixed adsorbent cartridges	300	Collection study	11

Table 2
PCB BLANK VALUES (ng) FOR SOLID ADSORBENTS

Adsorbent	Aroclor® 1016/1242	Aroclor® 1254	Total PCB	Cleanup[a]	Remarks	Ref.
PUF	9	5	14	A, H_2SO_4	Immediately after cleanup	16
	19	10	29	A, H_2SO_4	3—35 mo storage	16
		6		F		19
			<32	None		10
			<10	None		18
			≤2	H_2SO_4		4
			<3[b]	HPLC, S		20
XAD-2	8	7	15	A, H_2SO_4	Immediately after cleanup	16
	11	14	25	A, H_2SO_4	8—11 mo storage	16
Tenax-GC	15	9	24	A, H_2SO_4	Immediately after cleanup	16
	15	14	29	A, H_2SO_4	8—11 mo storage	16
Florisil			0.1	F		32
			<5	F		34

[a] Cleanup: A = alumina, F = Florisil, S = silica, H_2SO_4 = concentrated or 7% fuming sulfuric acid.
[b] Hexa- and heptachlorobiphenyls.

trap, it is assumed that breakthrough beyond the backup trap has not occurred. However, as the quantities on the two traps become similar, it is apparent that a certain fraction of the sample has also passed through the entire adsorbent bed. This uncollected fraction cannot be calculated simply from the trap-1/trap-2 ratio; a more detailed knowledge of the vapor front movement is needed, as will be discussed later.

Most workers include some estimate of collection efficiency in their reports. Sampling conditions and adsorbent quantities differ greatly among studies, so while quantitative vapor trapping is usually documented for a particular investigation, it is difficult to generalize

Table 3
BREAKTHROUGH VOLUMES (V_B) FOR PCB CONGENERS AND HEXACHLOROBENZENE (HCB) ON PUF AND SAMPLING VOLUMES (V_S) CORRESPONDING TO DESIGNATED COLLECTION EFFICIENCIES[a]

		V_S, m³ at collection efficiency	
Compound	**V_B, m³**	**90%**	**95%**
HCB	125	103	78
3,3′-DCB	720	590	446
2,4′, 5-TCB	1440	1180	893

[a] PUF column 7.6 cm diameter × 7.5 cm thick (N = 7.5), 20°C.

collection efficiencies from this literature. Several collection studies are listed in Table 1. Among these, solute volatility stands out as a major factor influencing collection. Several investigators have shown that the ability of adsorbents to trap PCB vapors increases with the number of chlorines on a homolog. Under the conditions usually employed for hi-vol sampling, the later-eluting components of Aroclor® 1242 and all the Aroclor® 1254 components are quantitatively trapped by polyurethane foam (PUF). Some of the dichlorobiphenyls may be lost, depending on the sampling temperature and total air volume. Tenax-GC, XAD-2, and Florisil collect the lower molecular weight PCBs more effectively than PUF.

Recently, some work has been done to express PCB vapor collection in a way that has some predictive capability for designing sampling systems. Senum[41] pointed out that the collection efficiency of an adsorbent bed is a function of the solute retention volume (V_R) and the number of theoretical plates (N) in the bed. The first measurements of V_R for PCB congeners in a hi-vol system were reported by Simon and Bidleman.[12] A PCB spike was introduced to a column of 15 1-cm thick × 7.6-cm diameter PUF slices and eluted with clean air. The PCB moved through the column as chromatographic bands, with the band penetration linearly related to air volume. Subsequently Burdick and Bidleman[13] extended this work to a frontal chromatography system, where vapors were continuously introduced to the PUF bed while sampling at 20°C. Breakthrough volumes (V_B) were obtained for two PCB congeners and hexachlorobenzene (HCB). In a separate study[42] V_S for four PAH as well as values of N for a PUF hi-vol sampler were determined.

Knowing V_B for the PCB homologs and N for the adsorbent bed, V_S can be calculated for desired collection efficiencies.[41,42] This is seen in Table 3 for a PUF column 7.6-cm diameter × 7.5-cm thick (7.6 g), the approximate dimensions of a PUF plug used in several field investigations.[9,15,16,18,19] These V_S apply to 20°C sampling temperature only. PCB vapor pressures increase by a factor of 1.8 to 1.9 for a 5°C rise in temperature,[43] and sampling at 25°C rather than 20°C should have nearly the same effect on breakthrough as doubling the air volume. In field studies, breakthrough of Aroclor® 1016 from front to backup PUF traps was better correlated with a temperature-weighted air volume than with air volume alone.[16]

In practice, the choice of adsorbent used for PCB sampling depends on the application, concentrations to be determined, necessary sampling times, and available equipment. Simultaneous collection of PCBs using two or three sampling systems has been carried out to test the comparability of different adsorbents.[15,16] Side-by-side collections in three cities using PUF, Tenax-GC, and XAD-2 showed that concentrations of Aroclor® 1016, Aroclor® 1254, and the pesticides chlordane and toxaphene measured with the three adsorbents agreed excellently. Average relative standard deviations for organochlorine levels measured with the different adsorbents ranged from 10 to 15% in most cases, and were within the precision of the analytical method. HCB concentrations measured with Tenax-GC and XAD-2 agreed well, and were several times higher than those obtained using PUF. HCB was not quantitatively collected by PUF in the air volume range studied (300 to 1600 m^3), since its V_B on PUF is only 125 m^3 (Table 3).

B. PCB Concentrations in the Air

Concentrations of airborne PCB depend on numerous factors, including proximity to local sources, source emission strengths, and meteorological variables. Levels as high as 57 μg/m^3 have been reported in air directly over a PCB spill in North Carolina.[44] Air in homes and offices typically contains several hundred ng/m^3 PCB.[24,32,44] By comparison, the NIOSH proposed criterion for PCB in workplace air over an 8 to 10 hr workday is 1.0 μg/m^3.[45] Other areas where elevated PCB levels have been measured are landfills,[14-16,46,47] capacitor and transformer manufacturers,[33,44,46,48] and incinerators.[33] Away from direct sources, PCB concentrations are highest near urban centers, 1 to 10 ng/m^3 representing the usual range for cities in the U.S. (Table 4). As expected, PCB levels decrease away from cities. Values from nonurban continental areas are typically in the range of 0.1 to 0.5 ng/m^3. In remote marine areas concentrations of Aroclor® 1254 are usually 0.01 to 0.02 ng/m^3 (Table 4).

C. The Atmospheric Reservoir of PCBs

Computation of the total amount of PCBs in the atmosphere is complicated by several factors. Relatively few data for atmospheric PCBs are available outside the U.S., and even less data have been collected in the Southern Hemisphere. In addition, the vertical distribution of PCBs in the atmosphere has not been measured. Finally, since the ban on open use of PCBs, there is some evidence for a decrease in environmental PCB concentrations,[52] though insufficient data are available to evaluate any decline in atmospheric PCB concentrations. Therefore, some simplifying assumptions and extrapolations are required to obtain a reasonable estimate of the total atmospheric reservoir of PCB.

Most data on atmospheric concentrations of PCB have been presented in terms of the commercial Aroclor® mixtures 1016/1242 and/or 1254, rather than as single congeners. Generally, in atmospheric samples, especially at low PCB concentrations, very light PCBs (monochloro) and very heavy (hepta- through decachloro-) PCBs are not included in the calculation of total atmospheric PCB. By including PCBs only as 1242 and 1254 in our calculation we may slightly underestimate the total PCB reservoir, but given the approximate nature of the calculation this should not lead to serious error.

Table 4 illustrates the high degree of variability in the reported PCB concentrations between urban, rural, and remote marine locations. Furthermore, examination of the original data sets reveals concentration variations which depend on meteorological factors and proximity to sources. Thus, calculation of the atmospheric PCB reservoir can be only approximate, perhaps within a factor of two or three. Also, there is variability in the apparent composition of atmospheric PCBs. Relative proportions of Aroclor® 1242 to Aroclor® 1254 range from 1:1 to 6:1. For the purposes of these subsequent calculations, we will use the proportions of 2:1 to represent the relative amounts of Aroclors® 1242 and 1254 in the atmosphere and to calculate total PCB when only concentrations of Aroclor® 1254 are given. Using this approximation to calculate total atmospheric PCBs from data in Table 4, we find that total

Table 4
CONCENTRATION OF PCBs IN THE AMBIENT ATMOSPHERE

	Av. conc. (ng/m^3) as Aroclor®			
	Total	1242	1254	Ref.
Urban				
Denver, Colo. (1980)	2.25	1.80	0.45	16
Columbia, S.C. (1977—80)	4.70	3.20	1.50	15, 16
Houston, Tex.	—	—	3.00	49
Minneapolis, Minn. (1978—79)	7.10	5.60	1.56	28, 29
Chicago, Ill. (1975—76)	8.00	6.70	1.28	17
Madison, Wisc. (1978)	7.50	6.50	1.05	27
Milwaukee, Wisc. (1977)	2.25	1.64	0.61	27
Ontario, several cities				
1979	5.90	—	—	34
1980	0.21	—	—	34
Stockholm, Sweden (1983—84)	0.34	0.23	0.11	50
Jacksonville, Fla. (1975)	4.70	—	—	14
Gainesville, Fla. (1977)	20.00	—	—	14
Rural/suburban/coastal				
Pigeon Key, Fla. (1978)	—	—	0.41	49
North Inlet Estuary, S.C. (1977—79)	0.44	0.25	0.19	51
Texas Gulf Coast (1976)	—	—	0.52	19
Texas Gulf Coast (1980)	—	—	0.067	35
College Station, Tex. (1979)	—	—	0.29	49
White Sands, N.M. (1981)	—	—	0.11	49
Ontario Province				
1979	5.40	—	—	34
1980	0.19	—	—	34
Rural Sweden (1984)	0.07	0.04	0.03	49
Rural Norway (1982)	—	—	0.02[a]	20, 21
Great Lakes				
Lake Michigan (1977)	0.87	0.65	0.22	27
Lake Superior (1978—80)	1.20	0.86	0.34	28, 29
Marine				
Gulf of Mexico (1976)	—	—	0.16	19
Newfoundland (1977)	—	—	0.096	18
Barbados (1977—78)	—	—	0.065	18
Enewetak Atoll (1979)	0.11	0.061	0.049	3, 49
American Samoa (1981)	—	—	0.012	49
Peruvian Coast (1981)	—	—	0.012	49
Bermuda (1976—77)	—	—	0.075	7
E. Indian Ocean (1980—81)	0.15	—	—	9
W. Pacific	0.34	—	—	9
Antarctic/Southern Ocean	0.11	—	—	4, 9
Arctic Ocean (1982—83)	—	—	0.02[a]	21

Note: —, Not measured or not reported.

[a] Pentachlorobiphenyls.

atmospheric PCBs for urban, rural, and marine areas is 6.0, 0.63, and 0.19 ng/m^3, respectively. An overall weighted average of continental (30%) and marine (70%) air yields a global average PCB concentration of 0.25 ng/m^3.

This average computed from reported data probably overestimates the "true average" concentration for several reasons. Data from rural areas are biased toward higher values because of proximity to industrialized areas of the U.S. and Europe. Concentrations of PCBs in large areas of the Asian, African, and South American continents are potentially lower than 0.6 ng/m^3, and are probably nearer "background" concentrations. Similarly, the data reported in the marine atmosphere are influenced by coastal population centers. From close examination of the data in Table 4, we subjectively estimate concentrations typical of open ocean regions to be ~100 pg/m^3 in the Northern Hemisphere and 30 pg/m^3 in the Southern Hemisphere. The average, 65 pg/m^3, is approximately 1/3 of the marine average reported above. A modified average computed from only 10% "rural" contribution (0.6 ng/m^3) and 90% "marine/remote" (0.065 ng/m^3) contribution is 0.12 ng/m^3, which may be closer to representing the PCB load currently in the global atmosphere. In any event, this more subjective average is within a factor of two of the earlier estimate based on all reported data.

To calculate the atmospheric reservoir of PCB, we will assume a uniform concentration of PCBs in the troposphere up to 6 km over the earth's surface. If PCB vapors behave as other stable halocarbons in the atmosphere, such as the freons, then this is a reasonable assumption. Using an area of 5.1×10^{14} m^2 for the earth's surface and the "high" average of 0.25 ng/m^3, we calculate a total PCB burden in the atmosphere of 7.7×10^5 kg. There is obviously some uncertainty in this calculation; however, the actual PCB burden in the atmosphere should lie in the range of 1 to 10×10^5 kg. This amount represents a small percentage of the total "mobile environmental reservoir" of U.S.-derived PCB estimated by the National Academy of Sciences (820×10^5 kg) in 1979.[7] This result is somewhat surprising considering the important role of atmospheric transport to the widespread distribution of PCBs in remote regions and suggests several features of the atmospheric reservoir of PCB. First, the atmosphere appears to be a relatively rapid conduit for transport of PCBs to remote areas. If the atmosphere slowly accumulated and retained PCBs we would expect a larger fraction of "mobile" PCBs to reside in the atmosphere. Second, we expect that the atmospheric reservoir must have been larger in the past, e.g., prior to 1970, to account for the accumulated PCBs in ocean water and sediments. This aspect is discussed more completely in a later section.

D. Vapor-Particle Partitioning of PCBs

The vapor/particle (V/P) ratio of airborne PCBs is operationally defined by the quantities retained in the adsorbent trap and on the filter. How closely adsorbent/filter (A/F) ratios found with hi-vol samplers approximate the true V/P distribution in the atmosphere is unknown, but is an active area of investigation since the V/P ratio is important in assessing the various mechanisms of deposition. Some A/F ratios for PCBs in various cities and over the ocean are given in Table 5.

Many investigators feel that A/F overestimates V/P because of the "blow-off effect" — stripping of organics from particles on the filter by the flowing airstream.[53-58] On the other hand, it is possible that the particle mass on the filter acts as an adsorbent and scavenges vapors, which leads to an underestimation of V/P.[59] Experiments in Atlas' laboratory suggest that airflow alone has little effect on PCB vapor adsorption or stripping during hi-vol sampling. In the experiments air samples collected at different face velocities showed the same PCB A/F ratios. Changes in temperatures, aerial concentrations, and particle chemistry probably have the greatest effect on A/F, and whether blow-off losses or adsorption gains are observed will depend on how and when these variables change during a collection period.

Table 5
AVERAGE A/F RATIOS FOR ATMOSPHERIC PCBs

Location and PCB	A/F	% Vapor	Ref.
Columbia, S.C. (Aroclor® 1254)	17.00	94	15, 16
Denver, Colo. (Aroclor® 1254)	0.56	36	16
New Bedford, Mass. (Aroclor® 1254)	6.00	86	16
Toronto, Ont.	1.3—6.3	56—86	60
Chicago, Ill. (Total PCB)	33.00	97	17
Milwaukee, Wisc. (Total PCB)	6.30	86	27
Madison, Wisc. (Total PCB)	0.33	97	27
Lake Michigan	8.30	89	27
North Atlantic (Aroclor® 1254)	>10.0[a]	90	3
North Atlantic (Aroclor® 1254)	>99.0[a]	99	61
Gulf of Mexico (Aroclor® 1254)	>99.0[a]	99	9

[a] Filter-retained PCB less than detection limit.

For example, what is deposited on the filter at night may be blown off again in the heat of day.

Despite these problems, some attempts have been made to correlate A/F ratios with temperature and airborne particle concentrations. Yamasaki et al.[62] demonstrated that the A/F ratio for PAH could be described by an equation derived from Langmuir adsorption theory:

$$\log(A/F) = mT^{-1} + b - \log(TSP) \qquad (1)$$

where m and b are empirical constants, TSP is the total suspended particle concentration ($\mu g/m^3$), and T is the average absolute sampling temperature. In an attempt to correlate PCB A/F ratios with temperature, a plot of Equation 1 was made for Aroclor® 1254 in Columbia, S.C., New Bedford, Mass., Denver, Colo., and Stockholm, Sweden.[50] Data from the 4 cities represented A/F ratios from 0.2 to 190 and average temperatures for the sampling periods of −10 to 26°C. For 34 data points, linear regression gave: $m = -4686$, $b = 19.428$, and $r^2 = 0.885$. Using this relationship, the apparent percentage of particulate Aroclor® 1254 in a city with TSP = 60 $\mu g/m^3$ would be estimated as 2.1% at 20°C and 25% at 0°C. Because of the lower TSP levels in the marine atmosphere, even lower particulate PCB fractions over the oceans would be expected. For example, if oceanic TSP is 7 $\mu g/m^3$ (mineral and sea salt),[63] at 0°C, 3.7% of the Aroclor® 1254 would be on particles. (This percentage is expected to be an upper limit because of the difference in particle size distributions and chemical characteristics of urban vs. marine aerosols.) At 20°C this percentage would fall to ~0.3%. Since Aroclor® 1242 has a vapor pressure ~5 times higher than Aroclor® 1254, the particle-bound Aroclor® 1242 should be ~1/5 of the above values. Thus, the low percentage of particulate PCB measured in the marine atmosphere is consistent with theoretical estimates.

III. ATMOSPHERIC REMOVAL PROCESSES FOR PCBs

A. Collection Methods for Aerial Deposition

Wet and dry deposition are two important processes for transferring PCBs from the atmosphere to the ocean surface. Precipitation sampling methods can be classified as "bulk" or "wet", depending on whether they collect dry deposition in addition to rain or snow. "Event" sampling refers to collections made during a single precipitation period. These and other terms associated with precipitation monitoring have been defined.[64]

One advantage of bulk collectors is that they are cheap, often consisting of nothing more than a jug and a funnel. Bulk collectors require no power, an important consideration for sampling at remote sites, and they provide an integrated measure of atmospheric input per unit area. A serious drawback is that these collectors are easily contaminated by bird droppings and insects. Moreover, they do not differentiate between input by precipitation and by dry deposition. An interesting way to measure total aerially derived pollutants is to analyze snow or ice cores. This bulk collection method, subject to the same contamination problems as mentioned above, has been used by several workers.

Wet samples are taken only during precipitation periods by manually exposing a collector, or by using an automated collector with a rain-tripped sensor.[64-67] Sample containers are usually made of glass or metal. Plastics other than polytetrafluoroethylene are not suitable for trace organic work because they may introduce contamination into the sample or cause losses by adsorption to the plastic surface.

Organic compounds deposited into precipitation collectors are subject to revolatilization losses and to chemical and microbial breakdown. Problems with transportation also discourage the collection of large water volumes. Some investigators have tried to circumvent these difficulties by drawing the collected rain sample through a plug of PUF,[68] a glass fiber filter-PUF combination,[17] or an XAD resin column.[66]

Dry deposition of trace organics occurs by gravitational settling, turbulent impaction of aerosol-bound materials, and adsorption of vapors to the deposition surface. Södergren[69,70] and Bengston and Södergren[71] coated a nylon mesh screen with silicone oil and exposed the nets for 2 to 3 months at a time to determine total aerial fluxes per unit area. The nets probably collected particles by filtration as well as impaction. McClure and LaGrange[72,73] collected dry deposition on mineral oil-coated glass plates, a technique that was also used by Young et al.[74,75] to collect DDT and PCB fallout samples in southern California. Heesen et al.[76] exposed sets of these collection plates over several days and found that the collection efficiency decreased over an 8-day period; i.e., residues on plates exposed for 8 days were less than the sum of residues recovered from plates exposed for 1-, 2-, or 4-day intervals over the same time period.

Murphy[77] criticized the use of oil-coated surfaces to sample organic dry deposition because of the possibility that the oil might scavenge vapors as well as particles. Murphy recommended the use of a polar fluid, and has used glass fiber filters coated with diols to collect PCB dry deposition.[78] Eisenreich et al.[28] collected PCB dry deposition on glass plates coated with glycerol. Christensen et al.[79] evaluated three surfaces for sampling PCB, DDT, and chlordane fallout. Pans filled with water or ethylene glycol-water to simulate a natural water surface were compared to glycerol-sprayed and dry pans of the same geometry. On the average, the wet surface pans collected 1.5 to 3 times as much organochlorine dry deposition as did the dry surface pans with no differences among the types of wet surfaces.

How well any of these surrogate surfaces approximate dry deposition to the surface of environmental interest (ocean, lake, plant foliage) is unknown. Removal of particles is strongly influenced by their size, meteorological factors, and surface properties of the collector. Reviews of dry deposition[80,81] reveal an incredible number of field experiments carried out over the last 25 years. Particle deposition velocities, V_d, vary by three orders of magnitude. A discussion of all factors affecting washout and dry deposition is beyond the scope of this chapter and the reader is referred to several articles on the subject.[80-85] For trace organics, deposition processes are further complicated by their fractionation between the particle and vapor phases.

A workshop held at Argonne National Laboratory in November 1979 evaluated methods to measure dry deposition.[86] A conclusion of the workshop was that surrogate surfaces are not accurate enough estimators of fallout fluxes to use in monitoring networks. Instead, research was recommended on using micrometeorological parameters to calculate dry dep-

Table 6
PCBs IN RAIN AND SNOW

Location and year	Collection method	Range (ng/ℓ)	Mean (ng/ℓ)	Ref.
Norway, 1973—76	Bulk rain and snow	1.2—7.6	3.7	89
Chicago and Lake Michigan, 1975—76	Event rain and snow, PUF extraction		193.0 119.0(P)[a]	17
British East Coast, 1975—76	Bulk rain, PUF	1.80—74.0	14.9	68
Great Lakes Eastern Canada, 1975—77	Event rain, snow cores		21—29	90,91
South Carolina coast, 1976—78	Bulk and event rain	<1.0—27.0	1.6	88
Lakes Michigan and Huron, 1977—79	Bulk and event rain, snow and ice cores		14—65(P)	78
Southern Ontario, 1975—77	Bulk rain		100	92
Lake Superior, South, Isle Royal, 1974—76	Snow cores		50—230	93
Ontario cities, 1979	Bulk rain and snow	<20.0—50.0	21—35	94
Pigeon Key, Fla., 1978	Event rain	7.1—8.3	7.7	49
Enewetak Atoll, North Pacific, 1979	Event rain		<0.6	3
Hamilton, Ontario	Rain, wet only	9.0—61.0	27.0	66
Sault St. Marie, Ontario (remote)	Rain, wet only	1.0—8.0	3.0	66
Antarctica, Mt. Olympus Washington, 1975	Snow cores	0.03—1.2	0.3(median)	95
Antarctica, 1981	Snow and ice cores	0.16—1.0	0.4	4

[a] P = precipitation volume-weighted mean. No symbol = arithmetic mean of concentrations, or method of mean computation, not specified.

osition fluxes from atmospheric concentrations. These techniques are summarized in reports.[86,87] Other workers disagree (the present authors included), and feel that more emphasis should be placed on field comparisons of fallout fluxes to surrogate and "real" surfaces.

B. PCB Levels in Precipitation and Dry Deposition

Compared to the number of studies on wet and dry deposition of inorganic substances, relatively little has been done to investigate removal processes for organic compounds. PCB concentrations in rain and snow are presented in Table 6, with an emphasis on recent literature. Higher PCB concentrations are found in rain near urban areas. Murphy and Rzeszutko[17] reported a precipitation-weighted mean PCB concentration of 119 ng/ℓ in Chicago and on Beaver Island, Lake Michigan. By comparison, only 11 out of 51 rain samples taken at a South Carolina estuary showed detectable PCB residues; the mean concentration in these 11 samples was 7.5 ng/ℓ.[88] Strachan and Huneault[66] collected rain 10 km downwind from a heavily industrialized area near Hamilton, Ontario and at a "remote" site 50 km from Sault St. Marie, Ontario. Average PCB concentrations were 27 ng/ℓ near Hamilton and 3 ng/ℓ at the remote site.

The few fluxes and deposition parameters that have been measured for PCBs are summarized in Tables 7 and 8. Table 7 shows that deposition fluxes to the Great Lakes are in the range of 10 to 100 μg/m²/year, while marine deposition may be somewhat lower. Removal parameters tend to be higher for the less volatile compounds. For example, washout ratios (W) and V_d are higher for Aroclor® 1254 than for Aroclors® 1016 and 1242 (Table 8). The same trend has been found for chlordane and the less volatile DDT.[48] A likely explanation is that a greater proportion of the heavier organics are particle-bound and are removed more efficiently by wet and dry processes. This behavior is modeled in the next sections.

Table 7
PRECIPITATION AND DRY DEPOSITION INPUTS OF PCBs TO WATER BODIES

Location and year	Input mode	Area (km^2)	Flux kg (months)	$\mu g/m^2/year$	Ref.
S. California Bight, 1973—1974	Dry deposition	—	1,000 (12)	—	74
Lake Michigan, 1975—1976	Precipitation and dry deposition	5.80×10^4	2,500—5,000 (12)	43.0—86	17,78
Lake Huron, 1977—1978	Precipitation and dry deposition	6.00×10^4	1,100 (12)	18.0	78
Lake Superior, 1974—1977	Precipitation and dry deposition	8.20×10^4	6,600—8,400 (12)	80.0—102	28
Saginaw Bay, Lake Huron, 1977—79	Precipitation and dry deposition	1.50×10^3	6.5—8 (2—3)	17.0—23	96
North Sea, 1975—1976	Precipitation and dry deposition	5.75×10^5	1,900 (12)	3.3	68
Five Great Lakes	Precipitation and dry deposition	2.40×10^5	29,000 (12)	121.0	2
Lake Michigan (calculated)	Total	—	—	5.4—10	97
Lake Superior (calculated)	Total	—	—	7.3—73	30

IV. ATMOSPHERIC INPUTS ON PCBs TO THE OCEAN

In this section we will attempt to model the deposition of PCBs to the ocean surface. Similar attempts at modeling the air/water flux of PCBs to the Great Lakes have been reported.[95,96] In a broad sense our approach here will be similar to other models, but some of the deposition parameters and assumptions necessarily will be modified for application to the marine environment. Also, where possible, our estimates will conform with data recently acquired for deposition of other substances, such as various trace metals to the ocean surface.[97] The main results from this model will be a comparison of the fluxes due to different deposition mechanisms of PCB and an estimate of the overall flux of PCB to the ocean.

Substances can be transferred from the atmosphere to the ocean by deposition of both particles and gases. Since PCB is partitioned between a vapor and particle phase, both mechanisms must be considered. Furthermore, both wet and dry deposition processes for each phase must be considered. Each process is considered separately below.

A. Dry Deposition of Particles

Particle deposition to a natural water surface is a highly complicated process. In a very simple mathematical form which masks many complexities, the flux of particles may be calculated from:

$$F_D^p = V_D C_p \tag{2}$$

where F_D^p = dry flux of particle-bound compound, V_D = particle deposition velocity, and C_p = concentration of the compound bound to particles. Large differences in particle deposition rates are expected as a function of particle size.[83] Measurements of V_D (total) described earlier (Table 8) may not be applicable to open-ocean areas. Data of Bidleman and Christensen[51] and Murphy et al.[78] suggest that dry deposition rates of PCB measured in continental areas may be dominated by a small percentage of PCB which is associated with large particles. Rapid deposition of these continentally derived large particles near the

Table 8
ATMOSPHERIC REMOVAL RATES AND DEPOSITION PARAMETERS FOR PCB

Compound	Location	Dry flux, ng/m²/day Range	Dry flux, ng/m²/day Mean	V_d, cm/sec[a]	W[b]	Ref.
Aroclor® 1242/1016	Columbia, S.C.	11—119	29	≤0.04		51, 79
	North Inlet Estuary, S.C.		<11	<0.06		51, 79
	Kingston, R.I.	<29—325	159	0.07		51, 79
	Enewetak Atoll				<1	3
	Chicago; Beaver Island, Lake Michigan				14	17
Aroclor® 1254	Columbia, S.C.	<24—368	118	0.43		51, 79
	North Inlet Estuary, S.C.	<2—74	21	0.16	94	51, 79
	Kingston, R.I.	61—360	237	0.11		51, 79
	La Jolla, Calif.			1.20		72
	Chicago; Beaver Island, Lake Michigan				103	17
	Pigeon Key, Fla.		2.4		18—32	49
Clophen® A50	Sweden[c]	20—350	110			69, 98
	Iceland[c]	<0.4—35	7			71
	Czechoslovakia[c]	4—113				70
	Kenya[c]	<0.4—0.7				70
Total (or unspecified)	Minneapolis, Minn.		630	0.13		28
	Chicago, Ill.					78
PCB	Waukegan, Ill.		121			78
	Lake Huron (towns)		17			78

[a] Deposition velocity for total PCB (particle and vapor).
[b] W = g/kg rain ÷ g/kg air.
[c] May include precipitation-derived residues.

source will lessen the effective deposition rate of PCB over most ocean areas. Since any particulate PCB present in the ocean atmosphere is mainly on small particles, it may behave similarly to other atmospheric compounds associated with small particles, such as Pb. A recent analysis of Pb deposition in the marine atmosphere[99] suggests a net dry deposition velocity of small-particle Pb on the order of 0.1 cm/sec.

Recent measurements of particle-borne PCBs in the marine atmosphere suggest that <1% is associated with particles. This result may be biased toward low values because measurements have typically been made near sea level, in temperate or tropical climates, and in relatively warm weather. As discussed earlier, temperature can affect the amount of PCB on atmospheric particles. For the purposes of our calculation we will assume that 2% of Aroclor® 1254 and 0.4% of Aroclors® 1016 and 1242 in the atmosphere are associated with particles. These percentages correspond to a temperature of ~4°C and should provide an upper limit of particulate PCBs for clean air conditions in the marine boundary layer.

B. Wet Deposition of Particles

Particles are efficiently scavenged from the atmosphere by precipitation. Data on particulate organics in precipitation[100,101] and on trace metal scavenging[102] suggest a value of 300 to 700 [ng/kg (rain)/ng/kg (air)]. For our calculation we will use a value of 500. The flux due to wet deposition of particulate PCB then is given by:

$$F_W^P = 500RC_p \tag{3}$$

where R = annual rainfall rate over the ocean (~1 m/year).

C. Dry Deposition of Gases

The magnitude (and direction) of the gas exchange of PCBs across the air/water boundary has been a subject of controversy in recent years, particularly with reference to the Great Lakes.[103,104] It has been shown from mass balance estimates and from water column measurements that the atmosphere must be an important source of PCB to the Great Lakes. However, physicochemical data on PCB vapor pressure, solubility, and Henry's Law constants suggest that the Great Lakes are supersaturated with respect to atmospheric PCB. Therefore, surface water should volatilize excess PCB and be a source, rather than a sink, of PCB. Some have suggested that the physical data are incorrect[105] or have pointed to assumptions regarding mass transfer rates and sorption of PCB as potential sources of miscalculation.[97]

A similar dilemma exists for gas transfer of PCB to the ocean surface. Direct measurements are not available to determine the net flux of vapor-phase PCB across the air/water interface. However, some indirect data are available to suggest the magnitude of the gas exchange flux. First, we will examine the equations and parameters used to estimate gas exchange.

The two-film resistance model of Whitman has been widely used to examine gas exchange in natural water systems.[106] In this model, the transfer of a gas is limited by diffusion across thin stagnant films of air and water adjacent to the interface. The overall transfer coefficient, K_{OL}, is based on the individual transfer coefficients across the liquid and gas films and on the Henry's Law constant of the compound. Thus:

$$1/K_{OL} = 1/K_L + RT/HK_G \tag{4}$$

where K_L and K_G are the individual mass transfer coefficients in the liquid (L) and gas (G) film, and (H/RT) = dimensionless air/water partition coefficient, H = Henry's Law constant (atm − m^3/mol), R = gas constant, and T = temperature. Depending on the magnitude of (RT/H), one of the two terms in Equation 4 may become large compared to the other and the exchange will be designated as either "liquid-phase" or "gas-phase" controlled. K_L and K_G are based on estimates of oxygen and water vapor mass transfer rates in the ocean. Typical values of 3000 and 20 cm/hr have been suggested for K_G and K_L, respectively.[106] These rates need to be corrected for the smaller diffusivity of PCB. Theoretical estimates[107,108] and laboratory experiments[109] suggest a correction factor of approximately 0.2. Thus, K_G ~600 cm/hr and K_L ~8 cm/hr for PCB.

Recent experiments by Atlas et al.[109,110] have shown that (H/RT) at 23°C for various PCBs in sea water is approximately 0.028 for Aroclor® 1242 and 0.014 for Aroclor® 1254. Using these values yields a K_{OL} of 5.4 cm/hr and 4.1 cm/hr for Aroclors® 1242 and 1254, respectively. These transfer coefficients show that both gas and liquid phase resistances are significant in the transfer of PCBs across the air/water interface.[111]

Given the mass transfer rate K_{OL}, the net flux F_D^G depends on the disequilibrium between air and water concentrations, or

$$F_D^G = K_{OL} (C_{eq} - C) \tag{5}$$

where C = the concentration of dissolved PCBs in water, and C_{eq} the aqueous concentration of PCBs in equilibrium with atmospheric PCB. If $C > C_{eq}$, the flux is from the water to the air, and vice versa. One assumption in prior estimates of gaseous PCB flux is that the ocean surface acts as a perfect sink for PCB, i.e., C = 0. Under this assumption the flux to the ocean is maximum and is given by $F = K_{OL} * C_{eq}$, where $C_{eq} = C_{air}/(H/RT)$. Using appropriate parameters yields a maximum air-sea flux of 5.0×10^{-6}g PCB/m²/year (2.9×10^{-6}g 1242 + 2.1×10^{-6}g 1254).

Reported concentrations of PCBs in the oceans are not zero, however. Aqueous concentrations of "dissolved" PCB of 50 to 500 pg/ℓ in surface ocean waters are at least 10 times the equilibrium concentration expected from Henry's Law calculations. The compilation of environmental air/water concentration ratios vs. theoretical estimates (Table 9) clearly illustrates this point. The data in Table 9 show that PCB partitions at least 10 times more into the aqueous phase (in marine and fresh water systems and in rain water) than is predicted from merely equilibrium with atmospheric PCB. Apparent supersaturation is evidently a generally widespread phenomenon in natural waters.

If we assume that the oceans are a sink rather than a source of atmospheric PCB, these observations suggest that the practical differentiation of particulate and dissolved PCBs in rain and other natural waters may not accurately reflect the actual physical state and chemical behavior of PCBs in solution. For example, PCBs sorbed to microparticulate material, which passes through glass-fiber filters, would be considered "dissolved"; consequently, the solution would appear supersaturated with respect to atmospheric PCB. There is little data available to test this hypothesis, however. Voice et al.[113] suggested the effect of microparticulate (=colloidal) PCB in laboratory experiments of PCB adsorption of sediments. Recently, Gschwend and Wu demonstrated this effect in a laboratory system.[114] Other studies have demonstrated the strong binding of synthetic organics to natural colloid material.[111,116] Available data on PCB distribution between air and water would be more easily explained if PCBs in natural waters were not in true solution, but rather associated with colloidal or microparticulate material. This does not suggest, though, that gas-phase PCB does not transfer to aqueous solution. Laboratory studies[109,110] on gas exchange of PCBs have demonstrated that gas-phase PCBs will reach equilibrium with aqueous solutions and subsequently can be purged from solution with clean air. Thus, gas deposition and volatilization of PCBs can occur; however, in natural systems this process may be masked by the presence of PCBs sorbed to very small particles. Data are presently unavailable to determine the extent to which the pool of particulate PCBs is involved in the equilibration of gas-phase PCBs with surface waters. Further work is required to determine the various mechanisms which control the air/water exchange processes in natural waters.

Table 9
MEASURED AND THEORETICAL AIR/WATER CONC. RATIOS OF PCBs [ng/m³ (air)/ng/ℓ (water)]

Location	PCB mixture (Aroclors®) 1242	1254	Ref.
Marine			
Antarctica[a]	3.4	3.4	9
Tropical Pacific[a]	1.7	1.8	9
Coastal China[a]	1.3	(14)	9
North Atlantic	—	2—6	6
Gulf of Mexico	—	1.2	19
Fresh water			
Lake Michigan surface water (total)	2.0	1.0	112
Microlayer	0.7	0.24	112
Lake Superior	1.6	0.4	28, 104
Average			
Marine	2.3	2.1	—
Fresh	1.8—2.3	0.7—0.9	—
Rainwater	0.13	0.03—0.17	Table 8
Theoretical			
Marine	28.5	14.1	110
Fresh	13.6	6.7	110

[a] Calculated from individual PCB isomer distributions.

Because of the uncertainty in the magnitude of "dissolved" PCBs, the actual saturation state of surface sea water with respect to atmospheric PCBs cannot be calculated directly and assumptions regarding the saturation state must be made. Two scenarios are considered. The first case assumes that processes removing PCBs from the "dissolved" form to colloidal or particulate matter are rapid compared to the gas exchange rate. This case (discussed above) suggests the ocean absorbs gaseous PCBs at the maximum rate. In the second case, gas exchange is rapid compared to other processes influencing aqueous PCB concentration. Since measurements of anthropogenic gases, such as freons, indicate that surface ocean waters are in near equilibrium with the atmosphere, it is probable that "dissolved" PCBs are also in equilibrium with atmospheric PCBs. If the ocean surface were stagnant there would be no net flux of PCBs from the atmosphere to the oceans. However, surface waters are being mixed constantly with old, deeper waters which are presumably undersaturated with respect to atmospheric PCBs. Thus, gas exchange of PCBs may be limited by the residence time of water in the mixed layer rather than by a diffusion-limited transfer across thin interfacial films. If, for example, we assume the 100 m surface layer of the ocean has a residence time, τ, of 2 to 4 years and is being mixed with water undersaturated in PCB, then the gas exchange flux becomes 100 m $\times$ C_{eq}/τ. This calculation results in a gas-exchange PCB flux of only 5% of the maximum rate.

D. Wet Deposition of Gases

This process involves equilibrium of rain with vapor phase PCBs in the atmosphere. Essentially, this mechanism has been discussed in the last section. The flux from wet deposition of gaseous PCBs is

$$F_W^G = RC_{eq} \tag{6}$$

Table 10
PARAMETERS FOR PCB FLUX CALCULATION

Parameters	PCB Aroclor® mixture 1242	PCB Aroclor® mixture 1254
Vapor conc. (pg/m^3)	126.0	63.0
Particulate conc. (pg/m^3)	0.5	1.3
Partition coefficient, m^3(air)/ℓ(water)		
Sea water	28.0	14.0
Fresh water	14.0	7.0
V_D (particles) cm/sec	0.1	0.1
Particle washout factor		
W_P	500.0	500.0
Rain rate (m/year)	1.0	1.0
Ocean area (m^2)	3.6×10^{14}	3.6×10^{14}

Table 11
FLUX OF PCBs TO THE OCEAN SURFACE

	Flux (ng/m^2/year)		
	Aroclor® 1242	Aroclor® 1254	Total
Particles			
Dry	16.0	41.0	57.0
Wet	250.0	650.0	900.0
Gas Phase			
Dry maximum	2,100.0	1,600.0	3,700.0
minimum	100.0	100.0	200.0[a])
Wet	9.0	9.5	19.0
Total	2375.0	2301.0	4676.0
Total flux to oceans			
($\times$ 10^8 g/year)	8.6	8.3	16.9

[a] Based on sea water composition ≃ 0.5 Aroclors® 1242 and 0.5 1254.

where the symbols used are those already described (Equation 5). As shown in Table 8 field measurements of PCB washout ratios are one to two orders of magnitude higher than those predicted based solely on dissolution of gas-phase PCBs in precipitation (Table 9). This suggests that PCBs in rainfall is obtained primarily from particulate matter and that gas-phase dissolution of PCBs is a relatively minor contribution.

E. Comparison of PCB Inputs to the Ocean

A summary of flux estimates is given in Tables 10 and 11. Under assumptions of maximum gas exchange, the flux of atmospheric PCB to the ocean is dominated by gas exchange processes (79%). Precipitation scavenging of the relatively small reservoir of particulate PCB is the second major mechanism for air/sea transfer of PCB (19%). On the other hand a smaller gas transfer flux would make these two processes comparable. Wet deposition of gaseous PCB and dry deposition of particulate PCB are less important processes in the marine environment. These estimates can be compared to other calculations of PCB input to the ocean. For example, ''total'' dry deposition velocity (particles and gas) have been measured for atmospheric PCB (see Table 8). A simple measure of ''total dry deposition'' is calculated from:

$$F^T = V_D^T C_A^T \tag{7}$$

where F = total dry deposition (gas and particle) flux, V_D^T = "total" dry deposition velocity, and C_A^T = total atmospheric PCB concentration. Using data from a pristine coastal area (Table 8), an estimated V_D for the marine environment is 0.15 cm/sec (Aroclor® 1254) and 0.03 cm/sec (Aroclor® 1242). Use of these factors yields a total PCB flux of 4.2 μg/m²/year (71% Aroclor® 1254, 29% Aroclor® 1242). This compares favorably with 3.8 μg/m²/year calculated in the maximum exchange model.

An independent measure of particulate PCB flux to the ocean may be obtained from recent measurements of dust flux to the ocean. Using an average flux of 50 μg/cm²/year of dust to the central Pacific[117,118] and a particulate PCB concentration of 0.25 μg/g (Table 10) and 7 μg/m³ particulate matter in marine air we calculate a net flux (not including gas exchange) of 0.125 μg/m²/year. This also agrees well with model calculations above (Table 11).

These calculations suggest an *upper limit* of PCB flux to the oceans is ~5 μg/m²/year, or a total annual flux of 1700 t. There are, of course, relatively large uncertainties in this estimate. The degree of gas exchange of PCBs between the atmosphere and ocean surface waters is the largest uncertainty. Effects of storms, wind speed, and temperature on air/sea exchange of PCB have been ignored in this model. Also, potentially important regional variations in particle size and concentration of atmospheric PCB have been ignored. A more refined model of PCB flux will require additional laboratory and field measurements of PCBs. In spite of the uncertainties, the flux calculated above is probably a reasonable upper limit of PCB input to the oceans over the last 10 to 15 years.

Calculations below will illustrate the importance of atmospheric inputs of PCBs by comparing the atmosphere-ocean input fluxes with other main inputs of PCBs to the ocean and with the size of the ocean reservoir of PCBs.

The second potential major input of PCBs to the ocean will be from rivers. We are aware of no data on PCBs in rivers with the largest flow, i.e., the Amazon, Congo, and Yangtze, which drain nonindustrialized regions. Mean PCB concentrations in the Mississippi River Delta area were reported as 2.5 ng/ℓ[6] and some rivers draining highly industrialized areas of Europe report PCB concentrations on the order of 10 to 20 ng/ℓ,[119] though much of this is associated with particulate material which will be deposited near the river mouth. Considering that the largest rivers drain rural or undeveloped lands, a reasonable upper limit for PCBs in "average" river water is 1 ng/ℓ. Using an annual runoff to the ocean of 3×10^{13} m³, we calculate an annual riverine flux of PCBs to the ocean of 30 t. This amounts to only 2% of the estimated atmospheric flux of PCBs to the ocean.

The ocean reservoir of PCBs is discussed in detail elsewhere in this volume,[120] and thus it will be considered only briefly here. The concentration of PCBs in ocean water is still relatively poorly defined. Data suggest PCBs are distributed relatively uniformly throughout the water column;[120] they are present even in very deep water. If we assume a uniform concentration of 0.1 μg/m³ (100 pg/ℓ)[121] to 4000 m depth in the world ocean (3.6×10^{14} m²), the amount of PCBs in the ocean is 1.4×10^{11} g. This does not include the ocean-sediment reservoir. This amount seems slightly high compared to the mobile environmental reservoir of 0.8×10^{11} g of U.S.-derived PCBs, but it is not unreasonable. At the *maximum* rate of atmospheric and riverine input, it would take approximately 82 years for the ocean reservoir to reach its present size. However, PCBs have only been produced in significant quantities over the last 40 to 50 years. Thus, even maximum estimated fluxes of PCBs cannot account for the size of the oceanic PCB reservoir. This suggests that atmospheric fluxes of PCBs to the ocean in the past (prior to 1970) may have been greater than present rates to account for PCBs accumulated in the ocean, or that there are still errors in estimating the reservoir sizes and exchange rates of PCBs in the environment. Indeed, some recent measurements of surface water PCB concentrations in the Pacific Ocean[8] are less than 100 pg/ℓ used in the above calculation, which would indicate a smaller oceanic PCB reservoir. A smaller reservoir of oceanic PCB would be more consistent with estimated rates of atmospheric inputs of PCB to the ocean.

V. SUMMARY

This chapter has examined the collection techniques, concentration, and deposition of PCBs from the atmosphere. Suitable collection and analytical techniques are available to measure low concentrations of PCBs in the ambient atmosphere. Direct determinations of wet deposition of PCB are rare, and there is still controversy concerning measurement and interpretation of dry deposition of atmospheric PCBs. Available data of PCB concentrations indicate an atmospheric reservoir size of approximately 7.7×10^5 kg. This reservoir is small compared to other reservoirs of "mobile" PCBs in the environment. However, the atmosphere is still a major transport mechanism for distributing PCBs throughout the environment. It is estimated here that up to 98% of the PCBs entering the oceans is currently being deposited from the atmosphere. A maximum atmospheric input of 1700 t/year of PCB is calculated.

The most important mechanisms for depositing PCBs from the atmosphere to the ocean surface are gas exchange and washout of particulate PCBs. Washout of gas-phase PCB is negligible. Dry deposition of particulate PCBs is only 5 to 10% of the wet deposition. However, there are still numerous uncertainties in the picture of PCB exchange with the ocean. The magnitude of PCB gas exchange with ocean surface waters is largely uncertain. Also, the model presented here is based on average properties and ignores temporal and regional differences in PCB transfer. Factors such as air temperature, wind speed, storm frequency, rainfall rate, and other meteorological factors will exert an important influence on the pattern and rates of PCB movement over the oceans and need to be considered in a detailed picture of PCB cycling between the oceans and the atmosphere. However, such a model will require more measurements of PCBs in the marine environment than are presently available. Finally, mass balance considerations should be applied to models of PCB cycling to set limits on observed and estimated concentrations and transport rates in the oceans.

REFERENCES

1. *Tropospheric Transport of Pollutants and Other Substances to the Oceans*, National Academy of Sciences, Washington, D.C., 1978.
2. **Eisenreich, S. J., Looney, B. B., and Thornton, J. D.,** Airborne organic contaminants in Great Lakes ecosystem, *Environ. Sci. Technol.*, 15, 30, 1981.
3. **Atlas, E. L. and Giam, C. S.,** Global transport of organic pollutants: ambient concentrations in the remote marine atmosphere, *Science*, 211, 163, 1981.
4. **Tanabe, S., Hidaka, H., and Tatsukawa, R.,** PCBs and chlorinated hydrocarbon pesticides in Antarctic atmosphere and hydrosphere, *Chemosphere*, 12, 277, 1983.
5. **Tanabe, S., Mori, T., Tatsukawa, R., and Miyazaki, N.,** Global pollution of marine mammals by PCBs, DDTs, and HCHs (BHCs), *Chemosphere*, 12, 1269, 1983.
6. **Giam, C. S., Chan, H. S., Neff, G. S., and Atlas, E. L.,** Phthalate esters plasticizers: a new class of marine pollutant, *Science*, 199, 419, 1978.
7. *Polychlorinated Biphenyls*, prepared by the Committee on the Assessment of Polychlorinated Biphenyls in the Environment, National Academy of Sciences, Washington, D.C., 1979.
8. **deLappe, B. W., Risebrough, R. W., and Walker, W., II,** A large-volume sampling assembly for the determination of synthetic organic and petroleum compounds in the dissolved and particulate phases of seawater, *Can. J. Fish. Aquat. Sci.*, 40(Suppl. 2), 322, 1983.
9. **Tanabe, S., Kawano, M., and Tatsukawa, R.,** Chlorinated hydrocarbons in the Antarctic, Western Pacific, and Indian oceans, *Trans. Tokyo Univ. Fish.*, 5, 97, 1982.
10. **Lewis, R. G., Brown, A. R., and Jackson, M. D.,** Evaluation of polyurethane foam for sampling of pesticides, PCB, and PCN in ambient air, *Anal. Chem.*, 49, 1668, 1977.
11. **Lewis, R. G. and Jackson, M. D.,** Modification and evaluation of a high volume air sampler for pesticides and semivolatile industrial organic chemicals, *Anal. Chem.*, 54, 592, 1982.

12. **Simon, C. G. and Bidleman, T. F.,** Sampling airborne PCB with polyurethane foam; a chromatographic approach to determining retention efficiencies, *Anal. Chem.,* 51, 1110, 1979.
13. **Burdick, N. F. and Bidleman, T. F.,** Frontal movement of PCB and HCB vapors through polyurethane foam, *Anal. Chem.,* 53, 1926, 1981.
14. **Stratton, C. L., Whitlock, S. A., and Allan, J. M.,** A Method for the Sampling and Analysis of PCBs in Ambient Air, EPA-600/4-78-048, U.S. Environmental Protection Agency, Research Triangle Park, N.C., 1978.
15. **Billings, W. N. and Bidleman, T. F.,** Field comparison of polyurethane foam and Tenax-GC resin for high volume air sampling of chlorinated hydrocarbons, *Environ. Sci. Technol.,* 14, 679, 1980.
16. **Billings, W. N. and Bidleman, T. F.,** High volume collection of chlorinated hydrocarbons in urban air using three solid adsorbents, *Atmos. Environ.,* 17, 383, 1983.
17. **Murphy, T. J. and Rzeszutko, C. P.,** Precipitation inputs of PCBs to Lake Michigan, *J. Great Lakes Res.,* 3, 305, 1977.
18. **Bidleman, T. F., Christensen, E. J., Billings, W. N., and Leonard, R.,** Atmospheric transport of organochlorines in the North Atlantic gyre, *J. Mar. Res.,* 39, 443, 1981.
19. **Giam, C. S., Atlas, E. L., Chan, H. S., and Neff, G. S.,** Phthalate esters, PCB, and DDT residues in the Gulf of Mexico atmosphere, *Atmos. Environ.,* 14, 65, 1980.
20. **Oehme, M. and Stray, H.,** Quantitative determination of ultra-traces of chlorinated compounds in high volume air samples from the Arctic using polyurethane foam as a collection medium, *Fresenius' Z. Anal. Chem.,* 311, 665, 1982.
21. **Oehme, M. and Manø, S.,** The long-range transport of organic pollutants to the Arctic, *Fresenius' Z. Anal. Chem.,* 319, 141, 1984.
22. **Doskey, P. V. and Andren, A. W.,** High volume sampling of airborne PCB with Amberlite XAD-2 resin, *Anal. Chim. Acta,* 110, 129, 1979.
23. **Lewis, R. G. and MacLeod, K. E.,** Portable sampler for pesticides and semivolatile industrial organic chemicals in air, *Anal. Chem.,* 54, 310, 1982.
24. **MacLeod, K. E.,** Polychlorinated biphenyls in indoor air, *Environ. Sci. Technol.,* 15, 926, 1981.
25. **Vannucchi, C. and Berlincioni, M.,** Sampling and analytical method to determine PCB concentrations in the air of the working areas of an electrical industry, *Am. Ind. Hyg. Assoc. J.,* 41, 352, 1980.
26. **Hollod, G. J. and Eisenreich, S. J.,** Collection of atmospheric PCB on Amberlite XAD-2 resin, *Anal. Chim. Acta,* 124, 31, 1981.
27. **Doskey, P. V. and Andren, A. W.,** Concentrations of PCB over Lake Michigan, *J. Great Lakes Res.,* 7, 15, 1981.
28. **Eisenreich, S. J., Hollod, G. J., and Johnson, T. C.,** Atmospheric concentrations and deposition of PCB to Lake Superior, in *Atmospheric Pollutants in Natural Waters,* Eisenreich, S. J., Ed., Ann Arbor Science, Ann Arbor, Mich., 1981, 425.
29. **Eisenreich, S. J., Looney, B. B., and Hollod, G. J.,** PCBs in the Lake Superior atmosphere 1978-1980, in *Physical Behavior of PCBs in the Great Lakes,* MacKay, D., Paterson, S., Eisenreich, S. J., and Simmons, M., Eds., Ann Arbor Science, Ann Arbor, Mich., 1983, 115.
30. **Eisenreich, S. J. and Looney, B. B.,** Evidence for the atmospheric flux of PCBs to Lake Superior, in *Physical Behavior of PCBs in the Great Lakes,* MacKay, D., Paterson, S., Eisenreich, S. J., and Simmons, M., Eds., Ann Arbor Science, Ann Arbor, Mich., 1983, 141.
31. **Andersson, K., Levin, J. O., and Nilsson, C. A.,** Sampling of polychlorinated aromatics in workroom air using Amberlite XAD-2 resin, *Chemosphere,* 10, 137, 1981.
32. **Giam, C. S., Chan, H. S., and Neff, G. S.,** Rapid and inexpensive method for detection of PCB and phthalates in air, *Anal. Chem.,* 47, 2319, 1975.
33. **Haile, C. L. and Baladi, E.,** Methods for Determining the Total PCB Emissions from Incinerators and Capacitor and Transformer Filling Plants, EPA-600/4-77-048, U.S. Environmental Protection Agency, Research Triangle Park, N.C., 1977.
34. **Singer, E., Jarv, T., and Sage, M.,** Survey of PCB in ambient air across the province of Ontario, in *Physical Behavior of PCBs in the Great Lakes,* MacKay, D., Paterson, S., Eisenreich, S. J., and Simmons, M., Eds., Ann Arbor Science, Ann Arbor, Mich., 1983, 367.
35. **Chang, L. W., Atlas, E., and Giam, C. S.,** Chromatographic separation and analysis of chlorinated hydrocarbons and phthalate esters from ambient air samples, *J. Int. Environ. Anal. Chem.,* 19, 145, 1985.
36. **Jackson, J. W. and Chang, D. P. Y.,** PCB recovery efficiency from stack gases and ambient air, *Air Pollut. Control Assoc. J.,* 27, 1006, 1977.
37. **Van Dyk, L. P. and Visweswariah, K.,** Pesticides in air: sampling methods, in *Residue Reviews,* Vol. 55, Gunther, F. A. and Gunther, J. D., Eds., Springer-Verlag, Basel, 1975, 92.
38. **Lewis, R. G.,** Sampling and analysis of airborne pesticides, in *Air Pollution from Pesticides and Agricultural Processes,* Lewis, R. G. and Lee, R. E., Eds., CRC Press, Cleveland, 1976, 51.

39. **Lamb, S. I., Petrowski, C., Kaplan, I. R., and Simoneit, B. R. T.,** Organic compounds in urban atmospheres: a review of distribution, collection and analysis, *Air Pollut. Control Assoc. J.*, 30, 1098, 1980.
40. **Simoneit, B. R. T. and Mazurek, M. A.,** Air pollution: the organic components, *CRC Crit. Rev. Environ. Control*, 11, 219, 1981.
41. **Senum, G. I.,** Theoretical collection efficiencies of adsorbent samplers, *Environ. Sci. Technol.*, 15, 1073, 1981.
42. **Feng, Y. and Bidleman, T. F.,** Influence of volatility on the collection of PAH vapors with polyurethane foam, *Environ. Sci. Technol.*, 18, 330, 1984.
43. **Westcott, J. W. and Bidleman, T. F.,** Determination of PCB vapor pressures by capillary gas chromatography, *J. Chromatogr.*, 210, 331, 1981.
44. **MacLeod, K. E. and Lewis, R. G.,** Measurement of contamination from PCB sources, in *Sampling And Analysis of Toxic Organics in the Atmosphere*, STP-721, Wheeler, J. B., Huersch, H. M., Mahy, H. P., and Kleinburg, A. S., Eds., American Society for Testing and Materials, Philadelphia, 1980, 56.
45. National Institute for Occupational Safety and Health (NIOSH), Recommended Criteria for Occupational Exposure to PCB, Publ. No. 77-225, U.S. Department of Health, Education, and Welfare, Washington, D.C., 1977.
46. **Weaver, G.,** PCBs in the area of New Bedford, MA, *Environ. Sci. Technol.*, 18, 22A, 1984.
47. **Hetling, L., Horn, E., and Tofflemire, J.,** Summary of Hudson River PCB Results, New York State Department of Environmental Conservation, Albany, N.Y., 1978.
48. **MacLeod, K.,** Sources of PCB Emissions into the Ambient Atmosphere and Indoor Air, EPA-600/4-79-022, U.S. Environmental Protection Agency, Research Triangle Park, N.C., 1979.
49. **Giam, C. S. and Atlas, E.,** Unpublished data.
50. **Bidleman, T. F., Billings, W. N., and Foreman, W. T.,** Vapor-particle partitioning of semivolatile organic compounds: estimates from field collections, *Environ. Sci. Technol.*, in press.
51. **Bidleman, T. F. and Christensen, E. J.,** Atmospheric removal processes for high-molecular weight organochlorines, *J. Geophys. Res.*, 84, 7857, 1979.
52. **Addison, R. F., Zinck, M. E., and Smith, T. G.,** PCBs have declined more than DDT-group residues in Arctic ringed seals *(Phoca hispida)* between 1972 and 1981, *Environ. Sci. Technol.*, 20, 253, 1986.
53. **Cautreels, W. and Van Cauwenberghe, K.,** Experiments on the distribution of organic pollutants between airborne particulate matter and the corresponding gas phase, *Atmos. Environ.*, 12, 1133, 1978.
54. **König, J., Funcke, W., Balfonz, E., Grosch, B., and Pott, F.,** Testing a high volume air sampler for quantitative collection of PAH, *Atmos. Environ.*, 14, 609, 1980.
55. **Broddin, G., Cautreels, W., and Van Cauwenberghe, K.,** On the aliphatic and polyaromatic hydrocarbon levels in urban and background aerosols from Belgium and the Netherlands, *Atmos. Environ.*, 14, 895, 1980.
56. **Grosjean, D.,** PAH in Los Angeles from samples collected on teflon, glass, and quartz fiber filters, *Atmos. Environ.*, 14, 895, 1980.
57. **Spitzer, T. and Dannecker, W.,** Membrane filters as adsorbents for PAH during high volume sampling of air particulate matter, *Anal. Chem.*, 55, 2226, 1983.
58. **Van Vaeck, L., Van Cauwenberghe, K., and Janssens, J.,** The gas-particle distribution of organic aerosol constituents: measurement of the volatilization artifact, *Atmos. Environ.*, 18, 417, 1984.
59. **Schwartz, G. P., Daisey, J. M., and Lioy, P. J.,** Effect of sampling duration on the concentration of particulate organics collected on glass fiber filters, *Am. Ind. Hyg. Assoc. J.*, 42, 258, 1981.
60. **Gilbertson, M.,** Background to the Regulation of PCB in Canada. Task Force on PCB. Report to the Environmental Contamination Committee of Environment Canada and Health and Welfare Canada, Toronto, Ontario, Canada, 1976.
61. **Harvey, G. R. and Steinhauer, W. G.,** Atmospheric transport of polychlorobiphenyls to the North Atlantic, *Atmos. Environ.*, 8, 777, 1974.
62. **Yamasaki, H., Kuwata, K., and Miyamoto, H.,** Effects of ambient temperature on aspects of airborne polycyclic aromatic hydrocarbons, *Environ. Sci. Technol.*, 16, 189, 1982.
63. **Prospero, J. M.,** Mineral and sea salt aerosol concentrations in various ocean regions, *J. Geophys. Res.*, 84, 725, 1979.
64. **Miller, J. and Summers, P. W.,** United States-Canada Memorandum of Intent on Transboundary Air Pollution, Report 2F-I, Atmospheric Sciences Review Subgroup, Work Group 2, 1982.
65. **Bogen, D. C., Nagourney, S. J., and Torquato, C.,** A field evaluation of the HASL wet-dry deposition collector, *Water Air Soil Pollut.*, 13, 453, 1980.
66. **Strachan, W. M. J. and Huneault, H.,** Automated rain sampler for trace organic substances, *Environ. Sci. Technol.*, 18, 127, 1984.
67. **Pankow, J. F., Isabelle, L. M., and Asher, W. E.,** Trace organic compounds in rain. I. Sampler design and analysis by adsorption/thermal desorption (ATD), *Environ. Sci. Technol.*, 18, 310, 1984.

68. **Wells, D. E. and Johnstone, S. J.,** The occurrence of organochlorine residues in rainwater, *Water Air Soil Pollut.*, 9, 271, 1978.
69. **Södergren, A.,** Chlorinated hydrocarbon residues in airborne fallout, *Nature (London)*, 236, 395, 1972.
70. **Södergren, A.,** Monitoring DDT and PCB in airborne fallout, *Environ. Qual. Saf.*, 3, 803, 1975.
71. **Bengston, S. A. and Södergren, A.,** DDT and PCB residues in airborne fallout and animals in Iceland, *Ambio*, 3, 84, 1974.
72. **McClure, V. E.,** Transport of heavy chlorinated hydrocarbons in the atmosphere, *Environ. Sci. Technol.*, 10, 1223, 1976.
73. **McClure, V. E. and LaGrange, J.,** Deposition of heavy chlorinated hydrocarbons from the atmosphere, *Bull. Environ. Contam. Toxicol.*, 17, 219, 1977.
74. **Young, D. R., McDermott, D. J., Heesen, T. C., and Jan, T. K.,** Pollutant inputs and distributions off Southern California, in *Marine Chemistry in the Coastal Environment*, Vol. 18, Church, T., Ed., ACS Symposium Series American Chemical Society, Washington, D.C., 1975, 424.
75. **Young, D. R., McDermott, D. J., and Heesen, T. C.,** Aerial fallout of DDT in Southern California, *Bull. Environ. Contam. Toxicol.*, 16, 604, 1976.
76. **Heesen, T. C., Young, D. R., and McDermott-Ehrlich, D.,** Evaluation of a technique for measuring dry aerial deposition rates of DDT and PCB residues, *Atmos. Environ.*, 13, 1677, 1979.
77. **Murphy, T. J.,** Evaluation of a technique for measuring dry aerial deposition rates of DDT and PCB — discussion, *Atmos. Environ.*, 15, 206, 1981.
78. **Murphy, T. J., Schinsky, A., Paolucci, G., and Rzeszutko, C. P.,** Inputs of polychlorinated biphenyls from the atmosphere to lakes Huron and Michigan, in *Atmospheric Pollutants in Natural Waters*, Eisenreich, S. J., Eds., Ann Arbor Science, Ann Arbor, Mich., 1981, 445.
79. **Christensen, E. J., Olney, C. E., and Bidleman, T. F.,** Comparison of dry and wet surfaces for collecting organochlorine dry deposition, *Bull. Environ. Contam. Toxicol.*, 23, 196, 1979.
80. **Sehmel, G. A.,** Particle and gas dry deposition: a review, *Atmos. Environ.*, 14, 983, 1980.
81. **Lunde, G., Gether, J., Gjos, N., and Stobet Lande, M. B.,** Organic micropollutants in precipitation in Norway, *Atmos. Environ.*, 11, 1007, 1977.
82. **Slinn, W. G. N., Hasse, L., Hicks, B. B., Hogan, A. W., Lal, D., Liss, P. S., Munnich, K. O., Sehmel, G. A., and Vittori, O.,** Some aspects of the transfer of atmospheric trace constituents past the air-sea interface, *Atmos. Environ.*, 12, 2055, 1978.
83. **Slinn, S. A. and Slinn, W. G. N.,** Predictions for particle deposition on natural waters, *Atmos. Environ.*, 14, 1013, 1980.
84. **Slinn, S. A. and Slinn, W. G. N.,** Modeling of atmospheric particulate deposition to natural waters, in *Atmospheric Pollutants in Natural Waters*, Eisenreich, S. J., Ed., Ann Arbor Science, Ann Arbor, Mich., 1981, 23.
85. **Scott, B. C.,** Modeling of atmospheric wet deposition, in *Atmospheric Pollutants in Natural Waters*, Eisenreich, S. J., Ed., Ann Arbor Science, Ann Arbor, Mich., 1981, 3.
86. **Hicks, B. B., Wesely, M. L., and Durham, J. L.,** Critique of Methods to Measure Dry Deposition — Workshop Summary, Report 600/9-80-050, U.S. Environmental Protection Agency, Washington, D.C., 1980.
87. **Dolske, D. A. and Gatz, D. F.,** A field intercomparison of methods for the measurement of particle and gas dry deposition, *J. Geophys. Res.*, 90, 2075, 1985.
88. **Harder, H. W., Christensen, E. J., Matthews, J., and Bidleman, T. F.,** Rainfall input of toxaphene to a South Carolina estuary, *Estuaries*, 3, 142, 1980.
89. **Lunde, G., Gether, J., Gjos, N., and Stobet Lande, M. B.,** Organic micropollutants in precipitation in Norway, *Atmos. Environ.*, 11, 1007, 1977.
90. **Strachan, W. M. J., Huneault, H., Schertzer, W. M., and Elder, F. C.,** Organochlorines in precipitation in the Great Lakes region, in *Hydrocarbons and Halogenated Hydrocarbons in the Aquatic Environment*, Afghan, B. K. and Mackay, D., Eds., Plenum Press, New York, 1980, 387.
91. **Strachan, W. M. J. and Huneault, H.,** PCB and organochlorine pesticides in Great Lakes precipitation, *J. Great Lakes Res.*, 5, 61, 1979.
92. **Sanderson, M. and LaValle, P. D.,** Surface loading from pollutants in precipitation in Southern Ontario: some climatic and statistical aspects, *J. Great Lakes Res.*, 5, 52, 1979.
93. **Swain, W. R.,** Chlorinated organic residues in fish, water, and precipitation from the vicinity of Isle Royale, Lake Superior, *J. Great Lakes Res.*, 4, 398, 1978.
94. **Davis, A. R.,** The Canadian Network for Sampling Organic Compounds in Precipitation, Inland Waters Directorate, Ottawa, Ontario, Canada, 1980.
95. **Risebrough, R. W., deLappe, B. W., and Walker, W.,** Transfer of higher molecular weight chlorinated hydrocarbons to the marine environment, in Marine Pollutant Transfer, Windom, H. and Duce, R. A., Eds., Lexington Books/D.C. Heath & Co., Lexington, Mass., 1976, 261.
96. **Murphy, T. J. and Schinsky, A. W.,** Net atmospheric inputs of PCBs to the ice cover on Lake Huron, *J. Great Lakes Res.*, 9, 92, 1983.

97. **Andren, A. W.,** Processes determining the flux of PCBs across air-water interfaces, in *Physical Behavior of PCBs in the Great Lakes,* MacKay, D., Paterson, S., Eisenreich, S. J., and Simmons, M., Eds., Ann Arbor Science, Ann Arbor, Mich., 1983, 127.
98. **Södergren, A.,** Transport, distribution, and degradation of organochlorine residues in a South Swedish lake ecosystem, *Vatten,* 2, 90, 1973.
99. **Settle, D. M. and Patterson, C. C.,** Magnitudes and sources of precipitation and dry deposition fluxes of industrial and natural leads to the North Pacific at Enewetak, *J. Geophys. Res.,* 87, 8857, 1982.
100. **Zafiriou, D. C., Gagosion, R. B., Peltzer, E. T., Alford, J. B., and Loder, T.,** Air-to-sea fluxes of lipids at Enewetak Atoll, *J. Geophys. Res.,* 90, 2409, 1985.
101. **Atlas, E., Sullivan, K., Madero, M., and Giam, C. S.,** Trace organic compounds in rural air and rainfall, in preparation.
102. **Arimoto, R., Duce, R. A., Ray, B. J., and Unni, C. K.,** Atmospheric trace and dry deposition, *J. Geophys. Res.,* 90, 2391, 1985.
103. **MacKay, D., Paterson, S., Eisenreich, S. J., and Simmons, M. S., Eds.,** *Physical Behavior of PCBs in the Great Lakes,* Ann Arbor Science, Ann Arbor, Mich., 1983.
104. **Eisenreich, S., Ed.,** *Atmospheric Pollutants in Natural Waters,* Ann Arbor Science, Ann Arbor, Mich., 1981.
105. **Doskey, P. V. and Andren, A. W.,** Modeling the flux of atmospheric PCB across the air-water interface, *Environ. Sci. Technol.,* 15, 705, 1981.
106. **Liss, P. S. and Slater, P. G.,** Fluxes of gases across the air-sea interface, *Nature (London),* 247, 181, 1974.
107. **Smith, J. H., Bomberger, D. C., Jr., and Haynes, D. L.,** Prediction of the volatilization rates of high-volatility chemicals from natural water bodies, *Environ. Sci. Technol.,* 14, 1332, 1980.
108. **Reid, R. C., Prausnitz, J. M., and Sherwood, T. K.,** *The Properties of Gases and Liquids,* 3rd ed., McGraw-Hill, New York, 1977.
109. **Atlas, E., Foster, R., and Giam, C. S.,** Air-sea exchange of high molecular weight organic pollutants: laboratory studies, *Environ. Sci. Technol.,* 16, 283, 1982.
110. **Atlas, E., Velasco, A., Sullivan, K., and Giam, C. S.,** A radiotracer study of air-water exchange of synthetic organic compounds, *Chemosphere,* 12, 1251, 1983.
111. **MacKay, D. and Wolkoff, A. W.,** Rate of evaporation of low-solubility contaminants from water bodies to atmosphere, *Environ. Sci. Technol.,* 7, 611, 1973.
112. **Rice, C. P., Eadie, B. J., and Erstfeld, K. M.,** Enrichment of PCBs in Lake Michigan surface films, *J. Great Lakes Res.,* 8, 265, 1982.
113. **Voice, T. C., Rice, C. P., and Weber, J. R.,** The effect of solids concentration on the sediment/water partitioning of hydrophobic pollutants, *Environ. Sci. Technol.,* 17, 513, 1983.
114. **Gschwend, P. M. and Wu, S.,** On the constancy of sediment-water partition coefficients of hydrophobic organic pollutants, *Environ. Sci. Technol.,* 19, 90, 1985.
115. **Means, J. C., Wijayaratne, R. D., and Boynton, W. R.,** Fate and transport of selected herbicides in an estuarine environment, *Can. J. Fish. Aquat. Sci.,* 40(Suppl. 2), 337, 1983.
116. **Means, J. C. and Wijayaratne, R.,** Role of natural colloids in transport of hydrophobic pollutants, *Science,* 215, 968, 1982.
117. **Uematso, M., Duce, R. A., Prospero, J. M., Chen, L., Merrill, J. T., and Donald, R. L.,** The transport of mineral aerosol from Asia over the North Pacific Ocean, *J. Geophys. Res.,* 88, 5343, 1983.
118. **Honjo, S., Manganini, S. J., and Cole, J. J.,** Sedimentation of biogenic matter in the deep ocean, *Deep-Sea Res.,* 29, 609, 1982.
119. **Duinker, J. C. and Hillebrand, M. T. J.,** Behavior of PCB, pentachlorobenzene, hexachlorobenzene, α-.HCH, β-.HCH, γ-.HCH, dieldrin, endrin, and p,p′-DDD in the Rhine-Meuse estuary and the adjacent coastal area, *Neth. J. Sea Res.,* 13, 256, 1979.
120. **Tanabe, S. and Tatsukawa, R.,** Distribution, behavior, and load of PCBs in the ocean, in *PCBs and the Environment,* Vol. 1, Waid, J. S., Ed., CRC Press, Boca Raton, Fla., 1986, chap. 8.
121. **Tanabe, S. and Tatsukawa, R.,** Vertical transport and residence time of chlorinated hydrocarbons in the open ocean water column, *J. Oceanogr. Soc. Jpn.,* 39, 53, 1983.

Chapter 5

SOLUBILITY AND SOIL MOBILITY OF POLYCHLORINATED BIPHENYLS

S. F. J. Chou and R. A. Griffin

TABLE OF CONTENTS

I. ABSTRACT

Aqueous solubilities of polychlorinated biphenyls (PCBs) are low, ranging from 2.7 ppb for Aroclor® 1260 to 3500 ppb for Aroclor® 1221. Water-soluble PCBs are reported to be richer in lower chlorinated isomers than were the original PCB fluids.

PCBs are strongly sorbed by earth materials. There is a very high direct correlation between the total organic carbon content of the soil and the amount sorbed. The sorption and retention of PCBs by soils and soil constituents are influenced by the number of chlorine atoms in the molecule, the more highly chlorinated derivatives being more tenaciously held. A linear regression relation for sorption of PCBs expressed on a unit carbon basis as a function of water solubility is presented. The correlation of sorption with compound *n*-octanol/water partition coefficients is also examined.

PCBs are not readily mobile in soils when leached with aqueous solutions such as water or sanitary landfill leachates but are highly mobile when leached with organic solvents. Mobility of PCBs is proportional to their solubility in the leaching solvent and to the soil organic matter content.

II. INTRODUCTION

PCBs are a class of chlorinated aromatic hydrocarbons that are thermally and chemically stable. They have been widely used as a dielectric fluid in transformers and as an impregnating agent for capacitors and condensers. The occurrence of PCBs in the environment and their consequences on environmental quality have been well emphasized.[1-4] The Toxic Substances Control Act (TSCA), Public Law 94-469, specifically prohibits production of PCBs within the U.S., regulates disposal of materials contaminated by PCBs, and restricts the use of any such materials already in service. The effect of these measures should be to eliminate further releases into the environment and eventually, to reduce quantities existing in the environment. However, because of the extreme stability of PCBs, environmental levels will not be reduced substantially for many years, and the problem of dealing with existing reservoirs of mobile PCBs will remain.

The disposal of PCBs and related materials in landfills is of great concern[5,6] because of the possibility of ground water contamination if these compounds are leached from landfills. The limited amount of available information indicates that PCBs have a strong affinity for soil[7-9] and are not readily leached by percolating water.[10] Briggs[11] reported that adsorption of nonionic organic compounds by soils was related to the organic matter content of soils and to their octanol/water partition coefficient. He predicted that PCBs would be immobile in soils. The water solubility, sorption, mobility, translocation, and plant uptake of PCBs will be discussed in this review. The correlation of the sorption constant of PCBs with soil and compound properties will also be examined.

III. AQUEOUS SOLUBILITY

The aqueous solubility of hydrophobic compounds such as PCBs is a valuable indictor of its environmental fate, for example, its tendency to sorb to soil or to bioaccumulate in organisms. A considerable volume of data has been published on PCB solubilities,[10,12-27] but discrepancies exist because of the experimental difficulty of generating and handling such dilute solutions. The aqueous solubilities of some individual PCB isomers and commercially available mixtures of PCBs are listed in Table 1. The soil sorption constant (K_{oc}) is also listed in Table 1. Aqueous solubilities of PCBs are very low ranging from 0.95 ppb for 2,4,5,2′,4′,5′-hexachlorobiphenyl to 5900 ppb for 2-chlorobiphenyl and 2.7 ppb for Aroclor® 1260 to 3500 ppb for Aroclor® 1221. The composition of Aroclors® is given in

Table 1
SOLUBILITY, K_{oc}, AND K_{ow} OF SEVERAL PCBs

Compound	Solubility (ppb)	log S	Ref.	K_{oc}	log K_{oc}[a]	Ref.	K_{ow}	log K_{ow}[b]	Ref.
Biphenyl	7,500	3.88	43	2,512	(3.40)		7,540	3.88	18
Monochlorobiphenyls									
2-	5,900	3.77	27	2,951	(3.47)		14,790	(4.17)	
3-	3,500	3.54	27	4,168	(3.62)		21,877	(4.34)	
4-	1,190	3.08	26, 27	7,943	(3.90)		79,400	4.90	50
Dichlorobiphenyls									
2,4-	1,400	3.15	27	7,244	(3.86)		41,686	(4.62)	
2,2′-	1,500	3.18	27	6,918	(3.84)		39,810	(4.60)	
2,4′-	1,260	3.10	23, 27	8,000	3.90	8	45,708	(4.66)	
4,4′-	80	1.90	27	42,658	(4.63)		346,736	(5.54)	
Trichlorobiphenyls									
2,4,4′-	85	1.93	27	40,738	(4.61)		323,593	(5.51)	
2′,3,4-	78	1.89	27	43,652	(4.64)		346,736	(5.54)	
Tetrachlorobiphenyls									
2,2′,5,5′-	36	1.56	23, 27	47,000	4.67	8	602,559	(5.78)	
2,2′,3,3′-	34	1.53	27	72,443	(4.86)		645,654	(5.81)	
2,2′,3,5′-	170	2.23	27	26,915	(4.43)		194,984	(5.29)	
2,2′,4,4′-	66	1.82	24, 25, 27	47,863	(4.68)		158,400	5.20	49
2,3′,4,4′-	58	1.76	27	52,480	(4.72)		436.555	(5.64)	
2,3′,4,5′-	41	1.61	27	64,565	(4.81)		562,341	(5.75)	
3,3′,4,4′-	180	2.26	27	25,633	(4.41)		186,208	(5.27)	
Pentachlorobiphenyls									
2,2′,3,4,5′-	22	1.34	27	95,324	(4.98)		870,693	(5.94)	
2,2′,4,5,5′-	31	1.49	27	76,948	(4.89)		691,830	(5.84)	
Hexachlorobiphenyl									
2,4,5,2′,4′,5′-	0.95	−0.02	23	1,200,000	6.08	43	5,248,000	6.72	15
Aroclors									
1221	3,500	3.54	19	4,123	(3.62)		12,300	4.09	25
1232	1,450	3.16	12	7,092	(3.85)		41,686	(4.62)	
Used capacitor fluid	698	2.84	19	10,725	4.03		70,794	(4.85)	
1016	332	2.52	10, 12, 14	17,684	(4.25)		202,000	5.31	14, 18

Table 1 (continued)
SOLUBILITY, K_{oc}, AND K_{ow} OF SEVERAL PCBs

Compound	Solubility (ppb)	log S	Ref.	K_{oc}	log K_{oc}[a]	Ref.	K_{ow}	log K_{ow}[b]	Ref.
1242	288	2.46	12, 13, 16, 17, 20	12,400	4.09	44	196,500	5.29	14, 18
1248	54	1.73	27	54,626	(4.74)		562,000	5.75	12
1254	42	1.62	15, 20, 21, 22	63,914	(4.81)		1,288,000	6.11	15
1260	2.7	0.43	12	349,462	(5.54)		4,073,800	(6.61)	

[a] Data in parentheses were estimated using the method of Hassett et al.[44]
[b] Data in parentheses were estimated using Equation 9.

Table 2
APPROXIMATE MOLECULAR COMPOSITION (%) OF AROCLORS® [24]

Empirical formula	Aroclor® no. 1221	1232	1016	1242	1248	1254	1260
$C_{12}H_{10}$	11	<0.1	<0.1	<0.1	ND	<0.1	ND
$C_{12}H_9Cl$	51	31	1	1	ND	<0.1	ND
$C_{12}H_8Cl_2$	32	24	20	16	2	0.5	ND
$C_{12}H_7Cl_3$	4	28	57	49	18	1	ND
$C_{12}H_6Cl_4$	2	12	21	25	40	21	1
$C_{12}H_5Cl_5$	<0.5	4	1	8	36	48	12
$C_{12}H_4Cl_6$	ND	<0.1	<0.1	1	4	23	38
$C_{12}H_3Cl_7$	ND	ND	ND	<0.1	ND	6	41
$C_{12}H_2Cl_8$	ND	ND	ND	ND	ND	ND	8
$C_{12}H_1Cl_9$	ND	ND	ND	ND	ND	ND	ND
Average mol wt	200.7	232.2	257.9	266.5	299.5	328.4	375.7

Note: ND means none detected.

Table 3
SUMMARY OF ISOMER COMPOSITION (%) OF WATER-SOLUBLE AROCLORS® [19]

Empirical formula	Aroclor® no. 1221	1016	1242	Capacitor fluid	1254
$C_{12}H_9Cl$	92	12	19	18	ND
$C_{12}H_8Cl_2$	7	34	35	32	ND
$C_{12}H_7Cl_3$	1	35	31	36	20
$C_{12}H_6Cl_4$	<0.1	19	14	13	56
$C_{12}H_5Cl_5$	ND	ND	1	1	24
$C_{12}H_4Cl_6$	ND	ND	ND	ND	<0.1
$C_{12}H_3Cl_7$	ND	ND	ND	ND	<0.1

Note: ND means none detected.

Table 2. The overall compositions of chlorobiphenyl isomers in the water-soluble fraction of these Aroclors® are listed in Table 3. Representative GC chromatograms of water-soluble Aroclors®[19] are shown in Figure 1. In comparing the composition of the water-soluble Aroclors® reported by Thurston[28] and Lee et al.,[19] the water-soluble fractions were found to be richer in the lower-chlorinated isomers than were the original PCBs. Aroclor® 1254 is reported to contain isomers from tetra- to heptachlorobiphenyl in the original fluid. The GC traces of hexane-soluble Aroclor® 1254 reported by several researchers[29,30] show some small peaks in the position of trichlorobiphenyl; this has also been confirmed by Lee et al.[19] Because trichlorobiphenyls are more water-soluble than the more highly chlorinated biphenyls, detection of trace amounts in water is reasonable. The hexa- and heptachlorobiphenyls are enriched in the Aroclor® 1254 fluid and do not dissolve readily in water. The low solubility of these highly chlorinated isomers in water is apparently the reason.

IV. SORPTION OF PCBs BY EARTH MATERIALS

Sorption of PCBs by earth materials is the major nondestructive physicochemical process

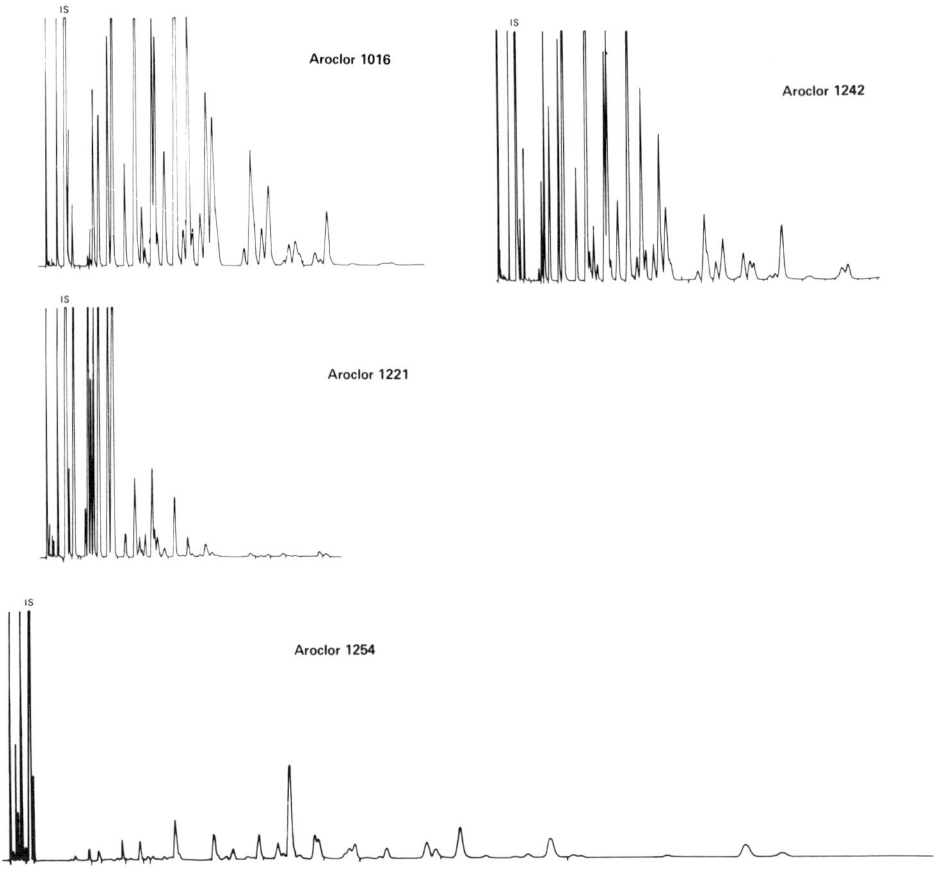

FIGURE 1. GC chromatograms of water-soluble Aroclors® [19] (IS means internal standard).

affecting PCB concentrations after introduction into the aquatic environment. The combination of low water solubility and high octanol/water partition coefficients (see Table 1) indicates that PCBs have a high affinity for suspended solids, especially those high in organic carbon.[31] This has been confirmed by a number of experiments which have shown that PCBs are rapidly sorbed and that the greatest amounts of PCBs are usually associated with the soils or sediments in soil-water systems.[7,14,32,33] Griffin and Chou[34] reported adsorption of water soluble PCBs by soil materials and coal chars. Data for PCB sorption by various earth materials are given in Table 4. Representative sorption isotherms are shown in Figures 2 and 3. The amount of PCBs sorbed by the earth materials was related to the equilibrium solution concentration of PCBs, and could be described by the simple linear relation[35]

$$x/m = KC \qquad (1)$$

where x = micrograms of compound sorbed, m = weight of adsorbent (g), C = the equilibrium concentration of the PCBs in solution (μg/mℓ), and K = the sorption constant (mℓ/g).

The PCB sorption data reported here are a special case of the Freundlich equation where $1/n = 1$, K_f, and K are thus identical and Freundlich K_f values reported in the literature for other compounds can be compared with the K values for PCBs reported in Table 4. The data presented in Table 4 illustrate the wide differences in sorption by the various earth materials. Sorption of PCBs followed the series Medium Temperature Coal Char > High

Table 4
EARTH MATERIALS USED IN SORPTION STUDIES, THEIR SORPTION CONSTANT (K), THEIR TOTAL ORGANIC CARBON (TOC) CONTENT, AND SURFACE AREA (SA)[35]

Sorbent	K	TOC (%)	SA (m^2/g)
Ottawa silica sand	22	<0.1	0.4
Montmorillonite clay	172	0.93	20.1
Montmorillonite clay (LTA)	145	0.13	20.2
Catlin silt loam	532	4.73	26.5
Catlin, 6 hr (LTA)	472	4.37	25.4
Catlin, 12 hr (LTA)	310	3.64	24.5
Catlin, 336 hr (LTA)	239	1.84	23.8
Medium temp. coal char 650°C	1,938	74.04	253
Medium temp. coal char 650°C (LTA)	1,432	64.00	214
High temp. coal char 980°C	1,220	76.62	44
High temp. coal char 980°C (LTA)	1,174	32.14	120

Note: LTA denotes low temperature ashed samples.

Temperature Coal Char > Catlin Soil > Montmorillonite Clay > Ottawa Silica Sand. Low-temperature ashing reduced the amounts of PCBs sorbed by all samples. For the samples of Catlin soil (CS), PCB sorption decreased as ashing time increased; this corroborates the observations of Briggs[11] that there is a relationship between organic matter content and sorption.

The sorption of isomers from water-soluble Aroclor® 1242 solutions by different sorbents was investigated by Lee et al.[9] They found that higher chlorinated isomers were sorbed more than the lower chlorinated isomers. Haque and Schmedding[8] indicated that the extent of sorption for the surfaces they studied followed the sequence hexachloro- > tetrachloro- > dichlorobiphenyl for the isomers studied. The sorption of the water-soluble isomers of Aroclor® 1242 by coal char was reported by Griffin and Chian.[35] The average sorption of all isomers in groups such as monochloro- or dichloro-isomers were considered together. Griffin and Chian[35] concluded that the higher chlorinated were preferentially sorbed over the lower-chlorinated isomers which agreed with the sequence found by Haque and Schmedding.[8]

The relationship between total organic carbon (TOC) content, surface area, and PCB sorption were investigated by Lee et al.[9] A highly significant (p = 0.001 level) correlation was found with a linear regression relation of the PCB sorption constant (K) and TOC of:

$$K = 255 + 18.5\ TOC$$
$$r^2 = 0.87 \qquad (2)$$

Thus, the PCB sorption constant (K) can be estimated from a knowledge of the TOC content of the earth material. A highly significant (p = 0.001 level) correlation was also found with a linear regression relation of the PCB sorption constant (K) with surface area (SA, measured by using CO_2 gas adsorption):

$$K = 230 + 6.64\ SA$$
$$r^2 = 0.82 \qquad (3)$$

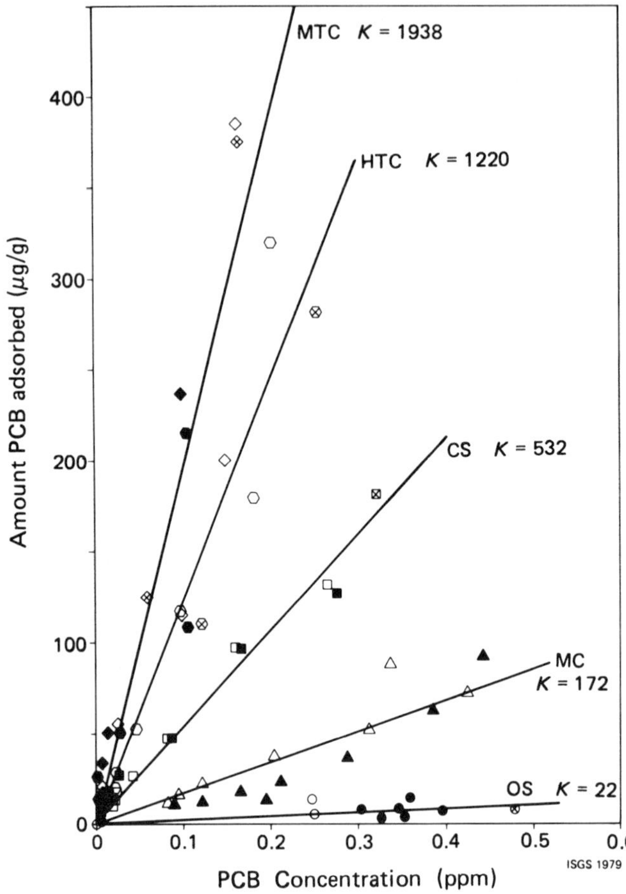

FIGURE 2. PCB adsorption by earth materials.[34] Solid symbols represent data for capacitor fluid, open symbols represent data for Aroclor® 1242, and Xs in open symbols indicate analysis using capillary column.

A three-variable regression analysis of the PCB sorption constant (K), TOC, and SA was investigated by Lee et al.[9] A very highly significant (p = 0.001 level) correlation was obtained with a linear regression relation of:

$$K = 188 + 3.36\ SA + 11.4\ TOC$$

$$r^2 = 0.94 \qquad (4)$$

The magnitude of the coefficients for SA and TOC indicate that TOC is the dominant factor in sorption. The best estimates of K were obtained by incorporating both SA and TOC; however, if only one earth material property must be chosen to estimate K, and TOC would be the most useful property.

The relation between the Aroclor® 1242 sorption constant, K, and the TOC content of the seven soil materials in Table 4 is illustrated in Figure 4. A linear relation was obtained and the slope of the line was used to estimate the K_{oc} value for Aroclor® 1242. The K_{oc} is the sorption constant normalized for the organic carbon content of soil and was found to have a value of 10,725. A three-variable regression analysis of the Aroclor® 1242 sorption constant, K, TOC, and SA was investigated using the data for soil materials (i.e., excluding coal chars) reported in Table 4 and is shown in Figure 5. A very highly significant correlation was obtained with a linear regression relation of:

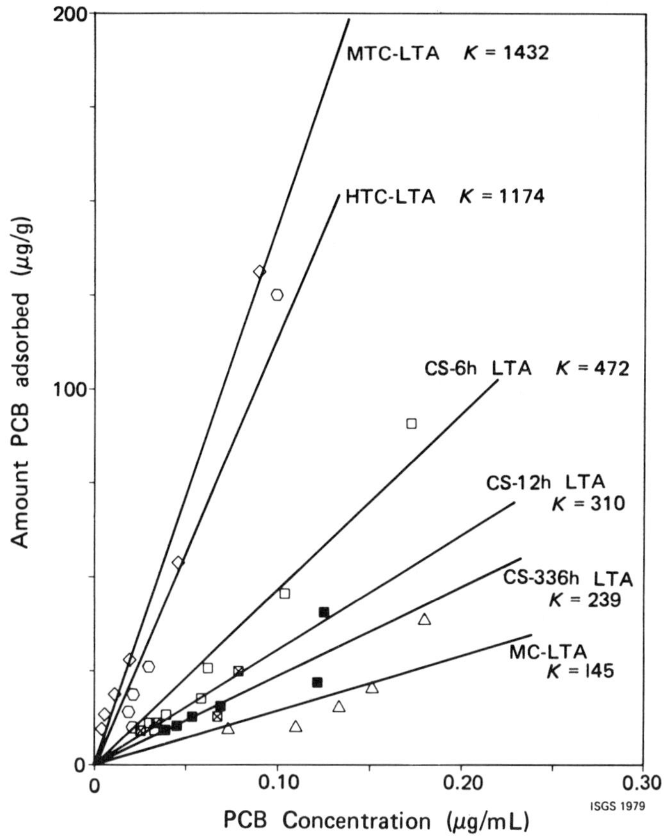

FIGURE 3. PCB adsorption by low-temperature ashes of earth materials.[34]

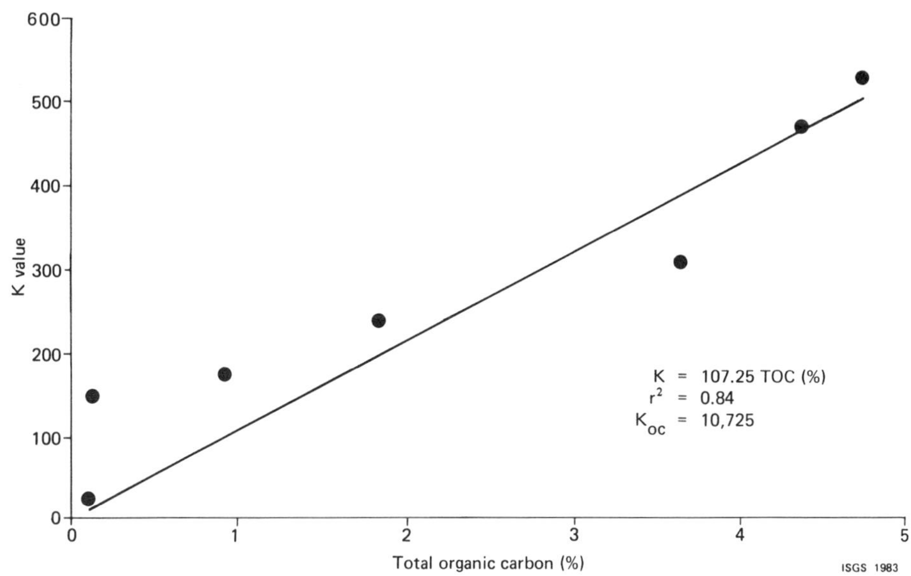

FIGURE 4. Aroclor® 1242 soil sorption constant (K) vs. soil material TOC content.

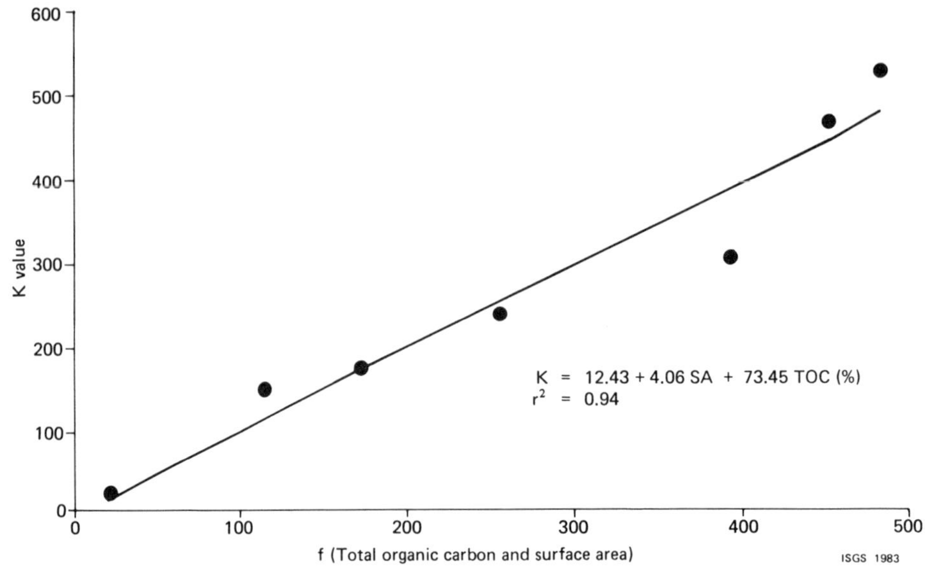

FIGURE 5. Multiple variable regression analysis of the Aroclor® 1242 soil sorption constant (K) plotted as a function of the TOC content and SA of 7 soil materials.

$$K = 12.43 + 4.06\ SA + 73.45\ TOC$$

$$r^2 = 0.94 \qquad (5)$$

The magnitude of the coefficients for SA and TOC indicates that TOC is the dominant soil property in sorption. The best estimates of K are obtained by a knowledge of both S and TOC; however, TOC is again, the single most important property controlling PCB sorption. When the PCB contamination level of soils or sediments becomes large, these materials may serve as a reservoir for desorption.[36] This is illustrated by Equation 1 which indicates that K is a ratio of x/m to C and as x/m becomes large, C also grows in relation to K. This fact has important ramifications to situations where PCBs are spilled or disposed in high concentrations because release of PCBs by soil materials can cause long-term pollution. For example, oysters still contained measurable levels of PCBs 7 years after a spill, even though the concentrations in the water were below detection limits.[37]

V. MECHANISM OF SORPTION

The sorption of PCBs by earth materials is suggested to be a hydrophobic sorption, which is the partitioning of the nonpolar solute out of the polar aqueous phase onto hydrophobic surfaces on the earth materials. The hydrophobic surfaces are primarily associated with soil organic matter but may also include ≡Si–O–Si≡ bonds at mineral surfaces.[38] The primary feature of hydrophobic sorption is the very weak interaction between the solute and the solvent. As the term hydrophobic suggests, the solvent water plays a major role in the interaction. Although not strictly the case, the interaction may be so weak as to acquire the nature of a repulsion of the PCB molecule from the water. The more hydrophobic the PCB molecule (lower in water solubility), the greater the repulsion from the water phase and the greater the sorption on a hydrophobic surface. In a stricter treatment, hydrophobic sorption is thought to be an entropy driven process governed by the basic thermodynamic relation:

$$\Delta G = \Delta H - T\Delta S \qquad (6)$$

where ΔG is the change in free energy of the system, ΔH is the heat of sorption (enthalpy), T is temperature, and ΔS is the change in entropy of the system. Sorption results when there is a negative free energy change. The primary driving force in hydrophobic sorption appears to be the large entropy change resulting from the removal of the organic molecule from solution and sorption onto a hydrophobic surface.[39,40] The entropy change is the primary driving force due to the destruction of the cavity occupied by the PCB molecule in the solvent and the destruction of the highly structured water shell surrounding the solvated PCB molecule.[40,41]

The role of the solute-solvent interaction in determining the degree of sorption of hydrophobic compounds was demonstrated by a soil thin-layer chromatography study using solutes and solvents (mobile phase) of different polarities.[42] When the mobile phase was 100% water, the polar compounds were weakly sorbed and moved with the solvent front, and the less polar compounds were strongly sorbed and showed little movement. As the mobile phase was made less polar by additions of ethanol, it became a better solvent for the nonpolar compounds and their sorption decreased. With ethanol additions, the mobile phase became a poorer solvent for the polar compound and its sorption increased and movement decreased. Similar results for Aroclors® 1242, 1254, and dicamba have also been reported.[35]

VI. CORRELATION OF K VALUES WITH EARTH MATERIAL AND COMPOUND PROPERTIES

Sorption experiments with nonpolar organic compounds and different earth materials produce a different K value for each earth material and organic compound. Hydrophobic sorption has been highly correlated with the organic carbon content of the earth material while at the same time being relatively independent of other sorbent properties.[43-46] When sorption of a hydrophobically sorbed compound is examined relative to the organic carbon content of the earth material, a constant, K_{oc}, is generated which is a unique property of the compound being sorbed. The K_{oc} can be determined graphically as shown in Figure 4 or computed as follows:

$$K_{oc} = K_f \times 100/(\%)TOC \tag{7}$$

where K_f is the Freundlich constant and TOC is the percent organic carbon in the respective soil material. K_{oc} values are compound properties, not soil properties, and have been related to other compound properties such as water solubility (S) and *n*-octanol/water partition coefficients (K_{ow}). Figure 6 shows the K_{oc}-S relation for the PCBs for which data are tabulated in Table 1. The linear regression equation obtained was

$$\log K_{oc} = 5.85 - 0.64 \log S \text{ (ppb)} \tag{8}$$

The data indicate that a good approximation of sorption of PCBs by soil materials can be made based on their water solubility. Water solubilities have been also shown to predict the K_{oc} values of a wide range of other hydrophobic compounds.[43,44,47]

The octanol/water partition coefficient (K_{ow}) measures the tendency of a compound to partition between an organic solvent and water. The octanol/water partition coefficient is usually obtained by measuring the concentration of the compound in equilibrium with *n*-octanol and water.[48] However, a satisfactory linear relationship is also observed between the log K_{ow} and log S, which extends over more than eight orders of magnitude in solubility (10^{-4} to 10^3 ppm) and four orders of magnitude in partition coefficient (10^2 to 10^6). The *n*-octanol/water partition coefficients of several PCBs are listed in Table 1. A plot of log

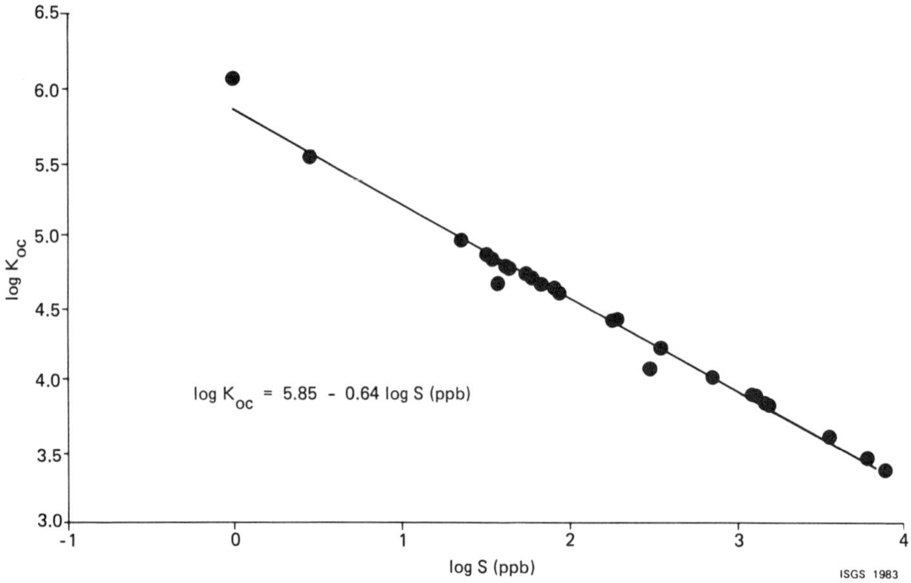

FIGURE 6. PCB soil sorption vs. water solubility.

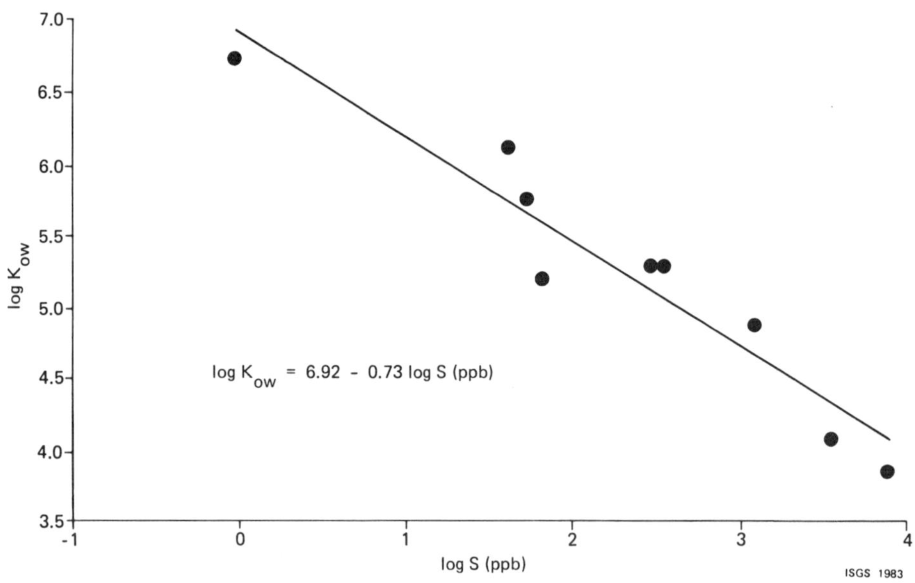

FIGURE 7. Relation of K_{ow} and the water solubility (S) of several PCBs.

K_{ow} vs. log S using the values cited in Table 1 is shown in Figure 7. The regression equation is

$$\log K_{ow} = 6.923 - 0.730 \log S \text{ (ppb)}$$
$$r^2 = 0.91 \qquad (9)$$

Equation 9 allows an estimation of the partition coefficient of a given compound from its aqueous solubility. The reliability of these relationships suggests that they may be very

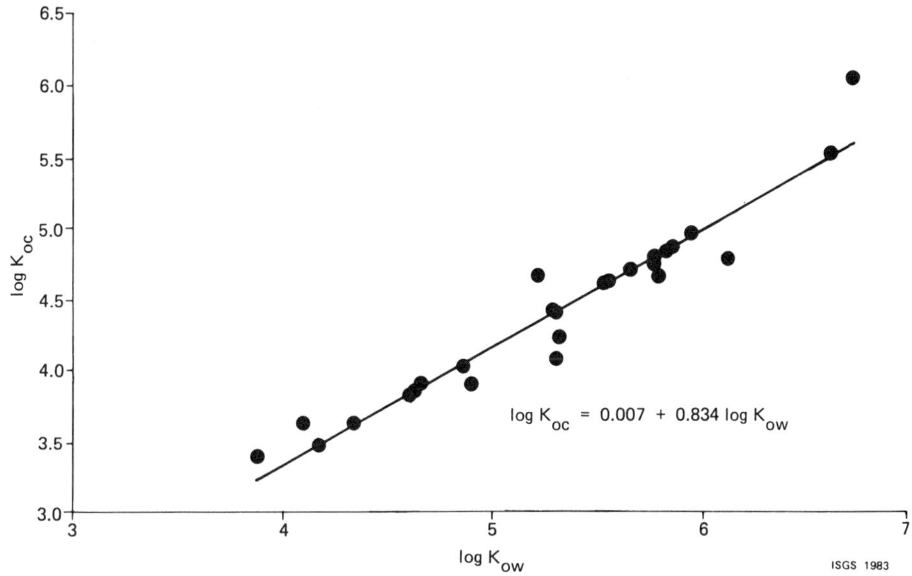

FIGURE 8. PCB K_{oc} vs. the K_{ow}.

useful in estimating the sorption of PCBs under many circumstances of soil and sediment conditions as well as bioaccumulation potential.

The sorption of nonionic organics has been related to the partitioning of an organic compound between an organic solvent (*n*-octanol) and water.[11,18,43,44,49] It follows therefore, that K_{ow} should be related to the tendency of the same compound to partition between water and soil organic matter. Figure 8 illustrates the relationship between K_{oc} and K_{ow} for PCBs. There is a high linear correlation between the two yielding the regression equation:

$$\log K_{oc} = 0.007 + 0.834 \log K_{ow} \tag{10}$$

VII. MOBILITY OF PCBs IN SOILS

The migration and vertical distribution of PCBs in soil profiles is a topic of much interest. A soil thin-layer chromatography (TLC) technique has been used to measure mobility of PCBs.[51] Soil TLC is a laboratory method of measuring the migration of a particular compound or element through a soil by using the soil as the sorbent phase and a developing solvent (water, landfill leachate, organic solvent, etc.) in a TLC system. The results are reported as R_f values, defined as the ratio of the distance the compound moved relative to the distance the solvent moved. The R_f value is a quantitative indication of the movement of a given compound in a soil-solvent system and a reproducible index of mobility.[52] The mobility of Aroclor® 1242 and Aroclor® 1254 in several earth materials was studied by Griffin and Chian[35] and expressed as frontal R_f values which are given in Table 5. The data indicate that under the conditions tested, Aroclor® 1242 and Aroclor® 1254 stayed immobile in these soil materials when leached with water and sanitary landfill leachates, but were highly mobile when leached with carbon tetrachloride. Similar results for polybrominated biphenyls (PBBs) and hexachlorobenzene (HCB) were also reported by Griffin and Chou.[53]

PCBs are nonpolar and are only very slightly soluble in polar solvents like water. Solubilities of Aroclor® 1242 and Aroclor® 1254 in distilled water have been determined to be approximately 288 and 42 ppb (see Table 1), respectively. However, PCBs are much more

Table 5
MOBILITY OF AROCLORS® 1242 AND 1254 IN SEVERAL SOIL MATERIALS LEACHED WITH VARIOUS SOLVENTS AS MEASURED BY SOIL-THIN-LAYER CHROMATOGRAPHY[35]

	R_f values					
	Water		Landfill leachate		Carbon tetrachloride	
Soil materials	1242	1254	1242	1254	1242	1254
Ava silty clay loam	0.02	0.02	0.02	0.02	1.00	0.96
Bloomfield loamy sand	0.03	0.03				
Catlin silt loam	0.02	0.02	0.04	0.04	1.00	1.00
Catlin loam	0.02	0.02	0.03	0.03	1.00	1.00
Cisne silt loam	0.03	0.02	0.03	0.02	1.00	1.00
Coal char (650°C)	0.03	0.03	0.04	0.04	1.00	1.00
Drummer silt loam	0.03	0.03			1.00	1.00
Flanagan silt loam	0.02	0.02	0.06	0.05	1.00	1.00
Ottawa silica sand	0.03	0.03	0.03	0.03	1.00	1.00

soluble in organic solvents such as acetone, methanol, benzene, or carbon tetrachloride. Griffin et al.[51] also tested the mobility of Aroclor® 1242 and Aroclor® 1254 in silica gel leached with acetone, methanol, benzene, carbon tetrachloride, and mixtures of water-acetone and water-methanol. They found the data consistent with the soil TLC data obtained by leaching with carbon tetrachloride; R_f values of 1.00 were obtained using the organic solvents. It is quite clear that mobility of PCBs in soil materials and silica gel is highly related to the solubility of PCBs in the solvent with which the TLC plates were being leached. In a soil column leaching study, Griffin and Chou[53] found that related compounds such as PBBs and HCB were not retained in the columns when ethanol was percolated through the soil columns. These data confirmed previous findings from soil TLC studies.

Leaching of water through soil containing PCBs may lead to the downward movement of PCBs, depending on the soil type and clay content. Moza et al.[54] incorporated ^{14}C-labeled 2,2′-dichlorobiphenyl in the top centimeter of a loamy sand at a concentration of approximately 1 ppm under field conditions. They found that the PCB dispersed to a depth of 30 cm during the first year and no radioactivity was detectable in the 30- to 40-cm layer. At the end of 1 year, when carrots grown on the test plots were harvested, concentrations of PCBs and their metabolites were highest in the 0- to 10-cm layer (0.24 ppm), intermediate in the 10- to 20-cm layer (0.17 ppm), and lowest in the 20- to 30-cm layer (0.03 ppm). Below 30 cm, the concentration of PCB and metabolites were less than 0.001 ppm. At the end of the second year, when sugar beets had been harvested, total PCB concentrations in the soil were 0.15 ppm in the top 10-cm layer, 0.04 ppm in the 10- to 20-cm layer, 0.02 ppm in the 20- to 30-cm layer, 0.008 ppm in the 30- to 40-cm layer, and 0.002 ppm in the leached water below the 40-cm depth. This study indicates that very small amounts of dichlorobiphenyls (or quite possibly a metabolite) moved below 40 cm in the soil; only 0.2% of the PCB-radioactivity applied to the soil appeared in leached water at the 40-cm depth over a period of 2 years.

Tucker et al.[10] studied the migration of PCBs in soil columns treated with Aroclor® 1016 and subjected to water percolation. The PCB application rate was 2.5% of soil weight, a very high rate for a land application system but more representative of landfill operations. The soils used were Norfolk sandy loam, Ray silty loam, and Drummer silty loam; the water application rates for each soil were 0.26, 0.53, and 0.32 ℓ/day, creating an essentially saturated flow. They concluded that soils containing more clay retained larger amounts of

Table 6
AROCLOR® 1016 FOUND IN PERCOLATING WATER FROM SOIL COLUMNS[10]

Norfolk sandy loam		Ray silty loam		Drummer silty clay loam	
Effluent vol. (ℓ)	PCBs (ppb)	Effluent vol. (ℓ)	PCBs (ppb)	Effluent vol. (ℓ)	PCBs (ppb)
1.3—8.1	ND	2.7—16.4	ND	1.6—9.9	ND
10.1	ND	20.7	65	12.5	ND
13.5	23	27.6	92	31.4	ND
48.1	63	51.9	153	59.2	ND

Note: ND means none detected.

Table 7
CLASSIFICATION OF SOIL MOBILITY POTENTIAL OF PCBs BY R_f, K_{ow}, OR K_{oc}[11,56]

Class	R_f	K_{ow}	K_{oc}
Very mobile	0.90—1.00	<1.2	0—50
Mobile	0.65—0.89	1.2—23	50—150
Medium	0.35—0.64	23—245	150—500
Low	0.10—0.34	245—6000	500—2000
Slight	—	—	2000—5000
Immobile	0.0—0.09	>6000	>5000

PCBs. The ease of leaching of Aroclor® 1016 from the different soil types (Table 6) was in the order: Norfolk sandy loam > Ray silty loam > Drummer silty clay loam. It should also be noted that this order of retention follows the organic matter content of these soils, as well as their clay content; the conclusion of the authors may be in error regarding the active fraction of the soil that caused the observed retention. Regardless of the mechanism involved, approximately 0.05% of total Aroclor® 1016 in the soil column was leached from the soil during the 4-month period when 50 to 100 ℓ of water had passed through the soil. This high percolation rate is equivalent to 15 to 30 m of rainfall, assuming no runoff or evaporation. Tucker et al.[10] also observed that the less chlorinated and more degradable species leached from soils more readily than highly chlorinated and more resistant PCBs. This was later confirmed by Suzuki.[55] Thus, the more resistant PCBs are retained in the surface soil layers.

The vapor loss of PCBs from soils depends strongly upon their adsorption characteristics by soils. Haque et al.[7] have shown that the vapor loss of Aroclor® 1254 from a sand surface was significant which could have been due to the poor sorption capacity of sand. When, similar experiments were carried out with a soil, the loss was very low.

The relative mobility of PCBs in soils can be classified using the parameters of R_f, K_{ow}, and K_{oc} discussed above. The soil mobility classes assigned to the various values of these three parameters are given in Table 7. The values of the various parameters assigned to each mobility class are somewhat arbitrary, but have been chosen from experience based on the relative mobility of these compounds in the environment.[11,56]

VIII. SURFACE RUNOFF

There are several ways by which PCBs get into the environment. One of these is by runoff from industrial wastes, dumps, and spills. Other sources are the points of manufacture and the plants where PCBs are processed into other products. PCBs can escape through the plant ventilation and exhaust system into the atmosphere and through its waste treatment system into sewers or directly into waterways. PCB input into fresh water by industries has been high in the past, and PCBs have accumulated in sediments.[57,58] Even if PCB input were stopped today, these sediments could continue to release PCBs into freshwater systems for years to come.

Nisbet and Sarofim[59] estimated that 4000 to 5000 tons/year of PCB were discharged into fresh and coastal waters in the peak year of PCB use. This estimate was based upon figures and industrial and disposal practices in effect at that time. Of the total PCBs, about 1000 tons may be assumed to have consisted of pentachlorobiphenyls. Analysis of waste waters and of surface water runoff entering the Southern California Bight since 1971 has shown that trichlorobiphenyls have constituted the majority of the total PCB residues. In this area, PCB input into the sea from waste waters is considerably higher than the input from surface runoff.

Martel et al.[66] reported that PCBs were found in the suburban watershed of Reston, Va. PCB concentrations in the Lake Anne basin increased going from water (<0.05 to 0.2 ppb), to bed sediment (<2.5 to 105 ppb), to fish (140 to 700 ppb) in an average ratio of 1:500:3000. The highest concentrations were observed in the lake itself, which is 10 years old, and receives no sewage or industrial waste discharges. Evidence indicates that the PCBs originated from diffuse sources associated with urban development and urban living and entered the hydrologic system through storm-water runoff. They also found that the PCB concentrations in sediment samples from two areas which were receiving rainfall runoff from two active construction sites were much higher than those from other streams feeding the lake. In the early 1970s, PCBs were commonly used in the formulation of plasticizers, adhesives, surface coatings, sealants, and fire-retardant compositions.[4,27] This suggests that building materials would be one possible source of PCBs.

In the summer of 1976, a serious spill of PCBs and chlorobenzenes was detected in Regina, Saskatchewan, Canada.[61] Approximately 6800 to 9100 ℓ of transformer oil containing PCBs (Aroclor® 1254) and chlorobenzene were spilled when an underground pipe broke at a transformer manufacturing plant. Large quantities of PCBs were found to have migrated both vertically and horizontally at the site. There was also strong evidence that the contaminants were being redistributed by surface processes. At the plant site, Roberts et al.[61] reported that PCB levels were up to 1000 ppm along some of the routes of surface drainage and zones of sediment accumulation after rainstorms. The sludge obtained from the catch basin in the street, which received the runoff, contained 20 ppm PCBs. The PCB levels at the surface in one area, which received the runoff from the highly contaminated area of the plant site ranged from less than 2 to 250 ppm.

A similar incident of a PCB spill occurred in a rural area near Kingston, Tenn. in 1973.[62] The spill resulted in the environmental contamination of two watersheds because the spill site was situated on the crest of a hill. Through the influence of rainfall, geological factors, and characteristics of the overlying stratum of soil, the chemical was subsequently dispersed through the soil both horizontally and vertically. The spread of the Askarel (a commercial mixture of Aroclor® and chlorinated benzenes) was also affected by the movement of contaminated surface water resulting from mass rainfall in the weeks immediately following the spill.

There are two potential means for PCBs to be transported at the surface: as a dissolved phase in water or as a sorbed phase on sediments. In the above cases of chemical spills, chlorobenzenes may play an important role in the movement of PCBs through soils. In previous soil-thin-layer chromatography studies,[34,35] it has been demonstrated that the mobilities of PCBs, PBBs, and HCB in soils were directly proportional to the solubility of the compounds in the leaching solvents.

Horn et al.[58] reported that measurable quantities of PCBs were leaching or eroding from some landfills and dredge spoil sites in the Upper Hudson River Basin. PCB losses to groundwater at the dredge spoil sites and at the landfills and dumps were approximately 44.4 kg/year and 172 kg/year, respectively. The results indicated that the landfills and dumps appeared to be larger contributors to the surrounding environment than the dredge-spoil areas. The erosion losses of PCBs from the same sites were 32.2 kg/year and 172 kg/year, respectively. The upper Hudson Basin delivered approximately 2.6 t of PCBs to the estuary in 1977.[58]

IX. PLANT UPTAKE

Since PCBs are strongly sorbed on carbonaceous materials and sediments, it is not surprising to observe that they are sorbed onto plant roots as well. Using Aroclors® 1242 and 1254, Suzuki[55] observed that soybean sprouts accumulated more lower chlorinated biphenyls than the higher chlorinated isomers during the first 2 weeks of exposure from a sandy soil containing 100 ppm PCBs. This may be due to the higher solubility and mobility of these isomers, thus allowing them to move to the plant root. A few gardeners in Indiana who used PCB contaminated sludge found that their vegetables had also picked up PCBs. Beets contained 0.6 ppm PCBs when grown in soil containing 4 ppm PCBs.[5] Grass clipped from a field with 8.5 ppm in the soil contained 1.16 ppm PCBs.[5]

Moza et al.[54] applied radioactive 2,2-dichlorobiphenyl to soils at a concentration of approximately 1 ppm. Carrot roots contained 0.24 ppm of 2,2′-dichlorobiphenyl while the phenolic metabolites from this PCB were found at a concentration of 0.012 ppm. In a similar study, Iwata et al.[63] found that 97% of the total absorbed Aroclor® 1254 remained on the carrot peel and very little was translocated into the plant tissue. Fries and Marrow[64] also reported that soybean plants grown on PCB contaminated soil would not be contaminated by root uptake and translocation, but some foliar contamination could occur from vapor sorption.

X. SUMMARY AND CONCLUSIONS

The aqueous solubility of PCBs is very low, ranging from 2.7 ppb for Aroclor® 1260 to 3500 ppb for Aroclor® 1221. Water-soluble PCBs were found to be richer in the lower-chlorinated isomers than were the original PCB fluids.

PCBs are strongly sorbed by soil materials. There is a very high direct correlation between the TOC content of the soils and the amount sorbed. The sorption and retention of PCBs by soils and soil constituents are influenced by the number of chlorine atoms in the molecule; the more highly chlorinated derivatives are more tenaciously held. The K_{oc} can be estimated from the water solubility of a PCB.

The higher-chlorinated isomers are preferentially sorbed to the organic matter of soils and suspended solids. Significant amounts of PCBs have been redistributed in the environment by surface-water runoff containing PCB contaminated soil particles. Sorption of PCBs onto suspended solids and their subsequent incorporation into sediments is considered the major mechanism for their immobilization in aquatic systems. However, the persistence of PCBs in sediments makes desorption a possibility for years after incorporation.

The vapor loss of PCBs from soils depends strongly upon their sorption to soils. The vapor loss of PCBs from a soil surface was very low but was significant from sand. The sorption of PCBs greatly influences their dynamics in soils. The diffusion, leaching, and vapor loss of PCBs in soils is retarded as the extent of sorption increases. The degree of sorption increases directly with an increase in organic matter content of a soil.

Highly chlorinated PCBs are not degraded or leached through soil by water to any significant extent. They are also apparently not taken up by plants and are thus not readily mobile in surface soil systems. However, PCBs do have a moderate vapor pressure and a likely path for redistribution or migration in soils may be by vapor phase transport through the unsaturated pores.

The above findings are significant to the land disposal of PCBs and related wastes. To prevent potential migration of PCBs from a landfill, PCB wastes and organic solvents should not be disposed of in the same landfill location, and should not be allowed to come in contact with leaching organic solvents.

REFERENCES

1. **Nelson, N., Hammond, P. B., Nisbet, L. C. T., Sarofim, A. F., and Drury, W. H.,** Polychlorinated biphenyls — environmental impact, *Environ. Res.*, 5, 249, 1972.
2. **Gustafson, C. G.,** PCBs — prevalent and persistent, *Environ. Sci. Technol.*, 4, 814, 1970.
3. **Gustafson, C. G., Ed.,** Proc. Symp. on PCBs — Still Prevalent — Still Persistent, 164th American Chemical Society Meeting, New York, September 1972.
4. **Peakall, D. B. and Lincer, J. L.,** Polychlorinated biphenyls: another long-life widespread chemical in the environment, *Biosphere*, 20, 958, 1970.
5. **Jordan, D.,** The town dilemma, *Environment*, 19, 6, 1977.
6. **Henderson, H.,** The dilemma of Wilsonville, *Ill. Times*, 4, 4, 1978.
7. **Haque, R., Schmedding, D. W., and Freed, V. H.,** Aqueous solubility, adsorption, and vapor behavior of polychlorinated biphenyl Aroclor 1254, *Environ. Sci. Technol.*, 8, 139, 1974.
8. **Haque, R. and Schmedding, D.,** Studies on the adsorption of selected polychlorinated biphenyl and isomers on several surfaces, *J. Environ. Sci. Health*, B11, 129, 1976.
9. **Lee, M. C., Griffin, R. A., Miller, M. L., and Chian, E. S. K.,** Adsorption of water-soluble polychlorinated biphenyl Aroclor 1242 and used capacitor fluid by soil materials and coal chars, *J. Environ. Sci. Health*, A14, 415, 1979.
10. **Tucker, E. S., Litschgi, W. J., and Mees, W. M.,** Migration of polychlorinated biphenyls in soil induced by percolating water, *Bull. Environ. Contam. Toxicol.*, 13, 86, 1975.
11. **Briggs, G. G.,** A simple relationship between soil adsorption of organic chemicals and their octanol/water partition coefficients, in *Proc. 7th British Insecticide and Fungicide Conf.*, British Crop Protection Council, London, 1973, 83.
12. Versar Inc., Water Related Environmental Fate of 129 Priority Pollutants. A Literature Search. V. Polycyclic Aromatic Hydrocarbons, PCBs, and Related Compounds, Office of Water and Waste Management, U.S. Environmental Protection Agency, Washington, D.C., 1979, 36-1.
13. Monsanto Co., Presentation to the Interdepartmental Task Force on PCB, Washington, D.C., May 15, 1972.
14. **Paris, D. F., Steen W. C., and Baughman, G. L.,** Role of the physicochemical properties of Aroclor 1016 and 1242 in determining their fate and transport in aquatic environments, *Chemosphere*, 7, 319, 1978.
15. **Chiou, C. T., Freed, U. H., Schmedding, D. W., and Kohnert, R. L.,** Partition coefficient and bioaccumulation of selected organic chemicals, *Environ. Sci. Technol.*, 11, 475, 1977.
16. **Wiese, C. S. and Griffin, D. A.,** The solubility of Aroclor 1254 in seawater, *Bull. Environ. Contam. Toxicol.*, 19, 403, 1978.
17. **Mackay, D., Mascarenhas, R., Shiu, W. Y., Valvani, S. C., and Yalkowsky, S. H.,** Aqueous solubility of polychlorinated biphenyls, *Chemosphere*, 9, 257, 1980.
18. **Kenega, E. E. and Goring, C. A. I.,** Relationship between water solubility, soil sorption, octanol/water partitioning, and concentration of chemicals in biota, *Aquatic Toxicology*, ASTM STP 707, Eaton, J. G., Parrish, P., and Hendricks, A. C., Eds., American Society for Testing and Materials, Philadelphia, 1980, 78.
19. **Lee, M. C., Chian, E. S. K., and Griffin, R. A.,** Solubility of polychlorinated biphenyls and capacitor fluid in water, *Water Res.*, 13, 1249, 1979.
20. **Chou, S. F. J. and Griffin, R. A.,** Adsorption of PCBs by cellulose fiber filteraids and carbonaceous adsorbents used for water treatment, *4th Annu. Conf. of Applied Research and Practice on Municipal and Industrial Waste*, University of Wisconsin-Extension, Madison, Wisc., 1981, 238.
21. **Lawrence, J. and Rosine, H. M.,** Adsorption of polychlorinated biphenyls from aqueous solution and sewage, *Environ. Sci. Technol.*, 10, 381, 1976.

22. **Nisbet, L. C. T. and Sarofim, A. F.,** Rates and routes of transport of PCBs in the environment, *Environ. Health Perspect.,* 1, 21, 1972.
23. **Haque, R. and Schmedding, D.,** A method of measuring the water solubility of hydrophobic chemicals: solubility of five polychlorinated biphenyls, *Bull. Environ. Contamin. Toxicol.,* 14, 13, 1975.
24. **Hutzinger, O., Safe, S., and Zitko, V.,** *The Chemistry of PCBs,* CRC Press, Boca Raton, Fla., 1980, 23.
25. **Tulp, M. M. and Hutzinger, O.,** Some thoughts on aqueous solubility and partition coefficients of PCB, and the mathematical correlation between bioaccumulations and physico-chemical properties, *Chemosphere,* 7, 849, 1978.
26. **Mackay, D. and Wolkoff, A. W.,** Rate of evaporation of low solubility contaminants from water bodies to the atmosphere, *Environ. Sci. Technol.,* 7, 611, 1973.
27. **Beeton, A. M. et al.,** *Polychlorinated Biphenyls,* Report of Committee on the Assessment of Polychlorinated Biphenyls in the Environment, National Academy of Sciences, Washington, D.C., 1979.
28. **Thurston, A.,** Quantitative analysis of PCBs, *PCB Newslett.,* July, No. 3, p. 4, 1971.
29. **Stalling, D. L. and Huckins, J. N.,** Gas-liquid chromatography mass-spectrometry characterization of polychlorinated biphenyls (Aroclors) and ^{36}Cl-labeling of Aroclor 1248 and 1254, *J. Assoc. Off. Anal. Chem.,* 54, 801, 1971.
30. **Sissons, D. and Welti, D.,** Structural identification of polychlorinated biphenyls in commercial mixtures by gas-liquid chromatography, nuclear magnetic resonance, and mass spectrometry, *J. Chromatogr.,* 60, 16, 1971.
31. **Hamelink, J. L., Waybrant, R. C., and Ball, R. C.,** A proposal: exchange equilibria control the degree chlorinated hydrocarbons are biologically magnified in lentic environments, *Trans. Am. Fish. Soc.,* 100, 207, 1971.
32. **Oloffs, P. C., Albright, L. J., and Szeto, S. Y.,** Fate and behavior of five chlorinated hydrocarbons in three natural waters, *Can. J. Microbiol.,* 18, 1393, 1972.
33. **Moein, G. J.,** Study of Distribution and Fate of PCBs and Benzenes after Spill of Transformer Fluid, EPA 904/9-76-014, U.S. Environmental Protection Agency, Washington, D.C., 1976, 128.
34. **Griffin, R. A. and Chou, S. F. J.,** Movement of PCBs and other persistent compounds through soil, *Water Sci. Technol.,* 13, 1153, 1981.
35. **Griffin, R. A. and Chian, E. S. K.,** *Attenuation of Water Soluble Polychlorinated Biphenyls by Earth Materials,* EGN 86, Illinois State Geological Survey, Champaign, 1979, 98.
36. **Veith, G. D. and Comstock, V. M.,** Sediments as a PCB source to aquatic systems, *J. Fish. Res. Bd. Can.,* 32, 1849, 1974.
37. **Wilson, A. J. and Forester, J.,** Persistence of Aroclor 1254 in a contaminated estuary, *Bull. Environ. Contam. Toxicol.,* 19, 637, 1978.
38. **Chen, N. Y.,** Hydrophobic properties of zeolites, *J. Phys. Chem.,* 80, 60, 1976.
39. **Nementhy, G. and Scherara, H. A.,** The structure of water and hydrophobic bonding in proteins. III. The thermodynamic properties of hydrophobic bonds in proteins, *J. Phys. Chem.,* 66, 1773, 1962.
40. **Horvath, C. and Melander, W.,** Reversed-phase chromatography and the hydrophobic effect, *Am. Lab.,* 10, 17, 1978.
41. **Frank, H. S. and Evens, M. W.,** Free volume and entropy in condensed systems. III. Entropy in binary liquid mixtures; partial molar entropy in dilute solutions: structure and thermodynamics vs. aqueous electrolytes, *J. Chem. Phys.,* 13, 507, 1945.
42. **Hassett, J. J., Banwart, W. L., Wood, S. G., and Means, J. C.,** Sorption of alpha-naphthol: implications concerning the limits of hydrophobic sorption, *Soil Sci. Soc. Am. J.,* 45, 38, 1981.
43. **Karickhoff, S. W., Brown, D. S., and Scott, T. A.,** Sorption of hydrophobic pollutants on natural sediments, *Water Res.,* 13, 241, 1979.
44. **Hassett, J. J., Banwart, W. L., and Griffin, R. A.,** Correlation of compound properties with sorption characteristics of nonpolar compounds by soils and sediments: concepts and limitations, in *Environment and Solid Wastes: Characterization, Treatment, and Disposal,* Francis, C. W. and Auerback, S. I., Eds., Butterworths, Reading, Mass., 1983, 161.
45. **Hassett, J. J., Means, J. C., Banwart, W. L., Wood, S. G., Ali, S., and Khan, A.,** Sorption of dibenzothiophene by soils and sediments, *J. Environ. Qual.,* 9, 184, 1980.
46. **Khan, A., Hassett, J. J., Banwart, W. L., Means, J. C., and Wood, S. C.,** Sorption of acetophenone by sediments and soils, *Soil Sci.,* 128, 287, 1979.
47. **Chiou, C. T., Peters, L. J., and Freed, V. H.,** A physical concept of soil-water equilibrium for nonionic organic compounds, *Science,* 206, 831, 1979.
48. **Means, J. C., Hassett, J. J., Wood, S. G., and Banwart, W. L.,** Sorption properties of energy-related pollutants and sediments. Polynuclear aromatic hydrocarbons, *3rd Int. Symp. on Chemistry and Biology-Carcinogenesis and Mutagnesis,* Jones, P. L. and Leber, P., Eds., Ann Arbor Science, Ann Arbor, Mich., 1979, 327.

49. **McDuffie, B.,** Estimation of octanol/water partition coefficients for organic pollutants using reverse-phase HPLC, *Chemosphere,* 10, 73, 1981.
50. **Branson, D. R.,** A new capacitor fluid: a case study in product stewardship, in *Aquatic Toxicology and Hazard Calculation,* ASTM STP 634, American Society for Testing and Materials, Philadelphia, 1977, 44.
51. **Griffin, R. A., Au, A. K., and Chian, E. S. K.,** Mobility of polychlorinated biphenyls and dicamba in soil materials: determination by soil thin-layer chromatography, Proc. 8th Natl. Conf. and Exhibit on Municipal Sludge Management, 1979, 183.
52. **Helling, C. S.,** Pesticide mobility in soils. I. Parameters of thin-layer chromatography, *Soil Sci. Soc. Am. Proc.,* 35, 732, 1971.
53. **Griffin, R. A. and Chou, S. F. J.,** *Attenuation of Polybrominated Biphenyls and Hexachlorobenzene by Earth Materials,* EGN 87, Illinois State Geological Survey, Champaign, 1980.
54. **Moza, P., Weisgerber, I., and Klein, W.,** Fate of 2,2′-dichlorobiphenyl-^{14}C in carrots, sugarbeets, and soil under outdoor conditions, *J. Agric. Food Chem.,* 24, 881, 1976.
55. **Suzuki, M.,** Translocation of polychlorobiphenyls in soil into plants, *Arch. Environ. Contam. Toxicol.,* 5, 343, 1977.
56. **McCall, P. J., Swann, R. L., Laskowski, D. A., Unger, S. M., Vrona, S. A., and Dishburger, H. J.,** Estimation of chemical mobility in soil from liquid chromatographic retention times, *Bull. Environ. Contam. Toxicol.,* 24, 190, 1980.
57. **Hetling, L., Horn, E., and Tofflemire, J.,** Summary of Hudson River PCB Study Results, New York State Department of Environmental Conservation, Bureau of Water Research, Albany, 1978.
58. **Horn, E. G., Hetling, L. L., and Tofflemire, T. J.,** The problem of PCBs in the Hudson River system, *Ann. N.Y. Acad. Sci.,* 1979, 591.
59. **Nisbet, L. C. T. and Sarofim, A. F.,** Rates and routes of transport of PCBs in the environment, *Environ. Health Perspect.,* 1, 21, 1972.
60. **Martell, J. M., Rickert, D. A., and Siegel, F. R.,** PCBs in suburban watershed, Reston, Va., *Environ. Sci. Technol.,* 9, 872, 1975.
61. **Roberts, J. R., Cherry, J. A., and Schwartz, F. W.,** A case study of a chemical spill: polychlorinated biphenyls (PCBs). I. History, distribution, and surface translocation, *Water Resources Res.,* 18, 525, 1982.
62. **Moein, G. J., Smith, A. J., Jr., Biglane, K. E., Loy, B., and Bennett, T.,** Follow-up Study of the Distribution and Fate of Polychlorinated Biphenyls and Benzene in Soil and Ground Water Samples after an Accidental Spill of Transformer Fluid, EPA 904/9-76-014, U.S. Environmental Protection Agency, Washington, D.C., 1976.
63. **Iwata, Y., Gunther, F. A., and Westlake, W. E.,** Uptake of a PCB (Aroclor 1254) from soil by carrots under field conditions, *Bull. Environ. Contam. Toxicol.,* 11, 523, 1974.
64. **Fries, G. F. and Marrow, G. S.,** Chlorobiphenyl movement from soil to soybean plants, *J. Agric. Food Chem.,* 29, 575, 1981.

Chapter 6

FACTORS CONTROLLING BIOACCUMULATION OF PCBs

G. R. Shaw and D. W. Connell

TABLE OF CONTENTS

I. INTRODUCTION

In recent years, there has been considerable interest in developing an understanding of the mechanisms leading to the distribution of contaminant chemicals in the environment. Such an understanding would not only improve our scientific knowledge of environmental processes but could be used in a variety of management situations.

An important segment of the environmental distribution process is the uptake of contaminant chemicals from the environment by biota. Persistent lipophilic chemicals, including PCBs have been shown to exhibit somewhat similar control mechanisms. A picture is gradually emerging of the influence of the physicochemical properties of lipophilic contaminants on their biotic behavior.

Before discussing the nature of mechanisms of transfer of chemicals from the environment to biota it is important to briefly consider the terminology used. A variety of descriptive terms are used in the literature for the various processes. In this discussion bioaccumulation or accumulation is used to describe uptake and retention of a chemical by any mechanism or pathway. Bioconcentration is used in a more restricted sense to represent uptake and retention of chemicals from the water mass through such tissues as gills or epithelial tissues.

II. MECHANISMS AND ROUTES OF BIOACCUMULATION

Biota could potentially acquire PCBs from three sectors of the environment — atmosphere, water, and food. With terrestrial organisms, uptake could occur by three mechanisms:

1. Absorption of PCBs in the atmosphere through the lung walls
2. Similar uptake from the atmosphere but through the epidermis
3. Consumption of food containing PCBs and passage through the stomach walls

The atmosphere route through the lungs and epidermis is of little significance due principally to the very low concentrations of PCBs in the atmosphere. The major route of acquisition of PCBs by terrestrial organisms is through contaminated food. Aquatic birds and mammals are also included in this category.

A similar set of potential uptake mechanisms could be visualized from aquatic organisms

1. Absorption of PCBs in the water mass through the gills
2. Similar uptake through the epidermis
3. Consumption of contaminated food

There is no significant information indicating that uptake through the epidermis is of general importance and the structure of the epidermis does not facilitate such a mechanism. On the other hand, there is sound evidence that all aquatic organisms studied in aquaria, can absorb PCBs directly from water.[1-7]Also, bioconcentration of PCBs by fish from water in natural systems has also been shown to occur.[8,9]

In fact, the main route of uptake by aquatic organisms has been shown to be absorption by the gills since the gills represent the active membrane surface for water exchange.[10-14] The gill membrane permeability for PCBs can be estimated by the calculation of their efficiency transfer coefficients, E.[15] The E values are estimated using uptake rate constants for PCBs, oxygen concentration in the water, the E value for oxygen, and the metabolic rate constant for the organism involved. Once absorbed by the gills, the PCBs are then partitioned into the blood and then from blood to tissues.[12,16]

An informative study of PCB levels in various organisms in Puget Sound, Wash. by Clayton et al.[17] demonstrated that all biota in that ecosystem had similar PCB levels nor-

Table 1
BIOCONCENTRATION FACTORS FOR INDIVIDUAL CHLOROBIPHENYLS

Compound	BCF
Trichlorobiphenyl	54
Tetrachlorobiphenyl	460
Pentachlorobiphenyl	1510

From Sanborn, J. R., Childers, W. F., and Metcalf, R. L., *Bull. Environ. Contam. Toxicol.*, 13, 209, 1975. With permission.

malized to lipid content. This indicates that food content was not the controlling factor. It has been suggested that the accumulation of PCBs by aquatic organisms is a physicochemical process and that uptake is predominantly governed by equilibrium partitioning of the PCBs between the organism and ambient water.[1,18,19] Nevertheless, the assimilation of PCBs from ingested food organisms occurs by partition across the lipoprotein membranes lining the gut into the bloodstream[16] but is usually of much less quantitative significance.

The uptake of PCBs by some benthic organisms may be controlled by different mechanisms to those associated with free swimming aquatic organisms. The association of PCBs with bottom sediments leads to contamination of the organisms[19-22] and the resultant concentration of PCBs are directly related to the concentrations in the sediment.[23-25]

III. PHYSICOCHEMICAL PROPERTIES CONTROLLING BIOACCUMULATION

Most information is available on the bioaccumulation of PCBs by aquatic organisms. It is usually measured as the ratio between the concentration in the animal to concentration in water, i.e., BCF (bioaccumulation factor = C animal/C water). A variety of factors have been shown to influence bioaccumulation as outlined below.

A. Degree of Chlorination

The degree of chlorination of the PCB molecule is known to have an effect on bioaccumulation of PCBs with the penta- and hexachloro isomers being accumulated to the greatest extent.[19,26] Table 1 lists bioconcentration factors for PCBs from water by the green sunfish *(Lepomis cyanellus)* as determined by Sanborn et al.[27] In addition, the halflives of various PCBs in organisms have been shown to be related exponentially to their degree of chlorination by weight.[28]

B. Water Solubility and Partition Coefficients

With lipophilic compounds in general, a linear inverse relationship has been shown to exist between water solubility and BCF such that log BCF = − A log (water solubility) + C, where A and C are constants.[29-31] The use of water solubility as a simple and rapid measure of a PCBs potential for bioaccumulation has great appeal. However, some chemicals with low water solubility, including PCBs, deviate significantly from this relationship.[29] Also, the numerous conflicting solubility values in the literature[19,31] and the lack of solubility data for many chemicals have led to the use of another parameter, the *n*-octanol/water partition coefficient (P_{ow}) for estimation of bioconcentration factors.[32] The approximately

Table 2
RELATIONSHIP BETWEEN LOG BCF AND LOG P_{ow}

Relationship	Ref.
$\log BCF = 0.56 \log P_{ow} + 0.124$	34
$\log BCF = 0.85 \log P_{ow} - 0.70$	38
$\log BCF = 1.119 \log P_{ow} - 1.579$	54
$\log BCF = 0.76 \log P_{ow} - 0.23$	9

inverse relationship between water solubility and P_{ow} has been known for some time.[33] It has been shown that bioaccumulation from water by aquatic organisms is correlated with lipophilicity for PCBs and other chemicals.[9,34-39] Table 2 presents relationships between BCF and P_{ow} for a wide variety of lipophilic compounds, including PCBs. However, PCBs exhibit some significant deviations from this relationship as outlined in the section on the effects of molecular stereochemistry.

The values of P_{ow} are usually not determined directly from a simple partition between *n*-octanol and water because the concentrations in water at the extremely low concentrations involved are difficult to measure accurately.[31] A variety of methods can be used as outlined below.

The Hansch et al.[40] method involves the calculation of P_{ow} using substituent π constants, where the substituent constant is defined as: $\pi = \log P_X - \log P_H$ where P_X is the partition coefficient of the derivative group and P_H that of the parent molecule. Rearranging the equation gives:

$$\log P_X = \text{substituent constants} + \log P_H$$

A large number of substituent constant values and log P_H values have been tabulated by Lee et al.[41] The substituent constant π is a free energy related constant and is a measure of the relative free energy change resulting from moving a derivative from one phase to another.[42] A different method of calculating partition coefficients using the hydrophobic fragmental constant, f, introduced by Rekker[43] is based on the same approach. However, with PCBs there are 209 possible compounds, varying in number and position of chlorine atoms and neither the Hansch or Rekker system offers an acceptable approach to calculate log P values, since in these calculations the π or f values used are the same for each chlorine atom in the molecule, regardless of its position.

Another approach to predicting BCFs is the use of i* and o* characters as described by Matsuo[44] and utilizing the following relationships:

$$\log BCF = -0.016\, i^* + 0.014\, o^*$$

$$\log BCF = -6.3 \frac{i^*}{o^*} + 5.4$$

The i* character is both additional and proportional to the strength of a polar group unit (e.g., -OH; -COOH) with polar interactions such as H-bonding or u-u interaction. Similarly, o* is associated with nonpolar interactions such as Van der Waals or dispersion forces. Matsuo[45] concludes that bioaccumulation of chemicals in fish is caused by the total interactions at polar and nonpolar areas between the PCB molecule and fish tissues. Similar shortcomings to those observed with the Hansch and Rekker systems are found when predicting bioaccumulation factors using this method.

The most widely used method involves high performance liquid chromatography using

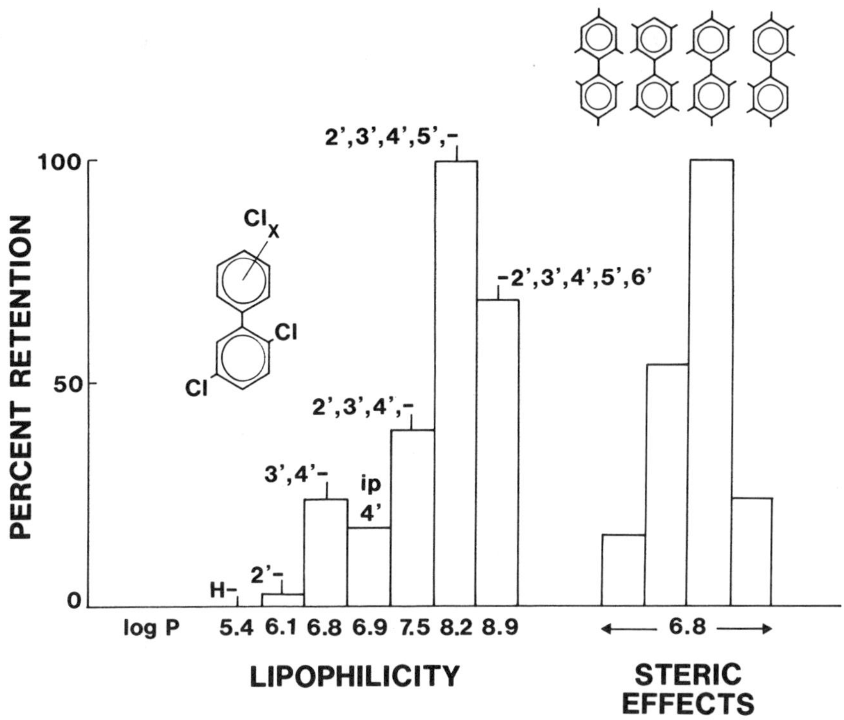

FIGURE 1. Relative retention of substituted biphenyls in abdominal fat of rats 2 weeks after single dose administration by stomach tube. Values are relative to the compound which is retained in highest amounts (= 100). Calculated log P values indicated at baseline. (From Hutzinger, et al. in *Aquatic Pollutants: Transformation and Biological Effects,* Hutzinger, O. et al., Eds., Pergamon Press, Oxford, 1978, 13. With permission.)

an aqueous phase, resembling water, and a permanently bonded C_{18} reversed phase, resembling *n*-octanol. With this system log retention time of lipophilic compounds has been shown to be linearly related to log P_{ow}.[38] This procedure is recognized as being the most appropriate because of the difficulties and inaccuracies in direct and calculation methods.[37,38,46,47]

C. PCB Stereochemistry

The correlation between BCF and log P_{ow} indicates that with increasing log P_{ow}, i.e., increasing lipophilicity of PCB molecules, there will be a greater tendency to bioaccumulate. However, there are some exceptions to this generalization as shown in Figure 1[37] which indicates the retention of PCBs in rat abdominal fat as influenced by lipophilicity and steric factors. The left part of Figure 1 indicates that there is not a linear relationship between lipophilicity and bioaccumulation (expressed as retention). The right part shows the different degrees of bioaccumulation exhibited by symmetrical hexachlorobiphenyls which have the same calculated log P_{ow} value (6.8). Similarly, this is consistent with the observations of Tulp and Hutzinger[31] who found that PCBs with comparable lipophilicities may exhibit different bioaccumulation factors. They concluded that there is an optimal steric configuration and molecular size for processes such as membrane passage, thereby affecting bioaccumulation. Similar results for PCBs have been found by other authors.[37,44,47-49]

More recent investigations have been conducted by Shaw and Connell[25,50,51] using polychaetes *(Capitella capitata)* and sea mullet *(Mugil cephalus)* in aquaria together with field investigations. A number of steric factors were found to influence the bioaccumulation of PCBs. A steric effect coefficient (SEC) was developed[50] and later revised[25] as shown in Table 3. These authors have suggested that bioaccumulation is controlled by both lipophilicity

Table 3
REVISED COEFFICIENTS FOR DIFFERENT CHLORINE SUBSTITUTION PATTERNS USED TO CALCULATE THE STERIC EFFECT COEFFICIENT (SEC)

Description	Structure	Coefficient
Three chlorines in the 2,2′,6,6′ positions	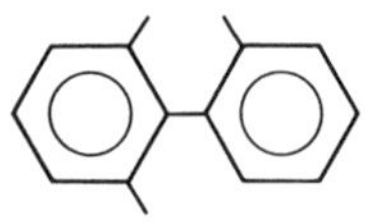	0.80
Four chlorines in the 2,2′,6,6′ positions	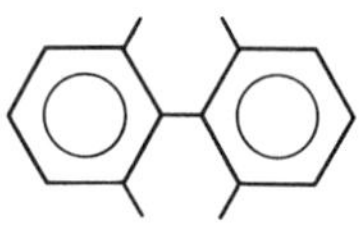	0.60
Two chlorines in the 2,6 or 2′,6′ positions	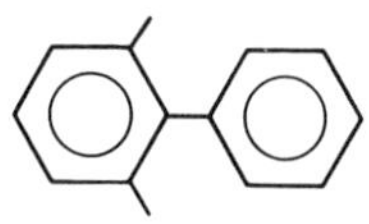	0.85
Four chlorines adjacent	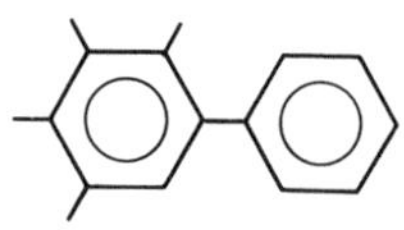	0.80 (1 ring) 0.60 (2 rings)
Three chlorines adjacent	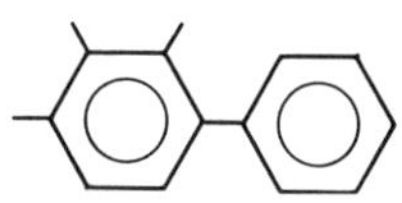	0.95 (1 ring) 0.90 (2 rings)
Chlorines in 3 or 5 positions if adjacent to chlorine in 2 or 6 positions	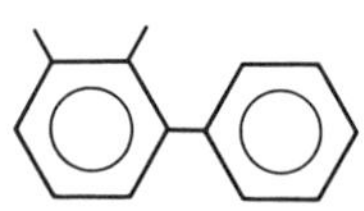	0.95 (1 in molecule) 0.90 (2 in molecule) 0.85 (3 in molecule) 0.80 (4 in molecule)
Chlorines in 3 or 5 positions not adjacent to chlorine	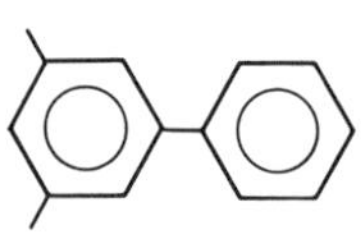	Subtract 0.02 from SEC for each chlorine in this configuration
No chlorines in the 2,2′,6,6′ positions	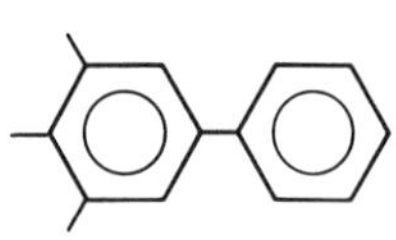	SEC is one irrespective of substitution pattern

From Shaw, G. R. and Connell, D. W., *Aust. J. Mar. Freshwater Res.*, p. 1057, 1982. With permission.

(P_{ow}) and steric effects (SEC) and a combination of these has a direct relationship to bioaccumulation (see Figure 2).

The stereochemistry of the PCB molecules affects the strength of adsorption of these compounds onto surfaces. Huckins et al.[52,53] have demonstrated that PCB molecules can be separated by chromatography on charcoal. The most planar molecules are most strongly absorbed and those with more irregular shapes, particularly the two aromatic rings out of plane, are comparatively weakly adsorbed. Chlorine substituents in the *ortho* positions give

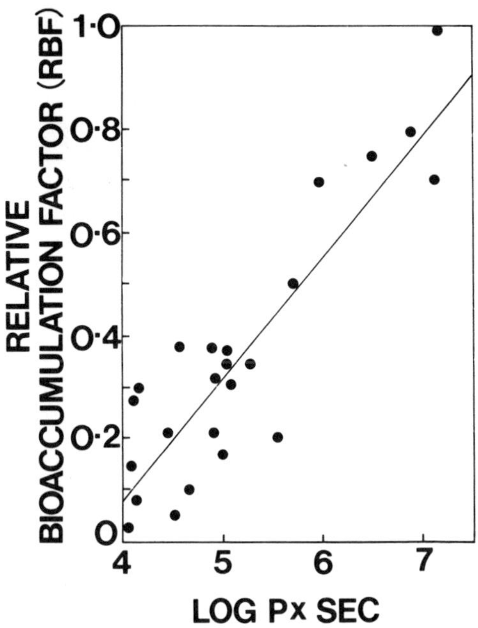

FIGURE 2. Relationship between relative bioaccumulation factor and the product of log P_{ow} and SEC. (From Shaw, G. R. and Connell, D. W., *Aust. J. Mar. Freshwater Res.*, p. 1057, 1982. With permission.)

this latter effect while chlorines in adjacent positions cause distortion of the aromatic rings (see Table 3). Consistent with these observations, the SEC developed by Shaw and Connell[51] is closely correlated with the elution time on charcoal and found that this combined with log P_{ow} provides a more accurate prediction for bioaccumulation (see Figure 3).

Thus, it has been shown that bioaccumulation of PCBs by aquatic organisms is influenced by both lipophilicity and adsorption characteristics of the PCB molecules. The adsorption of PCBs can be measured as elution volumes by chromatography on charcoal or by the SEC. The combination of lipophilicity and adsorption expressed as (log P $\times$ elution time) or (log P $\times$ SEC) is highly correlated with uptake. Shaw and Connell[51] have suggested that the bioaccumulation of PCBs by aquatic organisms involves adsorption of these substances onto membranes, such as the gills or stomach lining, and subsequent passage into the organism's lipid tissues.

IV. THERMODYNAMIC ASPECTS OF BIOACCUMULATION

In the process of bioconcentration, the organic molecules leaving the aqueous phase are adsorbed onto the membrane surface of the organism before penetration into the organism occurs.[42,54] Adsorption processes involve a decrease in free energy. Since the entropy change for adsorption must be negative, the system becoming more ordered with the surface binding, the enthalpy change must be positive. The heat of adsorption can thus give some index of the strength of binding. Smaller heats of adsorption (less than 10 kcal/mol) usually denote physical adsorption and involve Van der Waals interactions.[55]

The free energy related substituent constant, π, the hydrophobic fragmental constant, f, and the Hammett function, δ, used in structure activity relationships, are all measures of the relative free energy change resulting from moving a PCB molecule from one phase to another.[42,43] Matsuo[56] has calculated the free energy of accumulation of PCBs by fish from water as 2 to 6 kcal/mol using this expression:

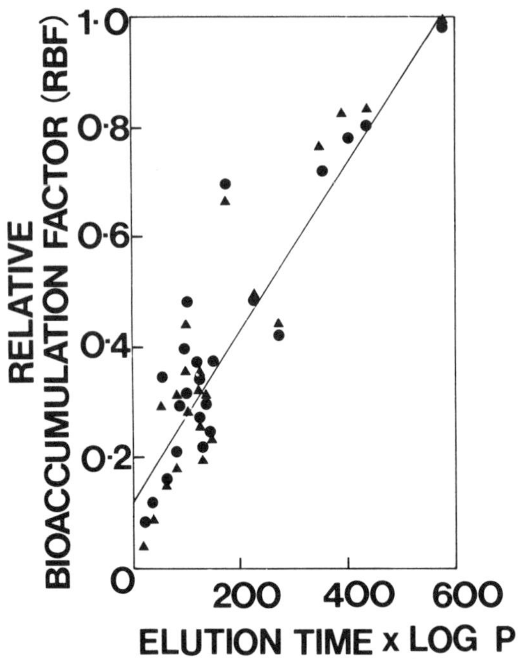

FIGURE 3. Relationship of relative bioaccumulation factor to the product of elution time (on charcoal chromatography) and log P_{ow}. (From Shaw, G. R. and Connell, D. W., *Aust. J. Mar. Freshwater Res.*, p. 1057, 1982. With permission.)

$$-\Delta\mu^\circ = 2.303RT \log (BCF)$$

where: $-\Delta\mu^\circ$ is the free energy of accumulation, R, the universal gas constant, and T, temperature. From the low values obtained, Matsuo[56] concluded that the driving force for accumulation originated in an energetic head drop corresponding to one hydrogen bonding or two or three Van der Waals forces.

When a nonpolar PCB molecule leaves the aqueous phase for the lipid phase an increase in entropy results.[40] The enthalpy and entrophy of bioaccumulation of PCB 1254 by fathead minnows *(Pimephales promelas),* green sunfish *(Lepomis cyanellus),* and rainbow trout *(Salmo gairdneri)* have been calculated.[56] To obtain enthalpy, ΔH°, and entrophy, ΔS°, R ln BCF was plotted against inverse absolute temperature change and the gradient and intercept on the ordinate were calculated by the method of least squares. These values were assigned to $-\Delta H^\circ$ and ΔS° respectively, according to the following equation:

$$R \ln (BCF) = -H^\circ/T + S^\circ$$

where; R is the universal gas constant; ΔH°, the enthalpy; S° the entrophy and T, temperature. Free energy is related to enthalpy and entrophy by the following expression:

$$\Delta\mu^\circ = \Delta H^\circ - T \cdot \Delta S^\circ$$

In the uptake of PCBs by fish, Matsuo[56] found that both ΔH° and ΔS° are positive. The positive enthalpy change means that the process of bioaccumulation is endothermic and requires heat. The positive and very large values of ΔS° obtained imply that the transfer of the PCB from water to fish increases the randomness of the PCB molecules to a large extent.

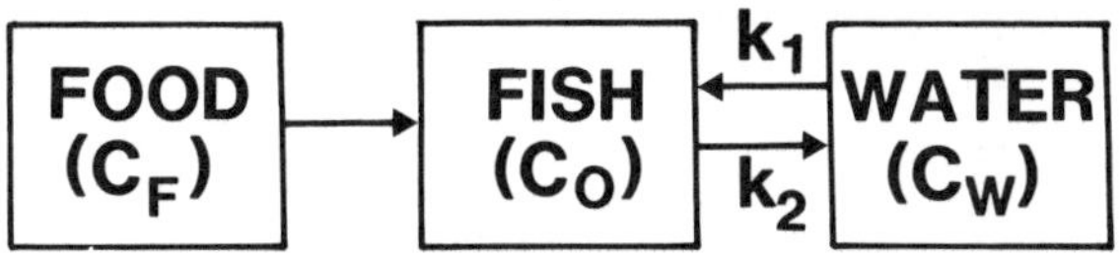

FIGURE 4. A model for the bioaccumulation of PCBs by fish from food and water.

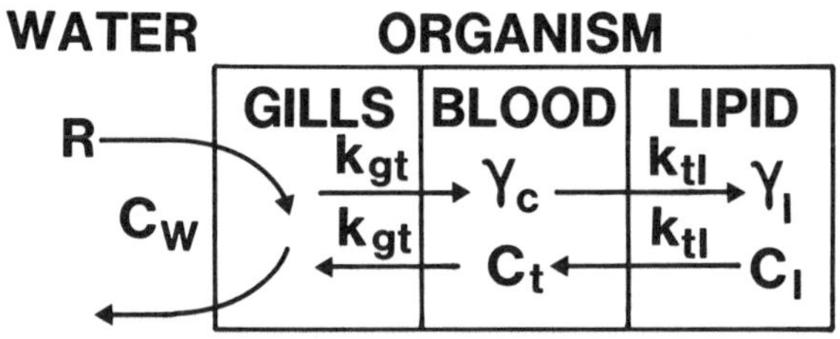

FIGURE 5. A compartmental model for the bioaccumulation of PCBs by an aquatic organism. (From Bruggerman, W. A. et al., *Chemosphere*, 10, 811, 1981. With permission.)

Matsuo[56] concluded that a large increase in entropy occurs because the water molecules in the vicinity of PCB molecules are arranged in a quasi-crystalline structure. Therefore, bioaccumulation of PCBs by fish is controlled by a large positive entropy change.

V. MODELS FOR BIOACCUMULATION

A variety of models have been developed to simulate uptake of lipophilic compounds by organisms. For example, Kerr and Vass[57] assumed that in fish and the larger more complex invertebrates, uptake occurs primarily through ingestion of contaminated food and direct absorption from water through the gills. Losses occur to the water through excretion in a chemically unchanged or a modified form. Figure 4 can be seen in accordance with this, the processes, and the following expression obtained assuming first order kinetics:[15,58]

$$\frac{dC_o}{dt} = \epsilon\, fC_F + k_1\, C_o - k_2\, C_w$$

where C_o is the concentration in the organism; ϵ is the absorption efficiency for the ingested chemical; f is the feeding rate; C_F is concentration in food; k_1 is the uptake rate constant; k_2 is the loss rate constant; and C_w is the concentration in water.

A more comprehensive model by Bruggerman et al.[15] assumes the body lipids are the storage compartment; the blood, the transport compartment, and the gills are the exchange or gate compartment, through which water is pumped. All compartments are considered homogeneous in relation to the concentration of the bioaccumulating chemical. The transfer rate from one compartment to another is proportional to the difference in thermodynamic activities, γ, and is dependent on the surface properties and area of the exchange barrier (such as a biomembrane), expressed together as the transfer coefficient, k. This model is diagrammatically represented in Figure 5 and is characterized by the equations below:

Net uptake in fish:

$$F\left(\frac{dC_f}{dt}\right) = R\,(C_w - C_g)$$

Net increase in the gate compartment:

$$G\left(\frac{dC_g}{dt}\right) = R\,(C_w - C_g) - k_{gt}\,(\gamma_w\,C_g - \gamma_t\,C_t)$$

Net increase in the transport compartment:

$$T\left(\frac{dC_t}{dt}\right) = k_{gt}\,(\gamma_w\,C_g - \gamma_t\,C_t) - k_{gt}\,(\gamma_t C_t - \gamma_l\,C_l)$$

Increase in lipid compartment:

$$L\left(\frac{dC_l}{dt}\right) = k_{tl}\,(\gamma_t C_t - \gamma_l C_l)$$

where C_g, C_t, and C_l are the concentrations of the chemical in each compartment; C_w is the concentration in ambient water; C_f is the concentration in fish; γ_w, γ_t, γ_l are activity coefficients of the chemical in the different compartments; F, G, T, and L are the mass of the different compartments; and k_{gt} and k_{tl} are transfer coefficients for exchange between compartments; and R is ventilation rate.

The capacity of the transport compartment plays an important role after a sudden change in the aqueous concentration, e.g., at the start of an experiment. In this case, the unloading of the transport compartment will initially dominate the depuration kinetics, followed later by the partitioning between the lipid and transport compartments, producing a two-phase clearance curve.[15] Other authors have found that clearance rate constants tend to decrease during clearance experiments if two or more compartments are involved which are cleared at different rates.[59] Some authors however, have attempted to describe this phenomenon by the use of concentration dependent rate constants and second order kinetics.[46] This implies that a lower dose of the chemical would result in a smaller clearance rate constant. However, assuming passive transport mechanisms for persistent, nonpolar lipophilic chemicals, first order kinetics are probably most accurate.[15]

VI. CONCLUSIONS

The most important routes of uptake of PCBs are via contaminated foods and contaminated foods with terrestrial organisms through the gills by aquatic organisms. With aquatic organisms the gill route is generally the most significant.

The physicochemical properties of the individual PCBs are the major factors controlling bioaccumulation. Two factors have been identified as of principal significance. First, is the P_{ow} which is a measure of the equilibrium partition between organism lipids, the site of PCB deposition, and the ambient water mass. Second, are the surface adsorption properties, which can be measured by chromatographic techniques. This is believed to represent the strength of adsorption of the PCBs on the gill membrane surface before uptake. A combination of these factors has a direct relationship to bioaccumulation of individual PCBs.

A number of models have been developed which provide a reasonably accurate representation of the bioaccumulation mechanism. The use of first order kinetics with a variety of different compartments within the organism probably provides the most accurate model.

REFERENCES

1. World Health Organization, *Polychlorinated Biphenyls and Terphenyls,* WHO, Geneva, 1976.
2. **Herbst, E., Scheuvert, I., Klein, W., and Korte, F.,** Uptake and conversion of 2,5,4-trichlorobiphenyl-^{14}C, 2,4,6,2′,4′-pentachlorobiphenyl-^{14}C and chloroalkylene-9-^{14}C by goldfish after a single water treatment, *Chemosphere,* 3, 221, 1978.
3. **Mayer, F. L., Mehle, P. M., and Sanders, H. O.,** Residue dynamics and biological effects of polychlorinated biphenyls in aquatic organisms, *Arch. Environ. Contam. Toxicol.,* 5, 501, 1977.
4. **Stalling, D. L. and Mayer, F. L., Jr.,** Toxicities of PCB's to fish, and environmental residues, *Environ. Health Perspect.,* 1, 159, 1972.
5. **Sanders, H. O. and Chandler, J. H.,** Biological magnification of a polychlorinated biphenyl (Aroclor 1254) from water by aquatic invertebrates, *Bull. Environ. Contam. Toxicol.,* 7, 257, 1972.
6. **Hansen, D. J., Parrish, P. R., Lowe, J. L., Wilson, A. J., Jr., and Wilson, P. D.,** Chronic toxicity, uptake and retention of Aroclor 1254 in two estuarine fishes, *Bull. Environ. Contam. Toxicol.,* 6, 113, 1971.
7. **Keil, J. E., Priester, L. E., and Sandifer, S. H.,** Polychlorinated biphenyl (Aroclor 1254): effects of uptake on growth, nucleic acids and chlorophyll of a marine diatom, *Bull. Environ. Contam. Toxicol.,* 6, 156, 1971.
8. **Shea, J. C., Simonin, H. A., Dean, H. J., Colquhoun, J. R., Spagnoli, J. J., and Veith, G. D.,** Bioaccumulation of Aroclor 1016 in Hudson River fish, *Bull. Environ. Contam. Toxicol.,* 22, 332, 1979.
9. **Veith, G. C., Macek, K. J., Petrocelli, S. R., and Carrol, J.,** An evaluation of using partition coefficients and water solubility to estimate bioconcentration factors for organic chemicals in fish, *Aquatic Toxicol.,* 707, 116, 1980.
10. **Phillips, D. J. H.,** *Quantitative Aquatic Biological Indicators — Their Use to Monitor Trace Metal and Organochlorine Pollution,* Applied Science, London, 1980.
11. **Nadeau, R. J. and Davis, R. A.,** Polychlorinated biphenyls in the Hudson River (Hudson Falls-Fort Edward, New York State), *Bull. Environ. Contam. Toxicol.,* 16, 436, 1976.
12. **Kenaga, E. E.,** Transport of insecticides by blood and body fluids, in *Toxicology of Insecticides,* Matsumura, F., Ed., Plenum Press, New York, 1975, 92.
13. **Gunkel, G. and Streit, B.,** Mechanisms of bioaccumulation of a herbicide (atrazine, S-triazine) in a freshwater mollusc *(Ancylus flureiatilio)* and a fish *(Coregonus ferajurine), Water Res.,* 14, 1573, 1980.
14. **Addison, R. F.,** Organochlorine compounds in aquatic organisms, their distribution, transport and physiological significance, in *Effects of Organic Pollutions on Aquatic Organisms,* Lockwood, A. P. M., Ed., Cambridge University Press, Cambridge, England, 1976, 126.
15. **Bruggerman, W. A., Martron, L. B. J. M., Kooiman, O., and Hutzinger, O.,** Accumulation and elimination kinetics of di-, tri-, and tetrachlorobiphenyls by goldfish after dietary and aqueous exposure, *Chemosphere,* 10, 811, 1981.
16. **Walker, C. H.,** Variations in the intake and elimination of pollutants, in *Organochlorine Insecticides: Persistent Organic Pollutants,* Morarity, F., Ed., Academic Press, New York, 1975, 77.
17. **Clayton, J. R., Pavlou, S. P., and Brietner, N. F.,** Polychlorinated biphenyls in coastal marine zooplankton. Bioaccumulation by equilibrium partitioning, *Environ. Sci. Technol.,* 11, 676, 1977.
18. **Pavlou, S. P. and Dexter, R. M.,** Distribution of polychlorinated biphenyls (PCB) in estuarine ecosystems. Testing the concept of equilibrium partitioning in the marine environment, *Environ. Sci. Technol.,* 13, 65, 1979.
19. **Kalmaz, E. V. and Kalmaz, G. D.,** Transport, distribution and toxic effects of polychlorinated biphenyls in ecosystems: review, *Ecol. Model.,* 6, 223, 1979.
20. **Langston, W. J.,** Accumulation of polychlorinated biphenyls in the cockle *Cerastoderma edule* and the tellin *Macoma balthica, Mar. Biol.,* 45, 265, 1978.
21. **Sodergren, A. and Larsson, P.,** Transport of PCB's in aquatic laboratory model ecosystems from sediment to the atmosphere via the surface microlayer, *Ambio.,* 11, 41, 1982.
22. **Halter, M. T. and Johnson, H. E.,** A model system to study the desorption and biological availability of PCB in hydrosoils, in *Aquatic Toxicology and Hazard Evaluation,* Mayer, F. L. and Hamelink, J. L., Eds., ASTM/STP 634, American Society for Testing and Materials, Philadelphia, 1977, 178.
23. **Nimmo, D. R., Wilson, P. D., Blackman, R. R., and Wilson, A. J., Jr.,** Polychlorinated biphenyl absorbed from sediments by fiddler crabs and pink shrimp, *Nature (London),* 231, 50, 1971.
24. **Murray, H. E., Ray, L. E., and Grain, C. S.,** Analysis of marine sediment, water and biota for selected organic pollutants, *Chemosphere,* 10, 1327, 1981.
25. **Shaw, G. R. and Connell, D. W.,** Factors influencing concentrations of polychlorinated biphenyls in organisms from an estuarine ecosystem, *Aust. J. Mar. Freshwater Res.,* P. 1057, 1982.
26. **Gruger, E. H., Karrick, N. L., Davidson, A. I., and Kruby, T.,** Accumulation of 3,4,3′,4′-tetrachlorobiphenyl and 2,4,5,2′,4′,5′- and 2,4,6,2′,4′,6′-hexachlorobiphenyl in juvenile coho salmon, *Environ. Sci. Technol.,* 9, 121, 1975.

27. **Sanborn, J. R., Childers, W. F., and Metcalf, R. L.,** Uptake of three polychlorinated biphenyls, DDT and DDE by the green sunfish *Lepomis cyanellus, Bull. Environ. Contam. Toxicol.,* 13, 209, 1975.
28. **Matthews, H. B. and Anderson, M. W.,** The effects of chlorination on the distribution and excretion of PCB's, *Toxicol. Appl. Pharmacol.,* 33, 183, 1975.
29. **Geyer, H., Viswanathan, R., Freitag, D., and Korte, F.,** Relationship between water solubility of organic chemicals and their bioaccumulation by the alga *Chlorella, Chemosphere,* 10, 1307, 1981.
30. **Haque, R. and Schmedding, D.,** A method of measuring the water solubility of hydrophobic chemicals: solubility of five polychlorinated biphenyls, *Bull. Environ. Contam. Toxicol.,* 14, 13, 1975.
31. **Tulp, M. Th. M. and Hutzinger, O.,** Some thoughts on aqueous solubilities and partition coefficients of PCB and the mathematical correlation between bioaccumulation and physicochemical properties, *Chemosphere,* 10, 849, 1978.
32. **Hollifield, H. C.,** Rapid nephelometric estimate of water solubility of highly insoluble organic chemicals of environmental interest, *Bull. Environ. Contam. Toxicol.,* 23, 579, 1979.
33. **Hansch, C., Quinlan, J. E., and Lawrence, G. L.,** Relationship of n-octanol/water coefficient to water solubility, *J. Org. Chem.,* 33, 347, 1967.
34. **Neely, W. B., Branson, D. R., and Blau, G. E.,** Partition coefficient to measure bioconcentration potential of organic chemicals in fish, *Environ. Sci. Technol.,* 8, 1113, 1974.
35. **Chiou, C. T., Freed, V. H., Schmedding, D. W., and Kohnest, R. L.,** Relationship between bioaccumulation and partition coefficient, *Environ. Sci. Technol.,* 11, 475, 1977.
36. **Mackay, D., Bobra, A., and Shiu, W. Y.,** Relationship between aqueous solubility and octanol-water partition coefficients, *Chemosphere,* 9, 701, 1980.
37. **Hutzinger, O., Tulp, M. Th. M., and Zitko, V.,** Chemicals with pollution potential, in *Aquatic Pollutants: Transformation and Biological Effects,* Hutzinger, O., Van Lelyveland, I. H., and Zoiteman, B. C. J., Eds., Pergamon Press, Oxford, 1978, 13.
38. **Veith, G. D., De Foe, D. L., and Bergstedt, B. V.,** Measuring and estimating the bioconcentration factor of chemicals in fish, *J. Fish. Res. Bd. Can.,* 36, 1040, 1979.
39. **Konemann, H. and Van Leeuwen, K.,** Toxicokinetics in fish. Accumulation and elimination of six chlorobenzenes by guppies, *Chemosphere,* 9, 3, 1980.
40. **Hansch, C., Leo, A., and Nikaitani, D.,** On the additive-constitutive character of partition coefficients, *J. Org. Chem.,* 37, 3090, 1972.
41. **Lee, A., Hansch, C., and Elkins, D.,** Partition coefficients and their uses, *Chem. Rev.,* 71, 525, 1971.
42. **Hansch, C. and Fujita, T.,** ρ-σ-π Analysis: method for the correlation of biological activity and chemical structure, *J. Am. Chem. Soc.,* 86, 1616, 1964.
43. **Rekker, R. F.,** *The Hydrophobic Fragmental Constant,* Elsevier, Amsterdam, 1977.
44. **Matsuo, M.,** The 1*/0 — characters to relate accumulation facors of some halogenated biphenyls in fish, *Chemosphere,* 9, 61, 1980.
45. **Matsuo, M.,** The i*/o — characters to describe ecological magnification of some organophosphorus insecticides in fish, *Chemosphere,* 7, 477, 1979.
46. **Ellgehausen, H., Girth, J. A., and Esser, H. O.,** Factors determining the bioaccumulation potential of pesticides in the individual compartments of aquatic food chains, *Ecotoxicol. Environ. Safety,* 4, 134, 1980.
47. **McDuffie, B.,** Estimation of octanol/water partition coefficients for organic pollutants using reverse phase HPLC, *Chemosphere,* 10, 73, 1981.
48. **Vreeland, V.,** Uptake of chlorobiphenyls by oysters *(Crasseostrea virginica), Environ. Pollut.,* 6, 135, 1974.
49. **Mizutani, T., Hidaka, K., Ohe, T., Matsumoto, M., Yamomoto, K., and Tajima, K.,** Comparative study on accumulation and elimination of hexachlorobiphenyls and decachlorobiphenyl in mice, *Bull. Environ. Contam. Toxicol.,* 25, 181, 1980.
50. **Shaw, G. R. and Connell, D. W.,** Relationships between steric factors and bioconcentration of polychlorinated biphenyls (PCB's) by the sea mullet *(Mugil cephalus), Chemosphere,* 9, 731, 1980.
51. **Shaw, G. R. and Connell, D. W.,** Physicochemical properties controlling PCB concentrations in aquatic organisms, *Environ. Sci. Technol.,* in press, 1985.
52. **Huckins, J. N., Stalling, D. L., and Smith, W. A.,** Foam-charcoal chromatography for analysis of polychlorinated dibenzodioxins in herbicide orange, *J. Off. Anal. Chem.,* 61, 32, 1978.
53. **Huckins, J. N., Stalling, D. L., and Petty, J. D.,** Carbon-foam chromatographic separation of non-o'chlorine substituted PCBs from Arochlor mixtures, *J. Assoc. Off. Anal. Chem.,* 63, 750, 1980.
54. **Kenaga, E. E.,** Predicted bioconcentration factors and soil sorption coefficients of pesticides and other chemicals, *Ecotoxicol. Environ. Safety,* 4, 26, 1980.
55. **Tinsley, J.,** *Chemical Concepts in Pollution Behaviour,* Wiley Interscience, New York, 1979, 86.
56. **Matsuo, M.,** A thermodynamic interpretation of bioaccumulation of Aroclor 1254 (PCB) in fish, *Chemosphere,* 9, 671, 1980.

57. **Kerr, S. R. and Vass, W. P.,** Pesticide residues in aquatic invertebrates, in *Environmental Pollution by Pesticides,* Edwards, C. A., Eds., Plenum Press, New York, 1973.
58. **Branson, D. R., Blau, G. E., Alexander, I. S., and Neely, W. B.,** Partition coefficient to measure bioaccumulation of organic chemicals, *Trans. Am. Fish. Soc.,* 104, 785, 1975.
59. **Konemann, H. and Van Leeuwen, K. V.,** Toxicokinetics in fish: accumulation and elimination of six chlorobenzenes by guppies, *Chemosphere,* 9, 3, 1980.

Chapter 7

FACTORS CONTROLLING PCBs IN FOOD CHAINS

G.R. Shaw and D.W. Connell

TABLE OF CONTENTS

I. INTRODUCTION

Early in the investigations of persistent chemicals in the environment it was recognized that often top predators may exhibit comparatively high concentrations of these substances. The process leading to this phenomenon has been described as "food chain magnification", "ecological magnification", "bioamplification", "bioaccumulation along food chains", "trophic contamination", and "trophic magnification". However, "biomagnification" is currently the most widely accepted descriptive term. Biomagnification can be described as:

1. The transfer of a chemical through a food chain to produce a high concentration in top predators
2. An increase of a given material at successive trophic levels within an ecosystem[3]
3. A process by which the tissue concentrations of bioaccumulated chemical residues increase as these materials pass up the food chain through two or more trophic levels[4]

The most convincing evidence for the occurrence of this process is provided by the chlorinated hydrocarbon insecticides, particularly DDT and related compounds.[1] The process can be seen as a series of bioaccumulation steps through a food chain and thus a compound which does not bioaccumulate would not be expected to be biomagnified. The PCBs have octanol to water partition coefficients and bioaccumulation factors higher than DDT[5] and thus have a greater tendency to bioaccumulate in individual organisms. The biomagnification process, being a series of bioaccumulation steps, takes a considerably longer period to develop and thus comparatively high persistence in organisms is necessary. The PCBs are generally more persistent than DDT and related compounds.

Thus, the PCBs have somewhat similar environmental properties to DDT and related compounds but with a greater tendency to bioaccumulate and persist. By analogy, biomagnification could be expected with the PCBs in natural ecosystems.

II. THEORETICAL CONSIDERATIONS

The biomagnification mechanism described above results from uptake of PCBs in the food of organisms. Of course, an alternative uptake route of considerable importance is directly from the environment, e.g., from water through the gills with fish. With bioaccumulation and biomagnification the individual organisms do not differentiate between PCBs and other food contaminants on uptake but on depuration the organism retains the PCBs due to its resistance to degradation compared to normal food components. Energy is lost at each level of a trophic pyramid as a result of respiration and nonassimilated energy, and the energy content at any level is approximately 90% less than the level below it.[6] It follows then that a persistent compound that enters one trophic level will be concentrated approximately ten times in the next trophic level. In fact, organisms can remove foreign compounds taken up and food is not the only route through which an organism can take up a persistent chemical, especially in the case of aquatic organisms.

A mathematical model can be developed for the transport of PCBs through different trophic levels.[7] The basic principle of mass conservation criteria can be applied at each trophic level, n, equating the inflow of PCBs to a level to the rate of ingestion of organisms from lower levels, i, and equating the outflow of PCBs to the rates of death and excretion. Using the conditions that biomass at any level is in a steady-state condition then the concentration of PCBs in the nth predator level as a function of time can be written as:

$$\left[\frac{M_n}{M_{d,n}}\right]\frac{dC_n}{dt} + C_n = \left[\frac{1}{M_{d,n}}\sum_{i-1}^{n-1} \sigma_i\, M_i\, C_i - (M_r)_{R,n} - C_\gamma M_{\gamma,n}\right]$$

where M is the biomass at the trophic level, M_d is the biomass death rate due to natural death or ingestion, $(M_r)_R$ is the rate of metabolism of the PCBs, M_γ is the rate of excretion, C_i is the concentration of PCBs in the nth predator level, C_γ is the concentration of PCBs in the excretions, and σ is an efficiency factor relating to the percent of ingested material which is retained in the trophic level.[7]

If it is assumed that the input of PCBs to the system is zero and the degradation rates are negligible then the equilibrium concentration for PCBs at the nth level can be written as:

$$C_n \text{ equilibrium} = \frac{T_n}{M_n}\left[\sum_{i-1}^{n-1} \sigma_i M_i C_i - C_\gamma C_{\gamma,n}\right]$$

where T_n is the life span of trophic level, n. This equation indicates that the equilibrium concentration of PCBs at any trophic level varies directly with the life span and inversely with the total mass, M_n, of the level.

Another approach[6] takes into account uptake from both food and water and assumes first order kinetics. Thus, for an individual organism consisting of a single homogeneous compartment the rate of uptake is

$$\frac{dQ}{dt} = C_o k_o$$

where Q is the amount of substance in the organism, C_o is the concentration of the substance in the external environment, and k_o is the uptake rate constant.

The loss of the substance by metabolism or excretion also follows first order kinetics; thus, overall the rate of bioaccumulation is

$$\frac{dQ}{dt} = C_o k_o - Qk$$

where k is the rate loss constant.

Differentiation and rearrangement of this expression give for the amount of substance in the organism at time t:

$$Q = \frac{C_o k_o}{k}(1 - e^{-kt})$$

thus:

$$C_a W = \frac{C_o k_o}{k}(1 - e^{-kt}) \tag{1}$$

where C_a is the concentration in the organism and W is its body weight.

This expression applies for uptake of the substance from the surrounding stable environment. However, with the biomagnification mechanism uptake will occur through contaminated food. Thus, taking both food and the external environment into account:

$$C_o\, k_o = Ck(\text{food}) + Ck(\text{water})$$

The intake from food with animals takes the general form of a W^{-b} where a and b are constants and W is its body weight. The intake from food can be written:

$$Ck(\text{food}) = a_1 W^{-b_1}\, r_1\, C_1$$

where a_1 and b_1 are constants, W is body weight, C_1 is the concentration in food, and r_i is the proportion of the concentration taken up. Since the volume of water passing over the gills is proportional to body weight a similar expression can be derived for uptake from water:

$$Ck(\text{water}) = a_2 W^{-b_2} r_2 C_2$$

where C_2 is the concentration of the substance in water and r_2 is the proportion taken up. Excretion may also be considered to vary with body size such that:

$$k = a_3 W^{\gamma - 1}$$

where a_3 and γ are constants.

Substituting the expressions above for Ck (food), Ck (water), and k in Equation 1 the following expression can be obtained for an organism at constant weight:

$$C_a = \frac{a_1 W^{-b_1} r_1 C_1 + a_2 W^{-b_2} r_2 C_2}{a_3 W^{\gamma}} 1 - e^{-a_3 W^{\gamma - 1} t}$$

This equation indicates a number of different aspects of biomagnification. The relative size of the coefficients preceding C_1 and C_2 is the principal factor indicating whether biomagnification is operating. If $C_1 \gg C_2$ then transfer in food is the major factor influencing the concentration of a substance in the organism and biomagnification is in operation.

Biomagnification can also be considered to result from the partition process.[8] At equilibrium the following relationship exists:

$$K_B = \frac{C_B}{C_M}$$

where K_B is the bioaccumulation factor, C_B is the concentration in the organism, and C_M is the concentration in the surrounding environment.

At the lowest trophic level, phytoplankton take up a pollutant in equilibrium with water according to the following relationship derived from the equation above:

$$C_{PP} = K_B C_W \qquad (2)$$

where C_{pp} is the concentration of pollutant in phytoplankton (or detritus), K_B is the bioaccumulation factor, and C_W is the pollutant concentration in water. When the phytoplankton are consumed by a herbivore the following relationship applies if (1) the food (phytoplankton) is the only source of pollutant; (2) the pollutant has not been chemically modified; (3) sufficient time has elapsed for equilibrium to be attained; and (4) K_B is constant and applicable to the herbivore to phytoplankton relationship:

$$C_H = K_B C_{PP} \qquad (3)$$

where C_H is the concentration in the herbivore.

Since food is the only source of pollutant, the equilibrium in this case will be between the stomach contents of the organism and the organism itself rather than the organism and the surrounding environment.

Substituting Equation 2 in Equation 3 the following is obtained:

$$C_H = (K_B)^2\, C_W$$

Similarly, the following general relationship can be established for the conditions outlined above:

$$C_{TR} = (K_B)^{TR}\, C_W$$

where C_{TR} is the concentration in organisms at a particular trophic level and TR is the number of transfers of the pollutant in the food web. It would be expected that concentrations would be somewhat lower than this due to losses from the organism by equilibrium establishment between the organism and the water mass.

Thus, a sequential increase in concentration could be expected in a food chain with organisms at the highest levels exhibiting the highest concentrations. However, in many natural food chains the conditions necessary for this to occur may not exist.

IV. OCCURRENCE OF PCB BIOMAGNIFICATION IN NATURAL ECOSYSTEMS

It is convenient to consider ecosystems in two classes: terrestrial and aquatic. Organisms in terrestrial ecosystems have only one significant source of PCBs, contaminated food, whereas in aquatic ecosystems both food and water are significant potential sources.

A. Terrestrial Ecosystems

It is generally accepted that biomagnification can occur with terrestrial ecosystems where food is the main source of PCBs for organisms.[9-11] For example, fish-eating birds and predatory raptors at the top of the food web are often found to contain high levels of PCBs.[4,12] In addition, Presst et al.[13] have established a correlation between PCBs in the livers of wild birds and their diets. Less than 1 mg/kg PCB was found with insectivorous birds while more than 70 mg/kg was found in the sparrow hawk, a top carnivore.

B. Aquatic Ecosystems

The significance of biomagnification of persistent residues as compared to direct uptake from water has been the subject of a considerable debate. Addison[14] has concluded that in invertebrates, both ingestion and contamination from water contribute to uptake but ingestion appears to predominate with increasing size, and is the main uptake route in fish.

A variety of other investigations have indicated that biomagnification of PCBs can occur in aquatic ecosystems. In a study of pollutant levels in marine ecosystems in southern California, Young et al.[15] found that increases in PCB 1254 concentrations were related to increases in the trophic level. The clearest relationship between PCB concentration and trophic position was usually obtained when the concentrations in wet weight were converted to a lipid basis. The authors attributed the biomagnification effect to the relatively long biological halflives of the PCBs. As a result of studies on the PCB levels of organisms in the Hudson River, Nadeau and Davis[16] have reported that a possible pathway for bioaccumulation exists between snails and larger game fish.

A comprehensive review of the literature information available on PCBs in diverse aquatic food chains has been carried out by Thomann[27] using a model food chain transfer number, f:

$$f = \alpha C/K + G$$

where α is the chemical absorption efficiency; C is the specific consumption; K is the excretion rate, and G is the net organism growth. Thomann[27] found that PCB levels in top predators are almost entirely due to biomagnification. In accord with this, the World Health Organization[4] has evaluated levels of PCBs found in marine ecosystems which concluded that the top predators in polluted areas generally exhibit the highest concentrations. This suggests that biomagnification may be generally occurring with the top predators. With fish Butler et al.[17] concluded that pollutant levels do not represent levels in the marine environment but are probably the result of storage and recycling in different links of the food web.

The alternative view proposed by Hamelink et al.[18] is that organochlorine levels in organisms depend on the physicochemical properties of the organochlorine and particularly those properties which control its tendency to distribute itself between lipoid and aqueous phases.

In an analysis of literature data, Moriarty[19] found that phytoplankton and zooplankton, at the lowest trophic levels, often contain PCB residues at higher levels than higher trophic level predators. Direct contamination from water and PCB-rich surface microlayers have been suggested as the most important uptake route.[9,20] Similarly, Shaw and Connell[26] found no evidence of biomagnification at lower trophic levels but carnivorous birds exhibited this phenomenon.

After extensive research into factors determining the bioaccumulation potential of chemicals in aquatic food chains, Ellgehausen et al.[21] concluded that biomagnification was of lesser importance than direct uptake from water. A similar conclusion has been reported by Bruggerman et al.[2] from studies with the goldfish. Investigations into ecosystems in the Baltic Sea[22] and an Oklahoma stream[23] did not reveal any relationship between trophic level and PCB concentration.

This and other data have led several authors[24] to conclude that, for pelagic biota at the lower trophic levels, food chain biomagnification is not a controlling factor in attaining observed PCB levels. Bioaccumulation is predominantly controlled by equilibrium partitioning of the chemical between the internal lipids of the biota and ambient water. In addition, Gruger et al.[25] have postulated that no matter how high the PCB concentrations in food are, the equilibrium level in the aquatic organism will be controlled by partitioning back into the water.

IV. CONCLUSIONS

The available evidence indicates that, as a general rule, biomagnification and food chain transfer are the major mechanisms for deposition of PCBs in the top carnivore members of terrestrial ecosystems. However, the overall significance of biomagnification in aquatic ecosystems is less certain. Direct uptake from water and sediments appears to be the dominant process for many persistent pollutants including PCBs in unstructured food webs and at lower trophic levels. Thus, equilibrium mechanisms between individual organisms and the water mass are the controlling factors. However, at higher trophic levels the information available suggests that biomagnification is generally the major mechanism of PCB deposition.

REFERENCES

1. **Rudd, R. L.,** *Pesticides and the Living Landscape,* University of Wisconsin Press, Madison, 1964.
2. **Bruggerman, W. A., Martron, L. B. J. M., Kooiman, D., and Hutzinger, O.,** Accumulation and elimination kinetics of di-, tri- and tetra-chlorobiphenyls by goldfish after dietary and aqueous exposure, *Chemosphere,* 10, 811, 1981.

3. GESAMP, Impact of Oil on the Marine Environment, No. 7, Joint Groups of Experts on the Scientific Aspects of Marine Pollution, Reports and Studies, GESAMP, 1977.
4. World Health Organization, *Polychlorinated Biphenyls and Terphenyls,* WHO, Geneva, 1976.
5. **Mackay, D.,** Correlation of bioconcentration factors, *Environ. Sci. Technol.,* 16, 274, 1982.
6. FAO/SIDA, Lectures presented at the Fifth FAO/SIDA Workshop on Aquatic Pollution in Relation to Protection of Living Resources, Food and Agricultural Organization, Rome, 1978, 268.
7. **Kalmaz, E. V. and Kalmaz, G. D.,** Transport, distribution and toxic effects of polychlorinated biphenyls in ecosystems: review, *Ecol. Model.,* 6, 223, 1979.
8. **Connell, D. W. and Miller, G. J.,** *Chemistry and Ecotoxicology of Pollution,* John Wiley & Sons, New York, in press, 1985.
9. **Peakall, D. B.,** PCB's and their environmental effects, *CRC Crit. Rev. Environ. Control,* 5, 468, 1975.
10. **Tinsley, J.,** *Chemical Concepts in Pollution Behaviour,* Wiley Interscience, New York, 1979, 86.
11. **Vermeer, K. and Peakall, D. B.,** Toxic chemicals in Canadian fish-eating birds, *Mar. Pollut. Bull.,* 8, 205, 1977.
12. **Bourne, W. R. P. and Bogan, W. A.,** Polychlorinated biphenyls in North Atlantic sea-birds, *Mar. Pollut. Bull.,* 3, 171, 1972.
13. **Presst, I., Jeffries, D. J., and Moore, N. W.,** Polychlorinated biphenyls in wild birds in Britain and their avian toxicity, *Environ. Pollut.,* 1, 3, 1970.
14. **Addison, R. F.,** Organochlorine compounds in aquatic organisms, their distribution, transport and physiological significance, in *Effects of Organic Pollutants on Aquatic Organisms,* Lockwood, A. P. M., Eds., Cambridge University Press, Cambridge, England, 1976, 126.
15. **Young, D. R., Mearns, A. J., Jan, T., Heesen, T. C., Moore, M. D., Eganhouse, R. P., Hershelman, G. P., and Gassett, R. W.,** Trophic structure and pollutant concentrations in marine ecosystems of Southern California, *Cal. COFI Rep.,* 21, 197, 1980.
16. **Nadeau, R. J. and Davis, R. A.,** Polychlorinated biphenyls in the Hudson River (Hudson Falls-Fort Edward, New York State), *Bull. Environ. Contam. Toxicol.,* 16, 436, 1976.
17. **Butler, P. A., Kennedy, C. D., and Schutzman, R. L.,** Pesticide residues in estuarine molluscs, 1977 versus 1972 — National Pesticide Monitoring Program, *Pest. Monit. J.,* 12, 99, 1978.
18. **Hamelink, J. L., Waybrant, R. C., and Ball, R. C.,** A proposal exchange equilibria to control the degree chlorinated hydrocarbons are biologically magnified in lentic environments, *Trans. Am. Fish. Soc.,* 100, 207, 1971.
19. **Moriarty, F.,** *Organochlorine Insecticides: Persistent Organic Pollutants,* Academic Press, New York, 1975.
20. **Nisbit, I. C. T. and Sarofim, A. F.,** Rates and routes of transport of PCB's in the environment, *Environ. Health Perspect.,* 1, 21, 1972.
21. **Ellgehausen, G., Guth, J. A., and Esser, H. O.,** Factors determining the bioaccumulation potential of pesticides in the individual compartments of aquatic food chains, *Ecotoxicol. Environ. Safety,* 4, 134, 1980.
22. **Clayton, J. R., Pavlou, S. P., and Brietner, N. F.,** Polychlorinated biphenyls in coastal marine zooplankton. Bioaccumulation by equilibrium partitioning, *Environ. Sci. Technol.,* 11, 676, 1977.
23. **Hunter, R. G., Randolph, J. C., and Carroll, J. H.,** Polychlorinated biphenyls in the sediments and fish of an Oklahoma stream, *Environ. Pollut. B,* 1, 233, 1980.
24. **Macek, K. J., Petrocelli, S. R., and Sleight, B. H.,** Consideration in assessing the potential for and significance of biomagnification of chemical residues in aquatic food chains, *Aquatic Toxicology,* ASTM/STP 667, American Society for Testing and Materials, Philadelphia, 1979, 251.
25. **Gruger, E. H., Karrick, N. L., Davidson, A. I., and Kruby, J.,** Accumulation of 3,4,3′,4′-tetrachlorobiphenyl and 2,4,5,2′,4′,5′- and 2,4,6,2′,4′,6′-hexachlorobiphenyl in juvenile coho salmon, *Environ. Sci. Technol.,* 9, 121, 1975.
26. **Shaw, G. R. and Connell, D. W.,** Factors influencing concentrations of PCB's in organisms from an estuarine ecosystem, *Aust. J. Mar. Freshwater Res.,* 33, 1057, 1983.
27. **Thomann, R. V.,** Equilibrium fate of micronutrients in diverse aquatic food webs, *Can. J. Fish. Aquat. Sci.,* 38, 280, 1981.

Chapter 8

DISTRIBUTION, BEHAVIOR, AND LOAD OF PCBs IN THE OCEANS

Shinsuke Tanabe and Ryo Tatsukawa

TABLE OF CONTENTS

I. INTRODUCTION

Polychlorinated biphenyls (PCBs) as well as DDT are the cause of much concern as manmade organic pollutants and many studies have been made of them in a wide range of environmental media and biota. The first indication of worldwide dispersion of PCBs was made in Antarctic birds.[1] Since then, the global distribution of PCBs has been demonstrated by their detection in the open ocean atmosphere,[2-7] water,[3,4,7-9] and marine organisms.[10-16]

In the early 1970s, Woodwell et al.[17] and the National Academy of Sciences[18] predicted that the open ocean environment serves as a vast reservoir for the persistent organochlorines. A recent report of the National Academy of Sciences[19] documented that the major portion of PCBs present in the environment of the United States was contained in North Atlantic waters, accounting for 50 to 80% of the total. This estimate suggests that open ocean water acts as a major sink for PCBs. There are some estimates of the PCB loads in terrestrial and coastal environments.[19,20] However, comprehensive and reasonable estimates regarding the PCB load in the open ocean environment have not been made. This seems to be due to the lack of PCB measurement data covering all the oceans, especially the Southern Hemisphere. In 1978, the National Academy of Sciences recommended that a major effort should be made to obtain monitoring data of organochlorines in the Southern Hemisphere.[21]

Since 1975, we have been surveying the Pacific, Indian, and Antarctic Oceans, and have measured PCBs and organochlorine insecticides in air, water, and biological samples.[4,7,13,16,22,23] In this chapter, our main concern is to estimate the PCB load in the open ocean environment. To understand the geochemical circulation and ultimate fate of PCBs, it is essential to estimate the PCB load in the ocean. Thus, we first describe the distribution and behavior of PCBs in open ocean environments including the atmosphere, hydrosphere, and biosphere. Based on our data and those of other workers, we attempt to estimate the general concentration level and load of PCBs in all environmental media and biota in the ocean. Finally, the prospects and ultimate fate of PCBs are discussed.

II. DISTRIBUTION AND BEHAVIOR OF PCBs IN THE OPEN OCEAN ENVIRONMENT

A. PCBs in the Atmosphere

There is not enough data about PCBs in the open ocean atmosphere. This is because the atmospheric concentrations are very low and therefore novel sampling devices and more accurate and sensitive analytical methods were not available. In the late 1970s, a suitable collection method using polyurethane foam was developed[24] and made it possible to isolate PCBs from large volumes of air. In addition, the introduction of capillary gas chromatography[5,6] and mass fragmentography[7] enabled us to determine the respective PCB isomers and congeners and to separate them from various interfering materials contained in environmental samples. Most of the recent studies have been carried out using these techniques and have enabled more detailed considerations of atmospheric transport processes, behavior, and flux of PCBs in the open ocean environment.

PCB concentrations in the open ocean atmosphere reported so far are summarized in Table 1. PCBs are widely distributed in the atmosphere over the oceans worldwide and atmospheric concentrations are mostly below 1 ng/m^3. A most interesting fact is their presence in the Southern Ocean and Antarctica, indicating that the impacts of industrial and human activities extend all over the world. The aerial concentrations of PCBs over the mid-latitudes of the Northern Hemisphere were found to be rather high, which is accounted for by their extensive use on land in this region. By contrast, relatively lower concentrations were observed in the Southern Ocean and the northern North Atlantic and Pacific Oceans. The atmospheric PCB levels over the North Pacific seem to be somewhat lower than those over the North Atlantic.

Table 1
PCB CONCENTRATIONS IN THE OPEN OCEAN ATMOSPHERE

Location	Year	N	PCB conc. (ng/m^3) Range	Mean	Ref.
North Atlantic					
Bermuda	1973	4	0.15—0.50	0.30	2
Bermuda	1973	8	0.21—0.65	0.51	3
Bermuda, U.S.	1973	4	0.72—1.6	0.99	3
Grand Banks (45°N, 52°W)	1973	5	0.05—0.16	0.086	2
Newfoundland	1977	6	0.042—0.15	0.12	6
Gulf of Mexico	1977	10	0.17—0.79	0.35	25
Barbados	1977-1978	17	<0.005—0.37	0.057	6
North Pacific					
Enewetak Atoll (12°N, 162°E)	1979	14	0.35—1.0	0.54	5
Western Pacific (3-35°N, 105-151°E)	1980-1981	7	0.089—0.74	0.25	7
Western Pacific (43-53°N, 154-172°E)	1981	2	0.041—0.061	0.051	26
Western Pacific (41-46°N, 144-174°E)	1982	5	0.022—0.095	0.043	26
Bering Sea	1981	3	0.026—0.059	0.041	26
South Pacific					
Western Pacific (1-46°S, 151-157°E)	1981	5	0.083—0.50[a]	0.27	7
Indian					
Eastern Indian (1-44°S, 104-125°E)	1980	5	0.066—0.33[a]	0.15	7
Western Indian (20-54°S, 48-57°E)	1982	4	0.060—0.24	0.16	27
Antarctic					
53-65°S, 125-161°E	1980-1981	5	0.056—0.18	0.091	7
54-68°S, 38-58°E	1982	4	0.076—0.11	0.091	27
Syowa Station (69°00′S, 39°35′E)	1981-1982	11	0.017—0.17	0.061	27

[a] Excluding the coastal regions.

This difference might be because smaller amounts of PCBs are used in Asian countries than in the U.S. and European countries. According to our recent survey in the western Pacific and eastern Indian Oceans,[7] rather high concentrations of aerial PCBs were found in the low latitudes of both the hemispheres. In another survey ranging from Syowa Station, Antarctica (69°00′S, 39°55′E) to Mauritius in the western Indian Ocean (21°10′S, 57°30′E),[27] an apparent increase of aerial PCB concentrations was observed towards the low latitudes. These observations may suggest that low latitude countries currently act as one of the PCB sources, as well as the developed countries. A recent rice bran oil incident in Taiwan[28] emphasizes the current use of PCBs without any restriction in low latitude countries.

Besides such geographical variations, seasonal variations have also been recognized in the aerial concentrations of PCBs. Figure 1 shows the atmospheric concentrations of PCBs and organochlorine insecticides at Syowa Station, Antarctica.[27] Although the seasonal variation was slightly different according to the variety of chemicals, rather high concentrations were generally found during the austral summer. Recently, Bidleman et al.[6] measured airborne organochlorines in the North Atlantic gyre. PCB concentrations in the Barbados atmosphere were apparently higher in summer than in winter, but organochlorine insecticides did not vary between the two seasons. Oehme and Stray[29] also noted seasonal variations of aerial organochlorines in the Arctic where higher concentrations of HCHs were found in summer. In Antarctica, snowfall prevails during the autumn and winter when relatively lower concentrations of atmospheric organochlorines were observed (Figure 1). Snowfall may assist the removal of atmospheric organochlorines in Antarctica. The seasonal usage of chemicals in countries located to the north of Antarctica may also contribute to such seasonal variations. The application of organochlorine insecticides on land most frequently occurs in

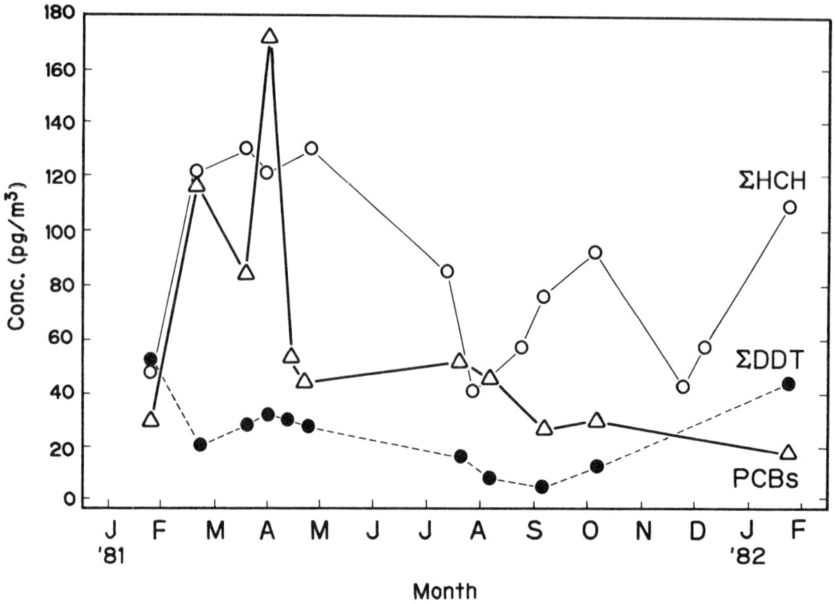

FIGURE 1. Seasonal variations of ΣDDT, ΣHCH, and PCB concentrations in the Antarctic atmosphere (Syowa Station, Antarctica; 69° 00′S, 39°35′E).[27]

summer. The dispersion of insecticides applied on land extends all over the world through the atmosphere, which may account for the high concentrations of HCHs and DDTs in the Antarctic atmosphere in the summer season. However, PCBs are unlikely to be used during a limited season, even though higher concentrations were found in summer. This might be a result of the active evaporation from various PCB sources on land due to high temperatures in the summer. The relatively high concentration levels of PCBs in the Antarctic atmosphere persisted during a short period of only two or three months. This fact suggests that the atmospheric lifetime of PCBs is rather short, probably less than 80 days.

Recent studies made clear the composition of PCBs in the open ocean atmosphere. Atlas and Giam[5] measured the aerial PCB concentrations at a land station on Enewetak Atoll and identified 30 PCB isomers and congeners which excellently matched the Aroclor® 1242 pattern. Bidleman et al.[6] also noted many lower chlorinated biphenyls in the Newfoundland and Barbados atmospheres by capillary gas chromatography. We also identified and quantified 53 PCB isomers and congeners ranging from di- to heptachlorobiphenyls using a gas chromatograph-mass spectrometer, and recognized relatively larger amounts of lower chlorinated biphenyls in the atmosphere over the western Pacific, eastern Indian, and Antarctic Oceans.[7] The enrichment of lower chlorinated biphenyls in the open ocean atmosphere is the result of their comparatively active evaporation and consequent preferential long distance transport. The behavior and lifetime of PCBs in the open ocean atmosphere are supposed to vary according to the isomers and congeners but this possibility remains to be studied.

B. PCBs in Water

A method to collect trace organics from a large volume of water has been developed using macroreticular resins.[30,31] This technique has also been applied to the isolation of PCBs and other organochlorines from open ocean water.[4,7,9,23,32,33] However, less information is available for PCBs in open ocean water as well as the overlying atmosphere. This is attributed to several difficulties such as sample storage, chemical analysis, contamination and so on. Some of these problems have not yet been examined sufficiently.

Table 2
PCB CONCENTRATIONS IN OPEN OCEAN SURFACE WATERS

Location	Year	N	PCB conc. (ng/ℓ) Range	Mean	Ref.
North Atlantic					
Sargasso Sea	1973	8	<0.9—3.6	1.0	3
Sargasso Sea, New York	1973	9		0.8	34
9—55°N, 9—73°W	1973-1975	39	0.4—8.0	2.9	9
North Sea and Scottish coast	1974	5	<0.15—0.52	0.23	33
South Atlantic					
11—36°S, 2—33°W	1975	8	0.3—3.7	1.0	9
North Pacific					
Western Pacific (22—35°N, 141—154°E)	1975	13	0.25—0.56	0.41	4
Western Pacific (32—42°N, 133—143°E)	1976	8	0.29—1.1	0.54	4
Western Pacific (12—33°N, 129—138°E)	1978	6	0.23—0.59	0.35	4
Western Pacific (29—34°N, 137—146°E)	1979	5	0.27—0.38	0.33	4
Western Pacific (5—31°N, 107—152°E)	1980-1981	9	0.039—0.15	0.089	7
Bering Sea	1981	3	0.073—0.13	0.10	26
South Pacific					
Western Pacific (2—41°S, 152—156°E)	1981	5	0.081—0.21	0.12	7
Indian					
Eastern Indian (4—45°S, 104—123°E)	1980	6	0.057—0.25	0.14	7
Antarctic					
48—65°S, 124—163°E	1980-1981	7	0.042—0.072	0.058	7
Syowa Station (69°00′S, 39°35′E)	1981-1982	3	0.035—0.069	0.053	27

PCB concentrations in open ocean surface waters reported so far are summarized in Table 2. There are some other reports on PCBs in North Atlantic water between 1971 and 1972,[32,35,36] but PCB concentrations are one or two orders of magnitude higher compared to those made since 1973 as shown in Table 2. No satisfactory and reasonable explanation has been made for this rapid decrease.

Concentration levels of PCBs in open ocean water are generally in the few nanograms per liter range or less. North Atlantic waters revealed relatively higher concentrations of PCBs than the other oceans. This seems to be due to the extensive use of PCBs in the U.S. and European countries. Much lower concentrations of PCBs were found in the Southern Ocean, ranging from 0.035 to 0.072 ng/ℓ. This concentration level is probably the lowest among the monitoring data on PCBs reported so far. According to our survey of the western Pacific and eastern Indian Oceans from 1975 to 1982,[4,7,26] PCB levels in surface waters were found to be apparently lower in the Southern Hemisphere than in the Northern Hemisphere and much higher concentrations were observed in the mid-latitudes of the Northern Hemisphere (Figure 2). Harvey and Steinhauer[9] reported higher levels of PCBs in North Atlantic surface waters than in the South Atlantic. Both observations strongly suggest progressive contamination by PCBs in the Northern Hemisphere. Harvey and Steinhauer further noted that higher PCB concentrations in the low (10°N to 20°N) and mid-latitudes (45°N to 55°N) of the North Atlantic might be caused by evaporation-precipitation processes. In the case of western Pacific and eastern Indian waters, we could not find any correlation between the latitudinal variation of PCBs and net annual precipitation. It seems that the extensive use of PCBs in the Northern Hemisphere, particularly in mid-latitude countries, is a major factor in this geographical distribution of PCBs.

The vertical transport of PCBs and DDTs in open ocean waters has been also demonstrated by their detection in deep sea fish,[12] sediments,[37] and subsurface water itself.[9,37-39] We also noted the presence of PCBs and organochlorine insecticides in North Pacific, eastern Indian,

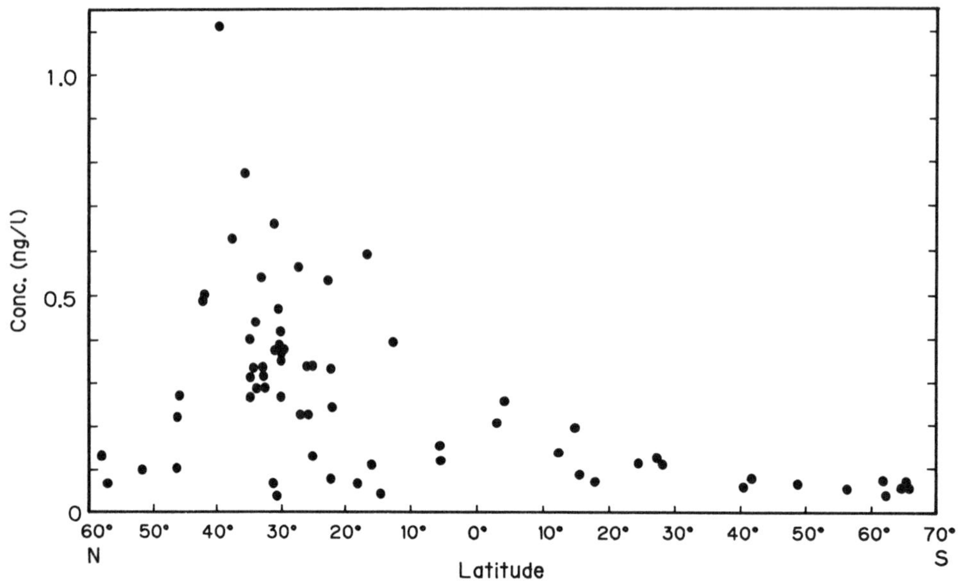

FIGURE 2. Latitudinal distribution of PCBs in open ocean surface waters from the western Pacific and eastern Indian Oceans during 1975 to 1982.

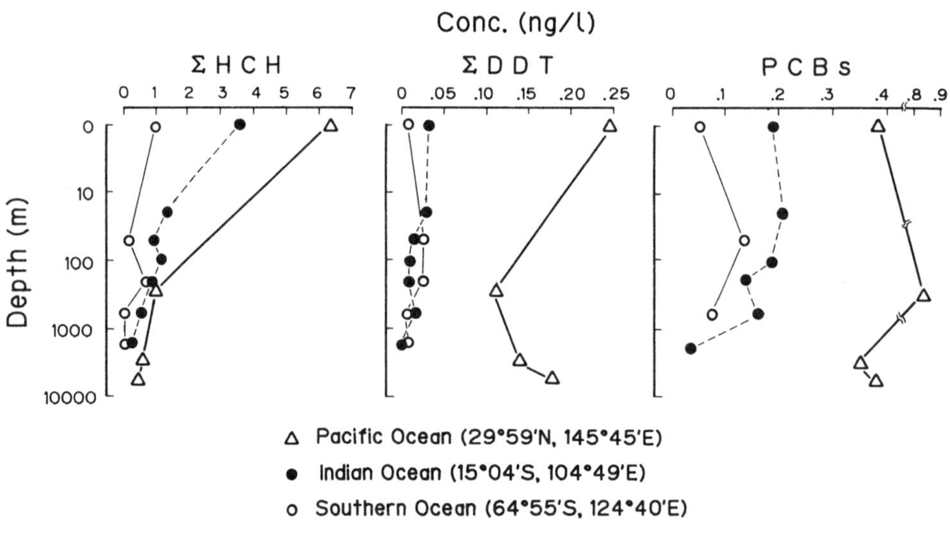

FIGURE 3. Vertical profiles of PCBs and organochlorine insecticides in open ocean water columns.

and Southern Ocean deep waters (Figure 3).[40] According to this report, vertical profiles of PCBs and ΣDDT (sum of p,p′-DDT, p,p′-DDE, and o,p′-DDT) concentrations were found to have small variations throughout the water column, whereas ΣHCH (sum of α, β, and γ isomers) concentrations decreased systematically with depth. PCB and organochlorine insecticide residues in deep water cannot be explained by the convectional and diffusional mixing of water, because it takes a long time for the surface and deep waters to mix and no more than 50 years has passed since PCB production and use began. The vertical transport of these chemicals seems to be strongly associated with sinking particles in the water column. Elder and Fowler[41] reported that fecal pellets from natural populations of euphausiids contained high concentrations of PCBs and also noted that such biogenic particles make a significant contribution to the vertical deposition of PCBs.

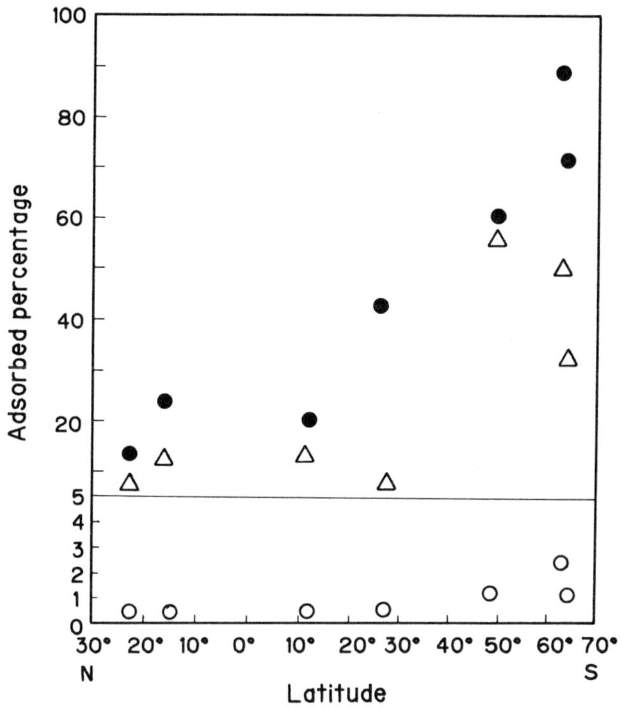

FIGURE 4. Latitudinal variations of the percentage of PCBs (△), ΣDDT (●), and ΣHCH (○) adsorbed on suspended solids in open ocean surface water.[40]

In order to elucidate the transport mechanism of organochlorines, it is necessary to know the forms in which they are present in open ocean water. Thus, we filtered large volumes of surface water and separated the organochlorines into dissolved and adsorbed fractions. Consequently, it became clear that a large portion of PCBs and DDTs was retained on particulate matter, while a major fraction of HCHs was present in filtered water (Figure 4). It was also apparent that there were prominent negative correlations in the relationships between the water solubilities of organochlorines and the concentration ratios of adsorbed to dissolved fractions (Figure 5). These observations strongly suggest that less water soluble organochlorines such as PCBs and DDTs are rapidly transported into deeper layers of water columns by sinking particles. By contrast, relatively water soluble organochlorines such as HCHs are slowly scavenged from surface to deeper layers. The different vertical profiles of organochlorines in open ocean water columns (Figure 3) are most likely because of their different affinity to particulate matter resulting from their water solubilities. Furthermore, the percentage of adsorbed fractions of organochlorines increased toward the high latitudes (Figure 4). It is well known that primary production and particle concentration in the oceans generally increase toward the high latitudes.[42,43] The latitudinal variations for the adsorbed percentage of organochlorines (Figure 4) appear to be related to particle concentrations in surface waters. According to the laboratory experiment by Biggs et al.[44] who examined the partition of ^{14}C-PCB between water and particles, 19 to 22% of ^{14}C-PCB were retained on particles at a particle concentration of 25 ppm, while at 100 ppm, 66 to 69% of ^{14}C-PCB were adsorbed on particles. This fact favorably accounts for the latitudinal variations of organochlorines in adsorbed fractions as shown in Figure 4.

PCB compositions in open ocean water consist mainly of lower chlorinated biphenyls and closely resemble those in the open ocean atmosphere.[7] Higher chlorinated biphenyls are also

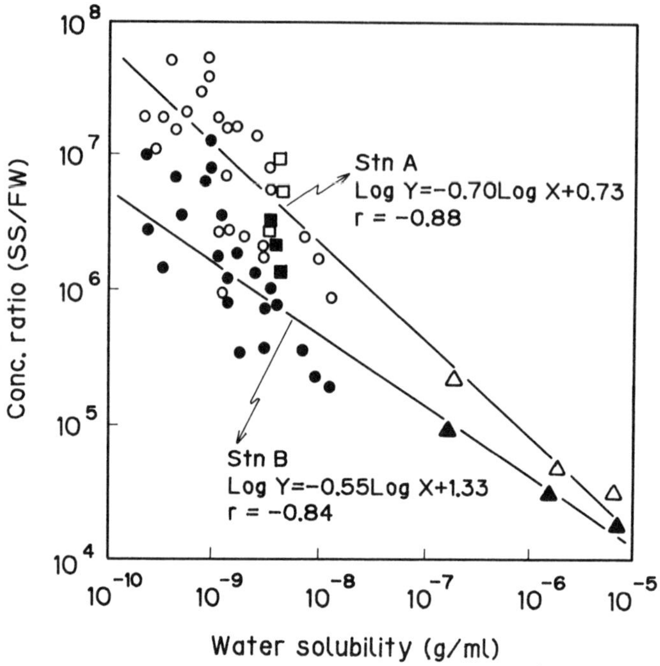

FIGURE 5. Relationship between the water solubilities of PCBs (○,●), DDT compounds (□,■), and HCH isomers (△,▲) and their concentration ratios in suspended solids to filtered water.[40] Stations A and B are located at 22°05′N, 145°02′E and 48°59′S, 127°31′E, respectively. The open and solid symbols indicate Stations A and B, respectively.

present in water, but in much smaller amounts. Such PCBs are mostly retained on particulate matter because of their relatively low water solubilities and high particle affinities, whereas the dissolved fraction mainly consists of lower chlorinated biphenyls (Figure 6). This discrepancy leads to the assumption that PCB compositions in deeper layers are relatively enriched with higher chlorinated biphenyls due to their comparable rapid deposition rates by sinking particles. Although no one has yet demonstrated this, studies of the PCB compositions in Antarctic[46] bottom living fish showed a relatively greater proportion of higher chlorinated biphenyls than surface living fish, may support the selective deposition of PCB isomers and congeners from surface to deeper layers in the water column.

C. PCBs in Organisms

Despite extensive literature on PCB concentrations in estuarine and coastal organisms, relatively few studies are available on open ocean organisms as shown in Tables 3 to 5. Among these organisms, marine mammals have been comparatively well studied all over the world (Table 5). This is due to the facility of PCB detection because of higher accumulation levels of PCBs in marine mammals than in other organisms. PCBs in open ocean fish have also been well monitored, but the portion of the fish body analyzed has been variable (Table 4). This makes it somewhat difficult to compare the PCB concentration levels in fish from different oceans. There is much less information on PCB monitoring data in open ocean plankton (Table 3). It is, therefore, most difficult to discuss the global viewpoint of PCB contamination in plankton samples.

In general, PCB concentrations in open ocean fish and mammals living in the Northern Hemisphere were found to be higher than those in the Southern Hemisphere. The highest range of PCB concentration was observed in the mid-latitudes of the Northern Hemisphere.

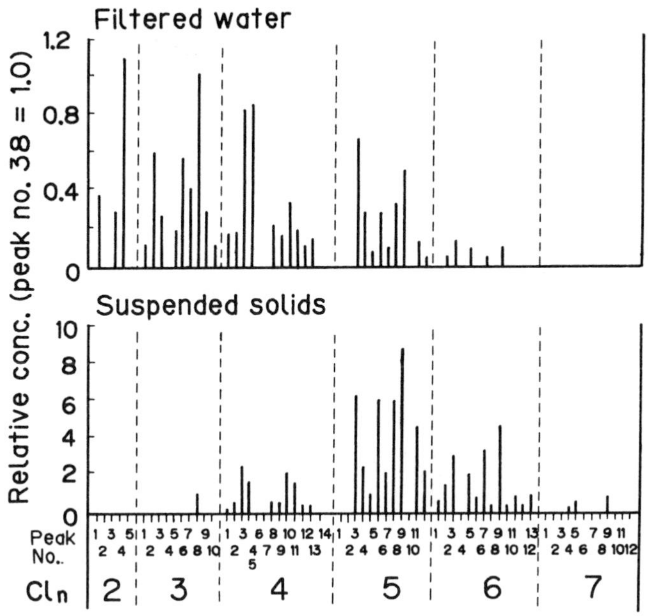

FIGURE 6. PCB compositions in filtered water and suspended solids from open ocean surface water (48°59′S, 127°31′E).[40] The notation Cl_n indicates the number of chlorine atoms substituted in biphenyls. Peak no. shows the order of retention time on the mass fragmentogram for PCBs with respect to substituted chlorine number. Each bar shows the relative concentration of individual PCB isomers and congeners measured by mass fragmentography. Details of each peak and its chemical structure are given in our previous report.[45] Mono-, octa-, nona-, and decachlorobiphenyls were below detection levels.

Table 3
PCB CONCENTRATIONS IN OPEN OCEAN PLANKTON

Location	Year	PCB conc (ng/g wet wt.) N	Range	Mean	Ref.
North Atlantic	1970		300—450	380	48
Northeast Atlantic	Before 1972	22	10—110		47
North and South Atlantic	1970-1972	53		200	11
South Atlantic	1971	4	18—640	200	48
Western North Pacific	1981	1		1.1	49
Bering Sea	1982	3	1.0—1.6	1.3	49
Western South Pacific	1981	3	1.2—2.3	1.7	49
Antarctic (Ross Sea)	1972	1		<3	50
Antarctic (50-65°S, 124°-126°E)	1981	3	0.2—1.0	0.5	49

By contrast, fish and mammalian samples from the northern North Pacific and the northern North Atlantic revealed rather low levels of PCBs. The Arctic and Antarctic mammals accumulated even less amounts of PCBs. The lowest concentration of PCBs in marine mammals was found in the Antarctic Weddell seal. The geographical distribution of PCBs in open ocean fish and mammals almost agreed with those in open ocean surface waters (Table 2). This suggests that the bioaccumulation of PCBs in open ocean organisms is strongly affected by their ambient water.

Table 4
PCB CONCENTRATIONS IN OPEN OCEAN FISHES

Location/species	Year	N	Analyzed portion	PCB conc. (ng/g fresh wt.) Range	Mean	Ref.
North Atlantic						
Flying fish	1970—1971		Whole		50	11
Flying fish *(Cypselurus exsilens)*	1970—1971		Muscle		1.4	11
Flying fish *(Prognichthys rondeletii)*	1970—1971		Muscle		4	11
Trigger fish *(Canthidermis maculatus)*	1970—1971		Muscle		1.9	11
Western North Pacific (off Japan)						
Emmelichthys struhsakeri	1980	3	Whole	12-75	44	49
Tropidinis amoenus	1979	3	Whole	17-71	35	49
Priacanthus boops	1979	3	Whole	12-73	45	49
Bering Sea						
Herring *(Clupea pallasi)*	1973	1	Muscle		80	51
Walleye pollock *(Theragra chalcogramma)*	1973	2	Muscle	40-40	40	51
Flatfish *(Limanda aspera)*	1973—1974	7	Muscle	20-130	50	52
Chum salmon *(Oncorynchus keta)*	1980	1	Whole		5.0	53
Sockeye salmon *(Oncorynchus nerka)*	1980	1	Whole		15	53
Chum salmon *(Oncorynchus keta)*	1982	3	Whole	5.3-9.8	7.3	49
Walleye pollock *(Theragra chalcogramma)*	1982	4	Whole	9.8-13	11	49
Eastern South Pacific (off Chile)						
Cheiloductylus sp.	1978	5	Muscle	1.2-2.2	1.5	13
Merluccius australis	1977	5	Muscle	0.3-0.6	0.4	13
Brama sp.	1978	5	Muscle	0.6-1.6	0.9	13
Neophrynichthys marmoratus	1978	4	Muscle	0.2-0.3	0.2	13
Coelorhyncus fasciatus	1978	5	Muscle	0.1-0.2	0.2	13
Micromesistius australis	1977	3	Whole	0.11-0.33	0.19	49
Coelorhynchus fasciatus	1978	4	Whole	0.06-0.11	0.09	49
North Indian (Arabian Sea)						
Argyrops spinifer	1976	2	Whole	0.74-1.4	1.1	49
Thryssa vitrirostris	1976	3	Whole	0.93-2.0	1.6	49
South Indian (off Australia)						
Coryphaena hippurus	1980	5	Whole	0.02-0.05	0.03	49
Antarctic						
Two whole fish	1972				2.0	50
Pagothenia borchgrevinki	1981	21	Whole	0.18-0.77	0.31	46
Trematomus berenacchii	1981	5	Whole	0.12-0.24	0.17	46
Trematomus hansoni	1981	4	Whole	0.28-0.59	0.48	46
Trematomus newnesi	1981	2	Whole	0.08-0.33	0.21	46

There was a marked variation of PCB concentrations in plankton samples taken from different oceans (Table 3). PCB concentrations in Atlantic plankton were one or two orders of magnitude higher than those in Pacific samples. All the PCB data on Atlantic plankton had been measured before 1972. Although no recent PCB data on Atlantic plankton could be found, the much higher concentrations before 1972 can probably be attributed to the same reasons as for the North Atlantic surface waters where an immediate decrease of PCB concentrations has been noted since 1973. As the monitoring data show that the discrepancy of PCB concentration levels between Atlantic and Pacific samples in respect of atmosphere, surface water, fish, and mammals were less than tenfold, the current PCB concentration in

Table 5
PCB CONCENTRATIONS IN THE BLUBBER OF MALE PINNIPEDS AND CETACEANS

Location	Species	Year	N	PCB conc. (µg/g fresh wt.) Range	Mean	Ref.
Arctic						
Canada	Ringed seal	1972	4	0.05—1.5	0.58	14
	Ringed seal	1972	15	1—6	4.1	54
North Greenland	Atlantic walrus	1975—1977	8	0.16—1.1	0.36	55
North Atlantic						
Newfoundland	Harp seal	1970	1		26	56
Gulf of St. Lawrence	Harp seal	1971	7	6—22	13	57
Nova Scotia	Atlantic white-sided dolphin	1972	1		37	58
Rhode Island	Striped dolphin	1972	1		39	58
Caribbean	Long-snouted dolphin	1972	1		5.0	59
South Atlantic						
Uruguay	Franciscana dolphin	1974	5	3.2—18	6.8	15
North Pacific						
Bering Sea	Dall's porpoise	1980	4	3.5—6.8	5.2	60
Japan	Striped dolphin	1978	3	22—23	22	61
	Finless porpoise	1968—1975	2	64—96	80	15
	Pilot whale		1		2.0	15
California	Common dolphin	1974—1976	10	80—300	120	15
	Pilot whale		1		14	15
Hawaii	Rough-toothed dolphin	1976	3	7.0—14	9.4	15
Eastern tropical	Striped dolphin	1973—1976	3	2.6—7.6	5.7	15
South Pacific						
Eastern tropical	Fraser's dolphin	1973—1976	1		5.2	15
	Striped dolphin		1		5.0	15
New Zealand	Dusky dolphin	1980	1		1.4	60
Antarctic						
Syowa Station	Weddell seal	1981	1		0.038	62

Atlantic plankton is assumed to be in the few nanogram per gram range on a wet weight basis.

PCB concentrations in open ocean organisms tend to be greater in the higher trophic levels. Open ocean plankton generally have concentrations of a few nanograms per gram or less. Fish samples accumulate nearly the same or one order of magnitude higher than this, while PCB concentrations in marine mammals are a further one or three orders of magnitude higher than those in fish. Such an amplification with trophic levels has been demonstrated in terrestrial and coastal aquatic environments.[63-65] We also found PCB amplification in open ocean ecosystems from the northwestern Pacific where bioconcentration factors (ratio of PCB concentrations in organisms to ambient water) were 10^3 for plankton, 10^4 to 10^5 for squid and myctophid, and 10^7 for striped dolphin.[66] In addition to the quantitative variation of PCBs with trophic level, qualitative variation regarding PCB isomers and congeners has also become a major interest recently.

According to detailed PCB determinations in various organisms taken from the Bering Sea,[53] apparent variations were recognized in their PCB compositions (Figure 7). Although the major fractions of PCBs in surface waters were lower chlorinated biphenyls with two and three chlorine atoms, plankton and fish samples retained not only the lower chlorinated biphenyls but also fairly large amounts of penta- and hexachlorobiphenyls. Furthermore, PCBs in Dall's porpoise were composed mainly of higher chlorinated biphenyls with more

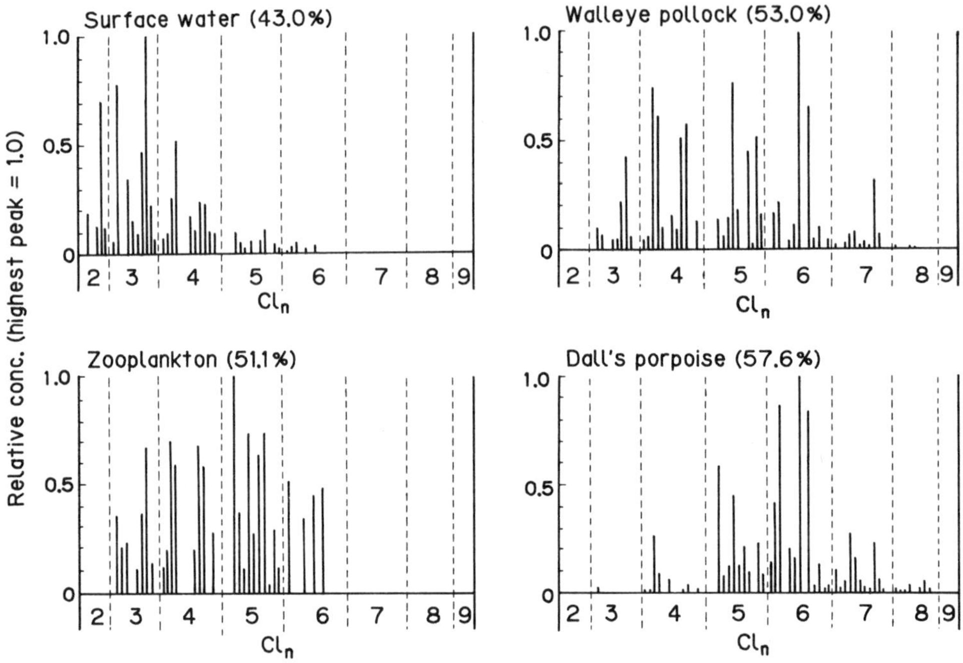

FIGURE 7. PCB compositions in surface water and organisms from the Bering Sea.[53] Parentheses show the chlorine contents. Details of each bar and Cl_n are the same as Figure 6. Dichlorobiphenyls in zooplankton could not be measured due to interfering materials.

than five chlorine atoms and a much lesser amount of lower chlorinated biphenyls. Jansson et al.[67] and Zell and Ballschmiter[68] also noted different PCB compositions in various biological samples such as fish, bird, seal, and human and they suggested that the metabolic potential inherent in organisms is a major factor for the variation of PCB compositions in different trophic levels. The PCB composition of Dall's porpoise (Figure 7) is most likely affected by a biological factor such as metabolism. However, plankton do not seem to have such an active metabolic potential. The different compositions of PCBs in plankton and surface water in the Bering Sea probably depend upon physicochemical factors. In fact, many authors support lipid/water partitioning and cell/water adsorption mechanisms as major uptake routes of PCBs by plankton,[69-72] while PCBs in those fish that are mesotrophic animals in marine food chains appear to be affected by both physicochemical and biochemical factors. It is known that the uptake of PCBs through the gills and digestive tracts in fish depends principally on lipid/water partitioning and molecular size,[73-76] whereas higher chlorinated biphenyls are accumulated more readily through gills but lesser amounts are absorbed through digestive tracts. Moreover, it is also noted that fish certainly have a potential to metabolize PCBs, but this is limited to lower chlorinated members.[76-78] The PCB composition of Bering fish (Figure 7) seems to basically reflect a result of physicochemical factors, but the relatively smaller proportion of lower chlorinated biphenyls in comparison with plankton is probably due to metabolism.

The ocean contains many species of organisms and their life spans vary considerably. These organisms more or less function as effective integrators of persistent organochlorines. The residues of these chemicals in plankton and fish which have rather short life spans reflect the present state or short-term contamination in the marine environment. By contrast, mammals, long-living animals, provide historical evidence of marine pollution in their bodies. Presently, we have no available information to trace the historical contamination by PCBs of the oceans. Organisms with long life spans may provide useful clues to the elucidation of the time course of PCB pollution in the ocean environment in the past and also probably in the future.

III. GENERAL CONCENTRATION LEVELS AND ESTIMATED LOAD OF PCBs IN THE OCEANS

Based on the measured data of PCB concentrations in various open ocean environments so far reported (Tables 1 to 5), PCB concentrations and loads in the world oceans were estimated and are shown in Table 6.

PCB monitoring data are not sufficient for the estimation of PCB concentration and load in the open ocean environment. In particular, there is much less data available for all environmental media and biota in the South Atlantic. Thus, we applied the PCB data of the South Pacific to the South Atlantic. Current data could not be found for North Atlantic waters. Thus, a median value of PCB data in the North Sea[33] and Sargasso Sea[34] was adopted as a general concentration level of PCBs in North Atlantic waters. There is only one paper available for estimating the PCB concentration and load in open ocean sediment, which was reported by Harvey and Steinhauer[37] and dealt with the North Atlantic Ocean. PCB concentrations in other oceanic sediments were estimated from this data and concentration levels in surface waters of the respective oceans. PCB monitoring data for plankton are extremely scarce for all oceans. Fish and mammalian data are also insufficient for the Southern Hemisphere. Moreover, we could not find any biomass data for the respective oceans, thus, there was no other way to estimate the PCB loads in organisms in each of the oceans but to describe them as a single or whole ocean.

Despite these limitations, this estimate (Table 6) provides sufficient information to understand the fate of PCBs. The total PCB load in the open ocean environment was estimated to be about 230,000 t. The major fraction of PCBs in the open ocean environment is in water, accounting for more than 99% of the total load. In spite of the fact that open ocean organisms accumulate PCBs in concentrations three to seven orders of magnitude higher than water, their PCB loads were extremely small. This is due to the relatively much smaller volume of biomass compared to the water mass. The open ocean environment in the Northern Hemisphere contains about 150,000 t of PCBs, accounting for nearly two thirds of the total load. The largest load was in the North Atlantic waters, which contained 90,000 t of PCBs. Recent estimates by the National Academy of Sciences[19] indicated that oceanic water over the North American Basin contained a maximum of 66,000 t of PCBs. This value compares favorably to our estimate of 90,000 t in the whole North Atlantic waters. Worldwide statistics regarding PCB production and use are not available. The total cumulative use of PCBs in the U.S. was estimated to be 61,000 t up to 1975.[19] Assuming the world PCB production to be twice this amount (1.2 million t), about 20% of the cumulative world production is retained in the open ocean environment. According to our estimate of PCB loads conducted in the Seto-Inland Area, Japan,[20] 12% of the total amount of the PCB used in this region was contained in the terrestrial and coastal environments. If this percentage is representative for all global terrestrial and coastal environments, it can be estimated that about 30% of the total PCB production in the world, an amount of approximately 400,000 t, has escaped into the open environment. Miller[82] noted that of the PCBs produced in the U.S., 3.6% has been degraded or incinerated. Thus, the destroyed portion of PCBs is negligible in any estimate of the world PCB load. Overall, it is estimated that about 800,000 t of PCBs are still in service, including the amounts used for electrical equipment and other products or deposited in landfills and dumps.

The open ocean atmosphere contains 790 t of PCBs at present (Table 6). If the following assumptions are allowed, (1) the land-stocked PCBs (800,000 t) disappear only through volatilization, (2) PCB concentrations in the atmosphere remain at their present level, (3)

Table 6
ESTIMATED CONCENTRATION AND LOAD OF PCBs IN THE OPEN OCEAN ENVIRONMENT

Compartment mass PCB conc., load	North Pacific	South Pacific	North Atlantic	South Atlantic	Indian	Antarctic	Total load
Compartment mass[a]							
Air ($\times 10^{16}$ m^3)[b]	70	76	38	35	58	46	
Water ($\times 10^{19}$ ℓ)	30	33	15	14	23	19	
Sediment (dry, $\times 10^{15}$ g)[c]	70	76	38	35	58	46	
Plankton (wet, $\times 10^{15}$ g)[d]			48 (whole ocean)				
Fish (fresh, $\times 10^{14}$ g)[d]			26 (whole ocean)				
Mammals (fresh, $\times 10^{12}$ g)[d]			85 (whole ocean)				
PCB conc.							
Air (ng/m^3)	0.3	0.2	0.5	0.2	0.2	0.1	
Water (ng/ℓ)	0.2	0.1	0.6	0.1	0.1	0.05	
Sediment (dry, ng/g)	0.4	0.2	1.0	0.2	0.2	0.1	
Plankton (wet, ng/g)	2.0	1.0	5.0	1.0	1.0	0.5	
Fish (fresh whole, ng/g)	10	2.0	30	2.0	2.0	0.2	
Mammals (fresh whole, μg/g)[e]	2.0	0.5	5.0	0.5	0.5	0.05	
PCB load (t)							
Air	210	150	190	70	120	50	790
Water[f]	60,000	33,000	90,000	14,000	23,000	10,000	230,000
Sediment	28	15	38	7	12	5	105
Plankton[g]			20—240 (whole ocean)				130[h]
Fish[g]			1—80 (whole ocean)				40[h]
Mammals[g]			2—200 (whole ocean)				100[h]
Total							231,165

[a] Data on the surface area and mean depth in respective oceans employed for the calculation of air mass, water mass, and sediment mass were adapted from Sugimura.[79]

[b] Calculated in troposphere (10 km height).

[c] Calculated in upper 1 mm sediment layer. Sediment was assumed to contain about 50% water and have a mass density of 2.

[d] Modified from Bowen[80] in consideration of the following moisture contents: plankton 95%, fish 75%, and mammals 65%.

[e] Values show the PCB concentrations regarding male specimens. PCB concentrations on whole body basis were calculated following the relationship of PCB concentrations between blubber and whole body obtained from the striped dolphin.[16]

[f] PCB loads in water were estimated on the assumption of their having vertically uniform concentrations in the water column as shown by their vertical profiles in the literature.[9,37]

[g] PCB loads in these organisms were estimated from the following probable concentrations: plankton 0.5—5 ng/g, fish 0.2—30 ng/g, and mammals (male) 0.05—5, μg/g. The PCB load in mammals took into account that female specimens generally have lower concentrations of PCBs in their bodies than males because of parturitional and lactational losses.[22,54,81]

[h] Median values.

Table 7
ESTIMATED RESIDENCE TIME OF PCBs IN THE OPEN OCEAN MIXING LAYER (UPPER 100 m OF WATER COLUMN)[40]

Ocean	Carbon productivity ($g/m^2/year$)[42]	Residence time (day)
Oligotrophic		
Pacific Ocean (22°05′N, 145°02′E)	50—100	130—280
Eutrophic		
Southern Ocean (64°42′S, 124°15′E)	150—250	26—44

the atmospheric residence time of PCBs is 60 days, and (4) atmospheric PCBs can become irreversibly transferred into open ocean water, then it is estimated that about 4800 t of PCBs are annually released from land-stocked PCBs and transferred into open ocean waters. This estimate suggests that, if further regulation is not enforced to ensure PCB disposal, more than 150 years are required for the complete disappearance of land-stocked PCBs and the contamination of PCBs in open ocean environments will continue during the same period. It should be noted that strict measures to prevent leakage or volatilization from land-stocked PCBs are needed and an effective technology to incinerate or destruct PCBs must be developed.

IV. PROSPECTS AND ULTIMATE FATE OF PCBs IN THE OCEAN RESERVOIR

Undoubtedly, open ocean water serves as a vast reservoir of PCBs. To predict the long-term contamination and ultimate fate of PCBs in the open ocean environment, it is necessary to know the removal rates of PCBs from the open ocean atmosphere and water. Bidleman et al.[6] reported that the atmospheric residence times of PCBs in the North Atlantic were in the range of 46 to 70 days. Recently, we reported PCB residence times in the open ocean mixing layer (upper 100 m of water column), based on our measured data of PCBs[40] and using an equation on the relationship between the organic carbon flux and the primary production rate in the world oceans reported by Suess.[83] The results are summarized in Table 7 and the detailed procedure used in the estimation is given elsewhere.[40]

The residence time of PCBs in the open ocean mixing layer was estimated to be in the range of 26 to 44 days in eutrophic oceans and 130 to 280 days in oligotrophic oceans. Supposing that the atmospheric residence time of PCBs is nearly the same over all oceans, ranging from 46 to 70 days, the residence time of PCBs in the eutrophic waters is shorter than in the atmosphere, while in oligotrophic waters it is rather longer. The different residence times between the atmosphere and the mixing layers indicate that, if the airborne PCB flux could be completely stopped, PCB concentrations in the mixing layers of both the eutrophic and oligotrophic oceans would rapidly decrease and eventually fall to zero within a rather short period. However, if the atmospheric flux of PCBs were to be kept in its present state, PCB contamination in the eutrophic ocean would be preserved at the present level, while in the oligotrophic ocean a gradual increase of PCB concentrations may occur according to the residual amounts of PCBs in the atmosphere and mixing layer. Mathematical simulation and more accurate monitoring surveys are required to validate this forecast.

Although the deposition rate of PCBs in the open ocean water column under the mixing layer is unknown, PCBs are certainly contained in deep waters and bottom sediments.[9,34,37,38,40] PCBs deposited into these environments are most unlikely to contribute to their geochemical and biochemical circulation, because the vertical mixing of deep water is extremely slow and the biomass there is very small. Thus, it can be concluded that open ocean deep water or bottom sediment serve as a final sink for PCBs.

The commercial production of PCBs has now ceased in most developed countries. However, as a rough estimate, we still use about 800,000 t of PCBs on land. This amount corresponds to more than three times the total PCB load in the open ocean environment. It is, therefore, justifiable to say that the large amount of land-stocked PCBs holds a crucial key for the future trend of PCB contamination in the open ocean environment as well as in the terrestrial and coastal environments. A major effort must be made to control the discharge of land-stocked PCBs and continuous surveillance is needed to pursue the trend of PCB pollution in the open ocean environment.

REFERENCES

1. **Risebrough, R. W. and Carmignani, G. M.,** Chlorinated hydrocarbons in Antarctic birds, in *Proc. Colloq. Conserv. Problems in Antarctica,* Lawrence, P. B. C., Ed., Allen Press, Lawrence, Kan., 1972, 63.
2. **Harvey, G. R. and Steinhauer, W. G.,** Atmospheric transport of polychlorobiphenyls to the North Atlantic, *Atmos. Environ.,* 8, 777, 1974.
3. **Bidleman, T. F. and Olney, C. E.,** Chlorinated hydrocarbons in the Sargasso Sea atmosphere and surface water, *Science,* 183, 516, 1974.
4. **Tanabe, S. and Tatsukawa, R.,** Chlorinated hydrocarbons in the North Pacific and Indian Oceans, *J. Oceanogr. Soc. Jpn.,* 36, 217, 1980.
5. **Atlas, E. and Giam, C. S.,** Global transport of organic pollutants: ambient concentrations in the remote marine atmosphere, *Science,* 211, 163, 1981.
6. **Bidleman, T. F., Christensen, E. J., Billings, W. N., and Leonard, R.,** Atmospheric transport of organochlorines in the North Atlantic Gyre, *J. Mar. Res.,* 39, 443, 1981.
7. **Tanabe, S., Kawano, M., and Tatsukawa, R.,** Chlorinated hydrocarbons in the Antarctic, western Pacific and eastern Indian Oceans, *Trans. Tokyo Univ. Fish.,* 5, 97, 1982.
8. **Williams, P. M. and Robertson, K. J.,** Chlorinated hydrocarbons in sea-surface films and subsurface waters at nearshore stations and in the north central Pacific Gyre, *U. S. NOAA Fish. Bull.,* 73, 445, 1975.
9. **Harvey, G. R. and Steinhauer, W. G.,** Transport pathways of polychlorinated biphenyls in Atlantic water, *J. Mar. Res.,* 34, 561, 1976.
10. **Giam, C. S., Richardson, R. L., and Wong, M. G.,** Polychlorinated biphenyls in Antarctic biota, *Antarctic J. U. S.,* 8, 303, 1973.
11. **Harvey, G. R., Miklas, H. P., Bowen, V. T., and Steinhauer, W. G.,** Observations on the distribution of chlorinated hydrocarbons in Atlantic Ocean organisms, *J. Mar. Res.,* 32, 103, 1974.
12. **Barber, R. T. and Warlen, S. M.,** Organochlorine insecticide residues in deep sea fish from 2500 m in the Atlantic Ocean, *Environ. Sci. Technol.,* 13, 1146, 1979.
13. **Tanabe, S., Tanaka, H., Tatsukawa, R., and Nakamura, I.,** Organochlorine compound residues in several species of fishes caught off Chile, *Bull. Jpn. Soc. Sci. Fish.,* 46, 763, 1980.
14. **Bowes, G. W. and Jonkel, C. J.,** Presence and distribution of polychlorinated biphenyls (PCB) in arctic and subarctic marine food chains, *J. Fish. Res. Bd. Can.,* 32, 2111, 1975.
15. **O'Shea, T. J., Brownell, R. L., Jr., Clark, D. R., Jr., Walker, W. A., Gay, M. L., and Lamont, T. G.,** Organochlorine pollutants in small cetaceans from the Pacific and South Atlantic Oceans, November 1968—June 1976, *Pest. Monit. J.,* 14, 35, 1980.
16. **Tanabe, S., Tatsukawa, R., Tanaka, H., Maruyama, K., Miyazaki, N., and Fukiyama, T.,** Distribution and total burdens of chlorinated hydrocarbons in bodies of striped dolphins *(Stenella coeruleoalba), Agric. Biol. Chem.,* 45, 2569, 1981.
17. **Woodwell, G. M., Craig, P. P., and Johnson, H. A.,** DDT in the biosphere: where does it go?, *Science,* 174, 1101, 1971.
18. Ocean Affairs Board, The ultimate fate of substances that relate to the quality of the marine environment, in *Marine Environmental Quality,* National Academy of Sciences, Washington, D.C., 1971, chap. 6.
19. Environmental Studies Board, PCB transport throughout the environment, in *Polychlorinated Biphenyls,* National Academy of Sciences, Washington, D.C., 1979, chap. 1.
20. **Tanabe, S. and Tatsukawa, R.,** Behavior of man-made organics in the coastal and estuarine environment, *Bull. Coast. Oceanogr.,* 19, 9, 1981.
21. Ocean Sciences Board, Halogenated hydrocarbons, in *The Tropospheric Transport of Pollutants and Other Substances to the Oceans,* National Academy of Sciences, Washington, D.C., 1978, chap. 6.

22. **Tanabe, S., Tatsukawa, R., Maruyama, K., and Miyazaki, N.,** Transplacental transfer of PCBs and chlorinated hydrocarbon pesticides from the pregnant striped dolphin *(Stenella coeruleoalba)* to her fetus, *Agric. Biol. Chem.,* 46, 1249, 1982.
23. **Tanabe, S., Tatsukawa, R., Kawano, M., and Hidaka, H.,** Global distribution and atmospheric transport of chlorinated hydrocarbons: HCH (BHC) isomers and DDT compounds in the western Pacific, eastern Indian and Antarctic Oceans, *J. Oceanogr. Soc. Jpn.,* 38, 137, 1982.
24. **Simon, C. G. and Bidleman, T. F.,** Sampling airborne polychlorinated biphenyls with polyurethane foam — chromatographic approach to determining retention efficiencies, *Anal. Chem.,* 51, 1110, 1979.
25. **Giam, C. S., Atlas, E., Chan, H. S., and Neff, G. S.,** Phthalate esters, PCB, and DDT residues in the Gulf of Mexico atmosphere, *Atmos. Environ.,* 14, 65, 1980.
26. **Tanabe, S., Tanaka, H., and Tatsukawa, R.,** Unpublished data, 1983.
27. **Tanabe, S., Hidaka, H., and Tatsukawa, R.,** PCBs and chlorinated hydrocarbon pesticides in Antarctic atmosphere and hydrosphere, *Chemosphere,* 12, 277, 1983.
28. **Chen, P. H., Gaw, J. M., Wong, C. K., and Chen, C. J.,** Levels and gas chromatographic patterns of polychlorinated biphenyls in the blood of patients after PCB poisoning in Taiwan, *Bull. Environ. Contam. Toxicol.,* 25, 325, 1980.
29. **Oehme, M. and Stray, H.,** Quantitative determination of ultratraces of chlorinated compounds in high-volume air samples from the Arctic using polyurethane foam as collection medium, *Fresenius' Z. Anal. Chem.,* 31, 665, 1982.
30. **Burnham, A. K., Calder, G. V., Fritz, J. S., Junk, G. A., Svec, H. J., and Wills, R.,** Identification and estimation of neutral organic contaminants in potable water, *Anal. Chem.* 44, 139, 1972.
31. **Junk, G. A., Richard, J. J., Greiser, M. D., Fritz, J. S., and Calder, G. V.,** Use of macroreticular resins in the analysis of water for trace organic contaminants, *J. Chromatogr.,* 99, 745, 1974.
32. **Harvey, G. R., Steinhauer, W. G., and Teal, J. M.,** Polychlorobiphenyls in North Atlantic ocean water, *Science,* 180, 643, 1973.
33. **Dawson, R. and Riley, J. P.,** Chlorine-containing pesticides and polychlorinated biphenyls in British coastal waters, *Estuarine Coast. Mar. Sci.,* 4, 55, 1977.
34. **Harvey, G. R., Steinhauer, W. G., and Miklas, H. P.,** Decline of PCB concentrations in North Atlantic surface water, *Nature (London),* 252, 387, 1974.
35. **Olney, C. E. and Wade, T. L.,** Chlorinated hydrocarbon in the marine atmosphere and sea surface microlayer, IDOE Workshop on Baseline Measurement, May 22, 1972, Brookhaven, N.Y.
36. **Bidleman, T. F., Rice, C. P., and Olney, C. E.,** High molecular weight chlorinated hydrocarbons in the air and sea: rates and mechanism of air/sea transfer, in *Marine Pollution Transfer,* Windon, H. L. and Duce, R. A., Eds., Heath & Co., Lexington, Mass., 1976, 323.
37. **Harvey, G. R. and Steinhauer, W. G.,** Biochemistry of PCB and DDT in the North Atlantic, in *Environmental Biochemistry,* Vol. 1, Niagu, J. O., Ed., Ann Arbor Science, Ann Arbor, Mich., 1976, 203.
38. **Scura, E. D. and McClure, V. E.,** Chlorinated hydrocarbons in sea water: analytical method and levels in the northeastern Pacific, *Mar. Chem.,* 3, 337, 1975.
39. **Jonas, R. B. and Pfaender, F. K.,** Chlorinated hydrocarbon pesticides in western North Atlantic Ocean, *Environ. Sci. Technol.,* 10, 770, 1976.
40. **Tanabe, S. and Tatsukawa, R.,** Vertical transport and residence time of chlorinated hydrocarbons in the open ocean water column, *J. Oceanogr. Soc. Jpn.,* 39, 46, 1983.
41. **Elder, D. L. and Fowler, S. W.,** Polychlorinated biphenyls: penetration into the deep ocean by zooplankton fecal pellet transport, *Science,* 197, 459, 1977.
42. **Koblentz-Mishke, O. J., Volkovinsky, V. V., and Kabanoba, J. G.,** Plankton primary production of the world ocean, in *Scientific Exploration of the South Pacific,* National Academy of Sciences, Washington, D.C., 1970, 183.
43. **Lal, D.,** The oceanic microcosm of particles, *Science,* 198, 997, 1977.
44. **Biggs, D. C., Powers, C. D., Lowland, R. G., O'Connors, H. B., Jr., and Wurster, C. F.,** Uptake of polychlorinated biphenyls by natural phytoplankton assemblages: field and laboratory determination of ^{14}C-PCB particulate-water index of sorption, *Environ. Pollut.,* 22, 101, 1980.
45. **Tanabe, S., Nakagawa, Y., and Tatsukawa, R.,** Absorption efficiency and biological half-life of individual chlorobiphenyls in rats treated with Kanechlor products, *Agric. Biol. Chem.,* 45, 717, 1981.
46. **Subramanian, B. R., Tanabe, S., Hidaka, H., and Tatsukawa, R.,** DDTs and PCB isomers and congeners in Antarctic fishes, *Arch. Environ. Contam. Toxicol.,* 12, 621, 1983.
47. **Goldberg, E. D.,** Halogenated hydrocarbons, in *The Health of the Oceans,* UNESCO Press, Paris, 1976, chap. 3.
48. **Risebrough, R. W. and Vreeland, V.,** PCB residues in Atlantic Zooplankton, *Bull. Environ. Contam. Toxicol.,* 8, 345, 1972.
49. **Tanabe, S., Maruyama, K., Hironaga, U., Matsumoto, M., and Tatsukawa, R.,** Unpublished data, 1983.

50. **Giam, C. S., Richardson, R. L., Wong, M. K., and Sackett, W. M.,** Polychlorinated biphenyls in antarctic biota, *Antarctic J. U. S.*, 8, 303, 1973.
51. **Murayama, H., Soma, S., Iida, A., and Takagi, M.,** PCB contents in several fish obtained in the Okhotsk Sea and the Bering Sea, *Proc. Hakodate Jr. Coll.*, 17, 1, 1974.
52. **Takagi, M., Murayama, H., and Soma, S.,** PCB contents in several species of flatfish collected in the eastern Bering Sea, *Bull. Jpn. Soc. Sci. Fish.*, 41, 685, 1975.
53. **Tanaka, H.,** Bioaccumulation of Chlorinated Hydrocarbons in Open Ocean Ecosystems, Master's thesis, Ehime University, Matsuyama, 1982.
54. **Addison, R. F. and Smith, T. G.,** Organochlorine residue levels in Arctic ringed seals: variation with age and sex, *Oikos*, 25, 335, 1974.
55. **Born, E. W., Kraul, I., and Kristensen, T.,** Mercury, DDT and PCB in the Atlantic walrus *(Odobenus rosmarus rosmarus)* from the Thule district, North Greenland, *Arctic*, 34, 255, 1981.
56. **Frank, R., Ronald, K., and Braun, H. E.,** Organochlorine residues in harp seal *(Pagophilus groenlandicus)* caught in eastern Canadian waters. *J. Fish. Res. Bd. Can.*, 30, 1053, 1973.
57. **Addison, R. F., Kerr, S. R., Dale, J., and Sergeant, D. E.,** Variation of organochlorine residue levels with age in Gulf of St. Lawrence harp seals *(Pagophilus groenlandicus), J. Fish. Res. Bd. Can.*, 30, 595, 1973.
58. **Taruski, A. G., Olney, C. E., and Winn, H. E.,** Chlorinated hydrocarbons in cetaceans, *J. Fish. Res. Bd. Can.*, 32, 2205, 1975.
59. **Gaskin, D. E., Smith, G. J. D., Arnold, P. W., Louisy, M. V., Frank, R., Holdrinet, M., and McWade, J. W.,** Mercury, DDT, Dieldrin and PCB in two species of odontoceti (cetacea) from St. Lucia, Lesser Antilles, *J. Fish. Res. Bd. Can.*, 31, 1235, 1974.
60. **Tanabe, S., Mori, T., Maruyama, K., and Tatsukawa, R.,** Unpublished data, 1983.
61. **Tanabe, S., Tanaka, H., Maruyama, K., and Tatsukawa, R.,** Ecology and bioaccumulation of *Stenella coeruleoalba:* elimination of chlorinated hydrocarbons from mother striped dolphins (Stenella coeruleoalba) through parturition and lactation, in *Studies on the Levels of Organochlorine Compounds and Heavy Metals in the Marine Organisms,* Fujiyama, T., Ed., Report for the Fiscal Year of 1980, Grant-in-Aid for Scientific Research, University of Ryukyus, Taiwan, 1980, 115.
62. **Hidaka, H., Tanabe, S., and Tatsukawa, R.,** DDT compounds and PCB isomers and congeners in Weddell seals and their fate in the Antarctic marine ecosystem, *Agric. Biol. Chem.*, 47, 2009, 1983.
63. **Woodwell, G. M., Wurster, C. F., and Issacson, P. A.,** DDT residues in an east coast estuary: a case of biological concentration of a persistent insecticide, *Science*, 156, 821, 1967.
64. **Robinson, J., Richardson, A., Crabtree, A. N., Coulson, J. C., and Potts, G. R.,** Organochlorine residues in marine organisms, *Nature (London)*, 214, 1307, 1967.
65. **Jensen, S., Johnels, A. G., Olsson, M., and Otterlind, G.,** DDT and PCB in marine animals from Swedish waters, *Nature (London)*, 224, 247, 1969.
66. **Tanabe, S., Tanaka, H., and Tatsukawa, R.,** Polychlorobiphenyls, ΣDDT, and hexachlorocyclohexane isomers in the western North Pacific ecosystem, *Arch. Environ. Contam. Toxicol.*, 13, 731, 1984.
67. **Jansson, B., Jensen, S., Olsson, M., Renberg, L., Sundström, G., and Vaz, R.,** Identification by GC-MS of phenolic metabolities of PCB and *p,p'*- DDE isolated from Baltic guillmot and seal, *Ambio*, 4, 93, 1975.
68. **Zell, M. and Ballschmiter, K.,** Baseline studies of the global pollution. III. Trace analysis of polychlorinated biphenyls (PCB) by ECD glass capillary gas chromatography in environmental samples of different trophic levels, *Fresenius' Z. Anal. Chem.*, 304, 337, 1980.
69. **Clayton, J. R., Jr., Pavlou, S. P., and Breitner, N. F.,** Polychlorinated biphenyls in coastal marine zooplankton: bioaccumulation by equilibrium partitioning, *Environ. Sci. Technol.*, 11, 676, 1977.
70. **Pavlou, S. P. and Dexter, R. N.,** Distribution of polychlorinated biphenyls (PCB) in estuarine ecosystems. Testing the concept of equilibrium partitioning in the marine environment, *Environ. Sci. Technol.*, 13, 65, 1979.
71. **Harding, L. W., Jr. and Phillips, J. H., Jr.,** Polychlorinated biphenyls (PCB) uptake by marine phytoplankton, *Mar. Biol.*, 49, 103, 1978.
72. **Hiraizumi, Y., Takahashi, M., and Nishimura, H.,** Adsorption of polychlorinated biphenyl onto sea bed sediment, marine plankton and other adsorbing agents, *Environ. Sci. Technol.*, 13, 580, 1979.
73. **Neely, W. B., Branson, D. R., and Blau, G. E.,** Partition coefficient to measure bioconcentration potential of organic chemicals in fish, *Environ. Sci. Technol.*, 8, 1113, 1974.
74. **Chiou, C. T., Freed, V. H., Schemedding, D. W., and Kohnert, R. L.,** Partition coefficient and bioaccumulation of selected organic chemicals, *Environ. Sci. Technol.*, 11, 475, 1977.
75. **Veith, G. D., Defoe, D. L., and Bergstedt, B. V.,** Measuring and estimating the bioaccumulation factor of chemicals in fish, *J. Fish. Res. Bd. Can.*, 36, 1040, 1979.
76. **Tanabe, S., Maruyama, K., and Tatsukawa, R.,** Absorption efficiency and biological half-life of individual chlorobiphenyls in carp (Cyprinus carpio) orally exposed to Kanechlor products, *Agric. Biol. Chem.*, 46, 891, 1982.

77. **Hinz, R. and Matsumura, F.,** Comparative metabolism of PCB isomers by three species of fish and the rat, *Bull. Environ. Contam. Toxicol.,* 18, 631, 1977.
78. **Melancon, M. J. and Lech, J. J.,** Isolation and identification of a polar metabolite of tetrachlorobiphenyl from bile of rainbow trout exposed to ^{14}C-tetrachlorobiphenyl, *Bull. Environ. Contam. Toxicol.,* 15, 181, 1976.
79. **Sugimura, Y.,** Personal communication, 1983.
80. **Bowen, H. J. M.,** Living organisms and elementary cycles, in *Trace Elements in Biochemistry,* Academic Press, New York, 1966, chap. 4.
81. **Gaskin, D. E., Frank, R., Holdrinet, M., Ishida, K., Walton, C. J., and Smith, L. M.,** Mercury, DDT and PCB in harbour seals *(Phoca vitulina)* from the Bay of Fundy and Gulf of Maine, *J. Fish. Res. Bd. Can.,* 30, 471, 1973.
82. **Miller, S.,** The persistent PCB problem, *Environ. Sci. Technol.,* 16, 98A, 1982.
83. **Suess, E.,** Particle organic carbon flux in the oceans — surface productivity and oxygen utilization, *Nature (London),* 288, 260, 1980.

Chapter 9

WHAT IS HAPPENING TO PCBs?

Elements of Effective Environmental Monitoring as Illustrated by an Analysis of PCB Trends in Terrestrial and Aquatic Organisms

Virginia F. Stout

TABLE OF CONTENTS

ABSTRACT

Since voluntary restriction of PCB usage began in 1970, PCB levels have fallen in some areas where they were previously highest but are increasing in previously clean areas and remaining more or less constant where low levels were already noted. These findings reflect the gradual spread of PCBs from earlier hot spots such as dumps, spills, estuarine disposal sites, and waste water discharges. Biological variability, environmental fluctuation, and evolving analytical techniques obscure the detection of PCB trends. Nonetheless, until the 680,000 t reservoir of PCBs produced worldwide since 1929 is destroyed or immobilized, PCB residues will continue to accumulate in aquatic and terrestrial organisms. Complexities of PCB analysis in trend assessment and issues in biological sampling are detailed. Studies of PCB trends in gulls, ducks, starlings, pelicans, mussels, and freshwater and marine fishes are reviewed.

The presence and danger of polychlorinated biphenyls (PCBs) in the environment is not yet a dead issue, but it is one that appears to be dying quite rapidly. — Thomas Maugh II, 1972[1]

22,000 kg of PCB have been buried at the [New Bedford municipal] landfill.....airborne PCB were 400—1300 ng/m^3 in June 1977.[2]

In 1977, there were individual goldfish.......[in the Hudson River].... which had 40,000 to 99,000 ppm PCB per gram of lipid...[3]

...between 32 and 54 metric tons of PCBs will be transported into the [New York] estuary during the next 20 years if no remedial action is taken.[4]

I. INTRODUCTION

Since polychlorinated biphenyls (PCBs) were first identified in Swedish fish in 1966,[5] they have been found throughout the world,[6,7] even in Antarctica.[8] * Initially they were observed as contaminants interfering with the quantification of DDT.[9,10] By the early 1970s, they were recognized as widespread problems in their own right. Before long, PCBs in high concentrations had been isolated from numerous lakes and rivers. Marine birds and fishes also contained amounts of PCBs quantifiable by early methods.[7,11] High levels in goatfish[11,12] *(Mulloidichthys auriflamma)*** confirmed that tropical islands such as Hawaii were not free of this new environmental pollutant.

* In this review, the term PCB is used to include all polychlorinated biphenyls regardless of the extent of chlorination. A specific degree of chlorination is indicated by the last two digits; thus, PCB 1242 contains 42% chlorine by weight. As PCBs weather, a residue which initially derived from a commercial mixture such as the Monsanto Company product Aroclor® 1221, 1242, or 1248 will come to resemble a more highly chlorinated substance such as Aroclor® 1254, 1260, or even 1268. As a reminder that no implication is intended that the higher chlorinated substances were ever used in any large amounts, I prefer the terms PCB 1254 and PCB 1260 to indicate the degree of chlorination observed in environmental samples. I have used the manufacturer's designation only in the case of Aroclor® 1016 which was intentionally dumped into the Hudson River. In this case, using the term PCB 1242 would be less accurate. Residues in the Hudson River are more appropriately related to Aroclor® 1016, a special formulation containing 42% chlorine but lacking the highly chlorinated components of Aroclor® 1242 itself. This chapter concerns PCBs, but not polar metabolites. Monitoring programs up to now have only considered the hydrophobic parent chlorinated hydrocarbons.

** Scientific names are included the first time each species is mentioned.

This chapter will discuss the issues involved in determining PCB trends over time. More than a decade has passed since quantitation of PCBs became routine. Techniques suitable for long-term assessment will be defined. Data from diverse laboratories will be summarized. These findings will be integrated into an overall picture of past and present PCB pollution. The prognosis for the future and recommendations for further research will be described.

A voluminous literature on PCBs has developed over the past 15 years. I have made an intensive and extensive search to find records suitable for this chapter. Numerous scientists have been generous in providing unpublished manuscripts and even raw data. Some reports are not readily available, while others may have been overlooked. Even with these limitations, distinct patterns have emerged that provide a clear picture of the present status of PCBs.

A. Experimental Design for Trend Assessment

Holden[13] described some of the parameters for an early international environmental monitoring program. These included determining the extent to which collaborating laboratories could both identify and quantify organochlorine residues. Protocols were also designed to assure representative samplings of marine, freshwater, and terrestrial ecosystems. Exact species were listed, although even within individual countries, no species was universally available. As far as possible, annual samples were to include 25 specimens of the same age, size, and sex, with the animals to be analyzed individually. Fishes were to be collected before spawning and birds before fledging. Holden even noted the importance of government support in mounting an effective monitoring program. Although sampling requirements were explicitly detailed, in practice the described protocol often was not followed exactly.

These early specifications are the foundation of modern monitoring programs, and they contain hints of difficulties in their implementation. General requirements for designing and implementing environmental studies were defined in 1980.[153] This subject still merits reflection, for again in 1981 Veith et al.[14] reiterated the need for stringent requirements in trend-monitoring programs including "precise methods for measuring small differences in concentration, and, to minimize biological variability, a fairly rigid procotol with respect to species and size". In the following section, some of the problems will be described in greater detail.

B. Chemical Analyses for Measuring Trends

1. The Art of PCB Analysis

Quantitation of PCBs will be discussed only insofar as it relates to trend assessment. Electron-capture gas chromatography (GC) has remained the standard tool for quantitation despite operational difficulties. PCB analysis started as an art with the analyst deciding how and what to quantify, e.g., how to draw baselines, what to eliminate because of interference, and what standard to use as a reference.

The complexity of PCBs challenged even the choice of standards and quantitation techniques. Commercial mixtures used as standards yielded numerous chromatographic peaks, from one to ten of which were chosen for quantitation. Either the height or area of peaks was measured. The PCB content would vary widely depending on the choice of peaks, baselines, and standards,[11] especially in weathered samples. Furthermore, patterns of peaks might differ from one gas chromatograph to another, environmental samples by no means provided consistent patterns, and the choice of peaks was changed sometimes in order to quantify unusual samples more accurately.

As computers were introduced, a more exact regimen was established. Arbitrary use of automation has not been a panacea, however, since it can also lead to inaccurate results when samples do not conform to a pre-established pattern. Capillary-column GC introduced another degree of sophistication, which gives hope of improved accuracy. Quantitation has not yet improved, however, according to an interlaboratory collaborative study. Even more discouraging, some components were *not identified correctly*.[15] After more than a decade, basic identification of pollutants in environmental samples is not assured.

In addition, both the proportion and quantity of components in an extract vary with extraction procedures. Diverse techniques are used for extracting PCBs from environmental samples because each type of sample presents its own specific analytical challenges. Methods have evolved also when standard procedures designed for surveillance of food products proved impractical for large-scale monitoring programs.[16]

International cooperative studies have measured the real effect of such analytical problems. An early study[17] reported a ± 30 to 60% coefficient of variation (C.V.) for wildlife samples requiring complex processing. In a subsequent study, 14 laboratories in 10 nations provided a total of 18 analyses on the same sample from a dead cormorant *(Phalacrocorax carbo)* found in the Netherlands in 1970. Seven standards were used for quantitation: PCB 1250, 1254, 1260, and 1264, mixtures of PCB 1250 and 1260, as well as DDE and dieldrin.[13] The wide variety of standards reflected early attempts to quantify complex mixtures and also batch variations of PCB formulations, sold by chlorine content rather than exact structure. Analysts chose the standard corresponding most closely to each sample, but matching of chromatographic profiles was a time-consuming process, sometimes compromised in routine monitoring. The C.V. for all results on the cormorant sample was $\pm 20.7\%$. This figure probably represents the best outcome up to that time since participants in such a study may well have been more skilled than other analysts, and collaborative study data reflect optimum performance resulting from extra care. A later oyster homogenate containing 0.3 ppm dry weight PCB 1254 showed a 55% C.V. among 14 laboratories,[18] even though data from 12 other laboratories were excluded because the levels were reported as below the detection limit or different, unknown, or mixed standards were used.

More recently, agreement between laboratories seems to have improved, though data may still vary up to 100-fold between laboratories.[19] Goldberg et al.[20] found a mean C.V. of 8.8% (range of 0 to 15%) for three samples. By 1980, C.V.s[21] for six samples had declined to 5.7% (range 1.5 to 9.7%). In contrast, an interlaboratory calibration of two urban air *extracts* showed C.V.s among nine participants ranging from 26 to 39%, even though the use of extracts had already eliminated one source of variability,[22] the extraction process itself. Improvement is selflimiting, however, since as analyte concentration decreases, the C.V. increases.[23] Perhaps quality assurance will benefit from the statistical evaluation described by Schmitt.[24]

2. *New Developments*

Chlorinated hydrocarbon studies at a remote site uncovered another problem in the late 1970s. Lake trout *(Salvelinus namaycush)*, fat lake trout *(S. namaycush siscowet)*, and lake whitefish *(Coregonus clupeaformis)* from Isle Royale contained elevated levels of several chlorinated hydrocarbon pollutants, compared to other Lake Superior sites closer to urban development. Initially, Isle Royale, located in the northern apex of Lake Superior, was chosen as a control site for environmental studies of Lake Superior because the island has been a national park since 1940[25] and it had never been logged nor contaminated by motor vehicles. Pollution of this supposedly clean site was attributed to atmospheric fallout until a better explanation surfaced. Extracts contaminated with toxaphene had been quantified inaccurately, perhaps as the result of automated analyses without careful examination of GC patterns.

This finding signaled a soon-to-be-discovered widespread difficulty, interference by multicomponent pollutants, including chlordane as well as toxaphene. (Although chlordane is a single compound, the commercial formulation contains several byproducts.) Once toxaphene and chlordane were rather uncommon pesticides, but more recently they have substituted for the banned DDT. Use of these compounds, which mimic PCBs in the usual separation procedures expanded without being noticed by analysts, because like PCBs they are mixtures appearing as a mass of mainly small peaks in GC traces. Ironically, PCBs themselves were identified in environmental samples because they interfered with DDT quantification. Now other interferences put a cloud on PCB trend assessment. The extent of interference, which undoubtedly varies from region to region has yet to be assessed. In the late 1970s, analysts were still searching for separation methods suitable for routine analysis. As PCB analysis was already expensive, each additional step would further constrain the already limited number of analyses that could be funded.

Another problem is the issue of evolving methodology. Capillary column GC brought indication of decreasing PCB levels in Lake Superior, but further study showed that the reduction was an artifact of a new technique, which resolved components previously assumed to be entirely PCBs.[26] Although it is impossible to freeze techniques for contaminant analysis, nonetheless it is difficult to compare current data to those obtained by older methods.

3. The Reality of PCB Data

In assessing trends, one must remember that PCB residues are complex mixtures. In the first place, PCBs are not exact substances, because the composition of each formulation varies with manufacturing conditions. Moreover, individual PCBs behave in different ways. For example, the lower molecular weight compounds evaporate more rapidly and also tend to react and degrade more readily. Even among the more highly chlorinated components, individual compounds do not behave and react uniformly.

To compare residue levels over the years is, in a sense, like asking whether we now have as many Winesap apples as we previously had MacIntosh apples. Numbers can be compared, but what do the answers mean?

As new methods are implemented, efforts are made to assure the comparability of data. Ideally we should be comparing the level of specific components today with levels in past years. Frequently that is not feasible, because most PCB data represent sums of several components. Even the purported identities of individual compounds have evolved with the development of techniques.

Assuming all the data are correct, interpretation is still an issue. If PCB 1260 starts appearing in fish somewhere, does it announce yet another PCB dump or a carefully hidden spill? Or does it mean that an old source of PCB 1242 or 1254 is slowly dissipating, leaving in its wake a tattletale, the high-boiling components which resemble PCB 1260?

I have described these analytical difficulties in order to support the restrictions I have placed on samples chosen for this review and to acknowledge imprecision in the process. It is vital to take stock of progress in an effort to rid the world of PCB pollution, but it is also important to spell out the limitations of trend assessment. Comparison between laboratories is usually a questionable process,[27] except in special cases where interlaboratory collaborative studies assure comparability.[28] Comparison between years within a single laboratory is more practical as long as a single quantitation technique is used and quality assurance programs are maintained.[27]

C. Issues in Biological Sampling

Over the years, numerous laboratories have analyzed for chlorinated hydrocarbons including PCBs. Many authors compare their data with published reports. Unfortunately, as Willford[29] has noted, "results frequently contradicted each other and sampling programs lacked the intensity, the uniformity of sampling procedures, and the continuity that permitted evaluation with time...." In this section, I shall address aspects of biological variability that affect the suitability of data for trend assessment.

1. Species

Organisms differ in the rate and extent of accumulation of chlorinated hydrocarbons.

Feeding habits, longevity, lipid content, metabolism, and migratory patterns are the kinds of factors that can contribute to marked intra- and interspecies differences. Twenty-year-old fish, for example, may reflect pollution over the entirety of their lives. Birds may migrate thousands of miles and accumulate pollutants wherever they live. Juvenile salmon, born from pea-sized lipid-filled eggs, carry forth a small contaminant burden from their mothers, while sole, born from tiny eggs, are nourished from their surroundings at a much earlier stage of development. Marine mammal pups receive a large dose of chlorinated hydrocarbons[30] in the fat-rich milk. Preening birds lose PCBs through the uropygial gland[31] but can also ingest PCBs from their feathers.

Over the years, a number of species have been chosen as indicator organisms. Butler[32] used mollusks, especially oysters, as sentinels of general pollution in estuaries. Oysters contain concentrations of organic contaminants as much as 50,000 times as high as those in the surrounding water, in which such compounds are nearly insoluble. Bioaccumulation expedites identification and quantification just because the higher levels can be measured so much more easily.

Mussels[20,28,33,34] have become common monitoring organisms because they are readily available in adequate numbers over a wide geographical range. Such sessile organisms have the advantage that they stay in one place. Perhaps for this reason sampling variance was considerably lower than analytical variance in the California Mussel Watch.[35] Compared to PCB determinations in water or sediment, PCB levels in mussels indicate bioavailability directly since contaminants are already in the living organism.[36] Mussels absorb and eliminate chlorinated hydrocarbons at known rates.[37,38] They reflect exposure over a distinct time interval. In contrast to fishes and crustaceans, mussels possess minimal enzymatic capability to metabolize contaminants. Perhaps for that reason, they can survive even in highly polluted environments. They can be transplanted readily and can be retained in bags or cages when suitable support is lacking.[39] They tolerate depths of at least 60 m.[40] These properties favor the use of mussels to monitor around sewer outfalls and in highly polluted harbors. Furthermore, consumers are directly concerned about the wholesomeness of wild and cultivated mussels, because these popular sport and commercial species live in potentially contaminated environments.

Estuarine contaminant data must be interpreted carefully. Intertidal animals sometimes are exposed to contaminant levels much higher than those present throughout the whole water column. Chlorinated hydrocarbons concentrate at the air/water interface in a lipid-rich monolayer.[41] Furthermore, at times of high runoff, organic compounds bound to particulate matter are carried in the surface layers where salinity is lowest. Consequently, mussels held below the surface may provide a more representative picture of PCB contamination. On the other hand, Richardson and Waid[21] found larger mussels collected 5 to 7 m below the surface contained much higher levels of PCBs than very small surficial specimens (0.12 to 0.21 ppm dry weight vs. <0.010 ppm). They suggested that size and rate of accumulation may have contributed to these differences. In the vicinity of a sewer outfall, however, PCB levels were highest at the greatest depth, i.e., closest to the point of their release.[42]

PCBs also concentrate in organic-rich sediment, which accumulates at the water/sediment interface. As a result, animals which live and feed on the bottom will receive greater exposure than animals living or foraging even a small distance above the sediment.

Numerous species of fishes have been chosen for monitoring free-ranging aquatic organisms. Often practical considerations are decisive, such as the sampling of commercially important species to ascertain their wholesomeness. Wide ranging species which are readily available in adequate numbers are taken in preference to more sedentary, but elusive or uncommon fishes. Several species classified only as bottom dwellers or predators[43] were selected for the U.S. National Pesticide Monitoring Program, because no single species is available at all 109 freshwater stations surveyed.

Fishes accumulate PCBs rapidly[44,45] but do not depurate them in measurable amounts.[46] Any apparent loss results from spawning or dilution through growth. Thus, the total body burden of PCBs reflects a time-averaged composite of exposure throughout life. To lessen the influence of long-past exposure, Butler and Schutzman[47] chose juvenile estuarine fish as indicator organisms. Fish less than 1-year-old contain contaminants from two possible sources: carry-over from parents and exposure during at most 1 year of life. For the most part, they reflect current exposure. One exception is juvenile dogfish *(Squalus acanthias):* this ovoviviparous species transfers a substantial burden to the next generation from its large eggs.[48]

Among terrestrial organisms, herring gulls *(Larus argentatus),* starlings *(Sturnus vulgaris),* and ducks were chosen because of availability and widespread distribution. Variability between specimens reflects individual feeding habits as well as the obvious ability of birds to travel. The gulls were also studied because they experienced reproductive difficulties on Lake Ontario.[49]

Brown pelicans were monitored in several areas mainly because they were disappearing in the late 1960s and early 1970s. The extent of sampling was, however, limited by the necessity of maintaining the dwindling stocks.

2. *Age and Size of Specimens*

The ideal sample would include specimens within specific size and age distributions. Unfortunately such samples rarely exist. Variations in weather, food supply,[50] pressure from predators, to name just a few factors, can contribute to fluctuations in size and abundance of animals. Aging in the field is not practical and analyzing enough specimens to assure the correct age distribution is too expensive. Samples of fishes have been restricted to fish less than 1 year old[47] and to a limited size range.[29] Alternatively, fish from a broad range of sizes and age have been compared[51,52] and size and age included in data comparison. Organic contaminants generally increase with age in fish[53] by a multiplicative factor.[54] Holden[13] suggested nestling birds as suitable indicators of terrestrial contamination. Alternatively, adult birds or their eggs have been studied.

Mussels are usually selected from a narrow range of shell lengths; aging is not currently possible.[20] In early mussel studies, no effects on contaminant levels were noted from divergent growth rates. Perhaps flow rate, water temperature, salinity, and available nutrients influenced growth only to a minor extent or canceled each other enough that they were not detected initially.

3. *Timing of Collection*

Farrington et al.[28] recently reported a two- to fourfold uncertainty in contaminant levels in mussels resulting from seasonal variability. Mussels from Narragansett Bay, R.I., and Bodega Head, Calif., exhibited different periodicities and magnitudes in contaminant levels. What causes these fluctuations? Changing biological and biochemical activities of mussels themselves or associated organisms, as well as changing concentrations and bioavailability of pollutants in the environment, contribute to seasonal fluctuation.

Weather can affect organisms in several ways. Some species reproduce only at specific light intensities or at critical temperatures. Climatic conditions alter the availability of food supplies. Heavy precipitation and rapid thawing alter water flow, salinity, and contaminant burden. Since the weather is not controllable, researchers can only remain mindful of its influences.

The life history of individual species determines the optimum period of collection. Both breeding status and weather influence PCB levels.[55] Regular sampling at the same stage of reproductive development increases the consistency of samples. Organisms which increase lipid stores preparatory to breeding transfer PCB-laden lipids to their spawn. The seasonal

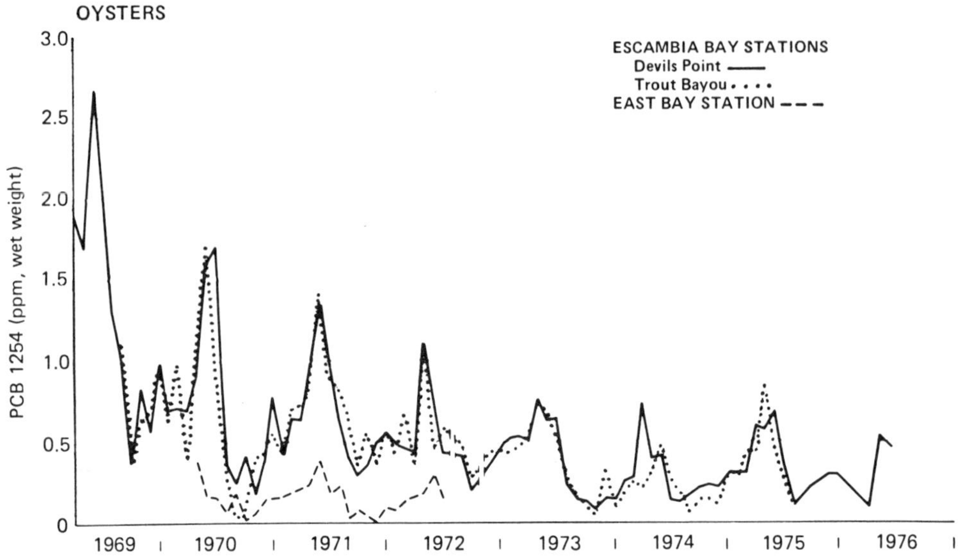

FIGURE 1. Annual fluctuation in PCB content associated with oyster *(Crassostrea gigas)* reproduction. In the spring, oysters accumulate lipid and lipophilic PCBs which are excreted again with the spawn. (Modified from Wilson and Forester.[56])

fluctuation associated with oyster reproduction[56] is shown in Figure 1. Because fishes eliminate chlorinated hydrocarbons mainly through reproduction, collection at different reproductive stages would reflect different types of contamination. Correcting for order of egg laying in common terns *(Sterna hirundo)* is necessary since the PCB content decreased 20% from the first to the third egg in a clutch.[57] Similarly, repeated spawning could alter PCB levels in fishes.

Beyond the more obvious factors which influence the consistency of sampling, the well-being of endangered species must also be considered. Brown pelicans and marine mammals along the California coast often abandon their young when disturbed. Only after the young have become independent are the rookeries approached to collect dead specimens. Thus, fresh eggs from the eastern brown pelican *(Pelecanus occidentalis carolinensis)* can be taken throughout the breeding season without ill effects on the young. In Texas[58] and in California, however, pelican *(P. o. californicus)* eggs are collected after the young have left the nest.[59] These samples are obviously atypical because they are confined to eggs which failed to hatch for lack of fertility, viability, or suitable incubation. Such specimens introduce further variability because loss of water and metabolic processes during incubation influence residue levels.

4. Sex

The sex of specimens frequently influences data.[60,61] Females of several species eliminate PCBs through reproduction. Salmon, pelicans, gulls, and terns lose chlorinated hydrocarbons through egg production. Marine mammals lose them by lactation.[30] Each species is unique, however, since chlorinated hydrocarbon levels in the roach *(Leuciscus rutilus)* showed no correlation with sex, age, or size in a set of 95 specimens.[55] The choice of sex, then, depends on the aim of individual projects.

5. Collection Site

Both pristine and polluted sites contribute useful data for trend assessment. Presumed "clean" sites in developed countries often receive unsuspected pollutants via aerial transport, runoff, clandestine dumping, or accidental spills. Cities, harbors, and in-shore areas contain high concentrations of PCBs from direct and indirect sources. Agricultural, sylvan, and open water areas reflect average burdens from numerous diffuse sources. Estuaries may be expected to receive contaminants in proportion to the intensity of human habitation.[62] They are monitored as nurseries for aquatic species[47] and as popular sports fishing areas.[63] Analyses of samples from truly pristine sites provide baseline data to be related to more polluted areas and to signal future increases in pollution. Data from highly polluted sites reflect the response of the environment to high levels of contamination, while studies at sites without intense or direct pollution are useful for predicting long-term progress.

Furthermore, the foraging and migratory habits of animals must be considered in choosing sampling sites and evaluating data. Only stationary plants[64] and sedentary or caged animals reflect pollutant levels at unique locations. Diurnal movement, for example, can expose organisms to significantly different levels of contaminants, and animals collected at different times of day or year may represent different populations. Tides and currents may disperse pollutants in unforeseen ways.[3,52] This brief discussion is intended to suggest the efforts necessary to obtain consistent samples — and to acknowledge that sometimes even these are unavailing.

6. Natural Biological Fluctuations

Short- and long-term fluctuations in fishery stocks and other wildlife populations[65] influence the levels of contaminants in complex but unpredictable ways.[66] For instance, the oil content of menhaden decreased markedly in recent years.[67] Chlorinated hydrocarbon concentrations in an organism may depend on lipid content,[55] partition equilibria, or extent of exposure. Assuming a specific total PCB content or burden, a change from 8 to 2% oil would quadruple the PCB concentration in the contained oil. Unless the mode of accumulation is defined, monitoring oil or individual tissues may not reflect environmental levels accurately. Even more of a problem, when animals disappear completely from their usual haunts, sampling fails and a prescribed sampling pattern is disrupted. The intrinsic variability of biological samples can be allowed for, but not entirely eliminated by appropriate design of monitoring programs. Biological variability is one of the main factors obscuring detection of new trends.

7. Tissues Sampled

The purpose of a study often dictates the type of sample. The National Pesticide Monitoring Program analyzed whole fish to assess the effect of contaminants on wildlife which consumed them.[68] Other studies[39,69] used muscle or fillets with skin in an effort to determine the safety to human consumers. Stout et al.[70] included whole fish and fish meal and oil to evaluate the impact of contaminants throughout the menhaden *(Brevoortia tyrannus* and *B. patronus)* fishery. Though sometimes done, comparing data from different tissues is not meaningful, unless the relationship for accumulation of contaminants in the various tissues has been documented. Preliminary data for fish muscle and liver suggested a complex relationship.[71]

8. Number of Specimens and Pooling

The number of specimens is dependent on the variability and the desired level of discrimination. Individuals in a single collection sometimes show nearly a 100-fold range in PCB content.[12,72] Furthermore, the distribution of organochlorine residues among individual specimens is often skewed.[17] To detect a 75% difference between residue levels, 10 specimens would suffice, but to detect a 25% difference, 50 specimens are necessary. Holden[13] recommended 25 specimens as an appropriate sample size. Gordon et al.[73] detailed the number of specimens required for varying degrees of resolution of several trace metal pollutants, data that can enhance the understanding of sampling problems for trace organic analysis as well. Butler[66] documented the combined variability of individual organisms and a river environment to have a standard error of 15 to 30%.

Jensen and Lassen[74] investigated problems of sample size and stratification of critical sample parameters. They calculated the effect of length stratification, i.e., choosing specific numbers of specimens in several length intervals, as compared to random sampling on the number of specimens required in a single sample. They did not advocate keeping biological variables constant because the influence of various parameters is not thoroughly documented. Instead they recommended continuing surveillance of the relationship between biological variable and contaminant levels. They determined that variance in chlorinated hydrocarbon values in one set of herring data contributed a 1.07- to 1.14-fold increase in the number of specimens needed in a random sample compared to one which took into account variations in length. In a sample of cod, the factor increased to 1.56 for PCBs in liver. Using these calculations, Jensen and Lassen[74] suggested that annual samples of 20 fish analyzed individually would detect a 10 to 20% annual change over a 5-year period. If the ratio for random vs. stratified-length variability increases substantially, the sample size would also need to increase to detect the specified trend. Likewise, if the trend is smaller, a larger sample size would be necessary to observe a statistically significant trend.

Not only statistical considerations, but also availability of specimens and cost of analyses influence the choice of these sampling parameters. Limited numbers of specimens of endangered species are available. Mussels, on the other hand, are plentiful in many regions of the world and easy to collect and handle in large numbers. Analysis of individual specimens, however, escalates the cost of monitoring. When limitations in funding restrict the number of analyses, compositing specimens becomes a necessary compromise. Nonetheless, the variability of individual specimens and their environment must be determined as frequently as practical at every site[73] to assure the accuracy of trend assessment. Where it is not practical to determine the variability of individuals, replicate pools at least provide pool variances. Whatever the experimental design, a consistent pattern of sampling and analysis simplifies statistical treatment of data.

9. New Approaches

Kaiser[75] has developed a new approach to trend assessment. He used interspecies contaminant quotients to compensate for background residue levels previously in biota.

$$\text{Quotient} = \frac{\text{PCB concentration in host}}{\text{DDT concentration in host}} \div \frac{\text{PCB concentration in parasite}}{\text{DDT concentration in parasite}}$$

He evaluated chlorinated hydrocarbon data from sea lamprey *(Petromyzon marinus)* and the associated host lake whitefish from Lakes Huron and Michigan. Trends in the original data were obscured because residue levels of individual specimens within a single catch varied widely (0.24 to 14 ppm* PCBs in whole fish) with a mean and standard deviation of 0.50 ± 0.30 in lake whitefish and 2.2 ± 4.2 ppm in lamprey. For the PCB:DDT ratio, the variation was much reduced to 1.9 ± 0.65 in lamprey. Kaiser[75] considered the use of ratios as a kind of normalization, akin to an internal standard. Using the new technique, he found that DDE and PCBs were both increasing in the two Great Lakes, while several other chlorinated hydrocarbons appeared to be decreasing.

* PCB concentration is reported on the basis of the tissue as received or wet weight, unless otherwise indicated. Dry weight means related to the weight after drying to remove water. Fat basis or lipid weight relates PCB residues to the fat content of the tissues. It is not meaningful to compare data reported on different bases. At the same time, data are not always available to convert to a common basis. Consequently, all data are reported as originally published. (See Section II.)

This concept requires further validation. All chlorinated hydrocarbon levels are not necessarily related to each other; the correlations of ratios may be a mere coincidence. Furthermore, since lampreys parasitize several species with as yet undefined organochlorine uptake ratios and interactions, interspecies ratios may not be as consistent as in this initial investigation. Nonetheless, data for lake trout and lake whitefish collected at nine sites in Lake Superior[75] did support the thesis. If validated, this technique will allow trends assessment with many fewer specimens by compensating for variability. Another application of contaminant ratios compared PCB levels of several species of plants at polluted sites with corresponding background PCB levels.[64] Perhaps this could be applied to stationary animals as well.

D. Statistical Aspects of Trend Assessment

What is the appropriate measure of central tendency? Some authors choose to report means based only on quantifiable values. Any or all values below either the quantitation limit or the detection level are ignored. Such a system overestimates the mean and to my mind misrepresents the picture, even when footnotes describe the methods of handling data. When this practice exaggerates the situation grossly, it leads to misinterpretation by the general public and does a disservice to environmental scientists.

A greater issue arises from assessing trends from averaged incidence rates alone. After collecting 38,000 fish and analyzing 1524 composites, the National Pesticide Monitoring Survey of estuarine fishes combined data for 144 estuaries into brief summaries for each state. New York State was reported to have a 63% incidence of quantifiable PCB residues based on sampling of three estuaries, while North Carolina had a 9% incidence based on 19 estuaries. Washington State reputedly had PCBs in 17% of its samples, even though none of the 128 samples from five estuaries contained PCBs and all 27 from one highly polluted river contained quantifiable levels. By the same logic, Washington State ranked highest in PCB level with a mean of 1.67 ppm, while *83% of all samples contained <0.05 ppm* PCBs.[47] Estuary-by-estuary analysis would be a more appropriate method of evaluating this massive body of data. Naming polluted estuaries and tallying the number in each state would be more meaningful than impugning whole states on the basis of single hot spots. In place of simple inspection, analysis of variance or some other statistical test is needed to determine any trend in this body of data.

Lassen[76] proposed substituting half the detection limit for observations below the detection level as long as they represent no more than 5% of the total number of values. When a major proportion of values was zero, he felt it was not meaningful to compute contaminant relationships. Alternatively, the National Pesticide Monitoring Program fresh water fishes study normalized lipid values by an angular transformation and concentration values by log transformation of the residue value + 1.0,[68] i.e., log (x + 1).

In their studies of pollutants in herring gull eggs from the Great Lakes, Gilman et al.[49] found not only skewed distributions of residue levels, but also a variation in direction of the skew for different contaminants. Consequently, they found the use of either arithmetic mean and confidence intervals or the assumption of log normality and geometric mean inappropriate as descriptions of central tendency. Furthermore, they avoided assigning absolute values to trace amounts because of the bias introduced by this process. They concluded that the median and range were accurate descriptive statistics regardless of the shape of the residue distribution and allowed direct inclusion of trace values. Nisbet and Reynolds[57] observed that the extent of the positive skew is not consistent as is evident from a comparison of the geometric and arithmetic means of tern eggs (Table 1). Individual data sets may determine the best descriptors, but lack of standardization then precludes comparison with other data.

Beyond these descriptive statistics, how have yearly data been evaluated? Schmitt[43] sum-

Table 1
COMPARISON OF GEOMETRIC AND ARITHMETRIC MEANS OF PCB RESIDUES IN COMMON TERN EGGS AT DIFFERENT SITES[57]

		Conc. in ppm, fresh weight			
Year	**No. of eggs**	**Geometric mean**	**Arithmetic mean**	**SD**	**Range**
1973	10	5.3	5.4	1.3	4.0—8.5
1975	11	19	20	7.9	12—40
1976[a]	8	15	22	17	2.1—53
1976[b]	8	4.0	8.9	15	1.5—45

Note: ppm: μg/g fresh weight; PCB 1254 + PCB 1260.

[a] Same station as 1975 data.
[b] Different station from preceding entry.

marized problems in applying statistical tests to contaminant residue data. He reminded the reader that no statistical approach is automatically correct. The objectives of a program and perspectives of those conducting the research influence the choice of statistical techniques. The National Pesticide Monitoring Program duck wing survey used *t*-tests for each ''data pair'', presumably each state, in a single flyway. Only pairs with detectable levels in 50% of all pools were compared.[77] Dropping data is a questionable procedure. Conclusions based on 40% of a data base cannot assess the whole picture.[77] The *t*-test is seldom applicable for trend assessment. With numerous stations and years, analysis of variance is preferable.

Nonparametric statistics allow testing of unequally sized samples and skewed data. Nisbet and Reynolds[57] and Gossett et al.[78] used the Spearman rank correlation coefficient to assess temporal trends in their data. The Mann-Whitney U-test,[79,80] the Kruskal-Wallis analysis of variance of ranks,[78] the Wilcoxon matched-pair signed-rank test,[81] the Friedman two-way analysis of variance,[82] and the Kramer multiple range test[83] have also been used for evaluating PCB trends. Scientists involved in trend assessment do not always understand the subtleties of statistical tests. For example, destructive analytical techniques such as PCB analysis cannot meet the requirements of the Wilcoxon test. It was designed for testing a single system under different conditions rather than similar populations in different years.

Schmitt[43] detailed a procedure for applying a partially nested factorial analysis of variance technique. He tested this method with 6 years of data from the National Pesticide Monitoring Program for fresh water fishes. He chose 15 stations which were sampled in each year and were consistently analyzed by a single method. Two- and three-way factorial techniques were compared with the partially nested factorial scheme. The effects of uneven replication and random vs. fixed location effects on a nationwide monitoring program were considered. Location and temporal differences were shown for PCB 1254 by all techniques, but were changing in a similar fashion at all 15 stations. Partially nested factorial analysis resolved species differences, but three-way factorial did not distinguish between predators and bottom-dwellers. These findings suggest differences between species at a single location, but not between species types.

This study[43] showed the complexities in analyzing a seemingly simple set of data. These included non-normality, unequal sample numbers, broad-ranging variance, multiple species, and inconsistent patterns at each site. The resolution of trends probably was enhanced by compositing specimens because individuals vary considerably and composites unavoidably underestimate variability. The data set used by Schmitt[43] comprised samples from sites with a history of pollution. To the extent that the data were influenced by point-source discharge, the detailed conclusions may not relate to sites with diffuse contamination. The method has also been applied to the whole 1976 to 1979 data set.[24] Partially nested factorial analysis represents a significant development in pollution assessment.

Multivariable linear regression (MLR) is another recent technique for evaluating contaminants in fishes. This procedure takes into account parameters such as length, weight, age, sex, season, area, and year as well as interactive effects such as season-year and length-season-year. A brief description of findings for metals will warn about the pitfalls of extrapolating data. A log-linear model appeared to yield a more normal distribution than a simple MLR model.[76] For cadmium, copper, and zinc in three species of fishes from the vicinity of a Greenland lead and zinc mine,[84] significant effects from year to year were observed, but no temporal trend was evident. Length, season, and year interacted but not consistently. Since interactive effects were difficult to quantify, time trends were hard to prove. Mercury from industrial pollution seemed to affect flounders differently near a point source than at greater remove. Near a former pesticide factory, no relationship between biological parameters and mercury levels in fish was observed, but year, length, and age relationships were evident further away.[85]

Scott et al.[86] compared rectilinear and interactive MLR with analysis of covariance. They were looking for less costly ways of determining time trends in resident fish. They tested predictor variables as a way to reduce variance. MLR was somewhat effective for metals but not for PCBs. Individual cod *(Gadus morhua)* did not respond to environmental contaminants in a consistent manner. MLR substantiated rather than eliminated the need for a range of sample sizes for effective monitoring.

These findings confirm the difficulties in extrapolating contaminant data from species to species, site to site, or substance to substance. They also show that statistical evaluation for trend assessment still needs refining.

II. FINDINGS

A. Small Projects

Studies limited in scope will be discussed first followed by more extensive programs. Comparison between laboratories remains uncertain and will not be attempted. Some reports depended on data from more than one laboratory. Such studies will help illustrate pitfalls and limitations of such interlaboratory comparisons.

A New Jersey salt marsh provides unusually early data. Clapper rails *(Rallus longirostris)* killed during a hurricane and contemporary samples of fishes and invertebrates were frozen until they were analyzed with samples collected 6 years later. Moisture determinations showed no dehydration during storage. Fauna in this salt marsh exhibited no PCB trend in the whole ecosystem. Instead, PCB levels in the rails increased 160% between 1967 and 1973, whereas levels decreased more than 86% in three species of fishes. In all three sets $p < 0.05$, according to two-tailed Wilcoxon two-sample tests.[87] The changes in both directions are large enough that discounting them would seem unjustified. They may result from transfer of PCBs from lower to higher trophic levels. Whatever the cause, these findings emphasize the complex characteristics of any natural site.

The presumption in discussing trends in PCBs in biota is that, eventually, as PCB usage and disposal cease, the levels in biota will decline. Often wildlife respond variously to PCB pollution. Three species of seabirds from eastern Canada showed no clear trend from 1972 to 1976. Only three of seven sets of egg data showed changes ($p < 0.01$): two decreases and one increase, one for each species.[88]

Wilson and Forester[56] studied the fate of Aroclor® 1254 after an accidental leak from an industrial site along the Escambia River in 1969. The PCBs dispersed more than 20 km downstream into Escambia Bay near Pensacola, Fla., where they accumulated up to 2.7

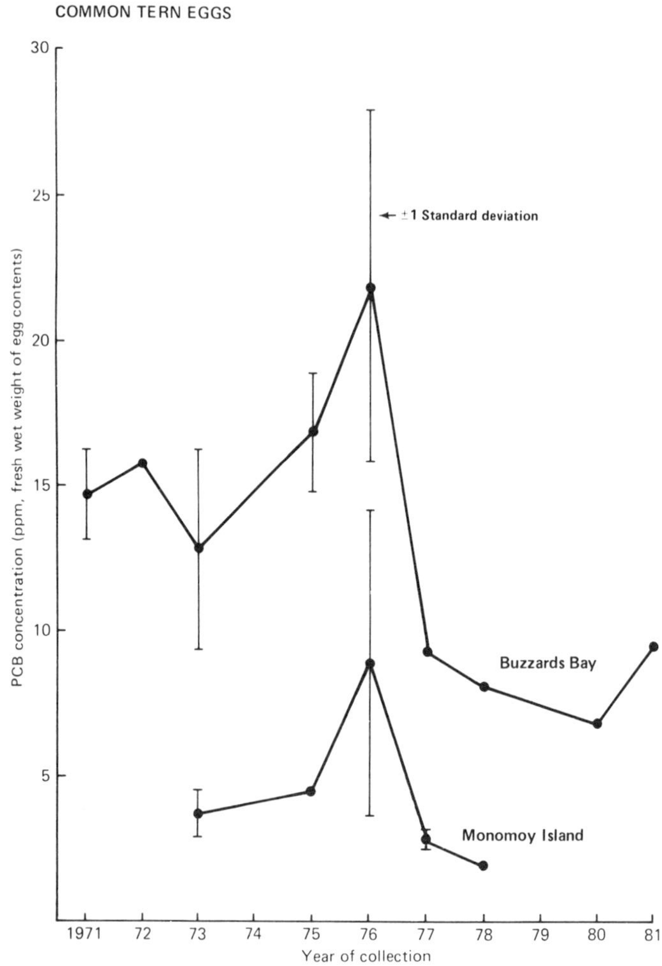

FIGURE 2. PCBs in common tern *(Sterna hirundo)* eggs from Buzzards Bay and Monomoy Island, a site relatively remote from known contamination. (From the data of Nisbet and Reynolds.[57])

ppm in oysters *(Crassostrea virginica)*. Monthly monitoring for 7 years showed that when the leak was stemmed, PCB levels declined initially. Although PCBs nearly disappeared from the water at the spill site by 1974, in Escambia Bay, oysters still contained up to 0.5 ppm Aroclor® 1254 in May 1976. The levels fluctuated annually; they increase during the spawning season when lipid contents increase temporarily (Figure 1). Presumably, the contaminated sediments provide a persistent reservoir for continuing pollution of the bay.

In Massachusetts, New Bedford Harbor and nearby Buzzards Bay have attracted attention because of the high levels of PCBs found in sediments, fishes, and shellfishes.[57,89] PCBs declined significantly between 1971 and 1981 in eggs of common terns from Buzzards Bay.[57] The Spearman rank correlation coefficient was -0.67 ($p < 0.05$). At Monomoy Island, 80 km eastward and relatively remote from known sources of contamination, PCB concentrations were lower but followed a similar temporal trend. In both areas, maximum levels were observed in 1976 (Figure 2). The authors noted that samples collected in that year were analyzed by a different laboratory than in other years. They did not eliminate the possibility of analytical variability between the laboratories. Assuming the maxima are not artifacts, these sets of data provide a vivid example of long-range aerial transport from the urban, industrial center of New Bedford to the eastern shore of Cape Cod, 80 km away.

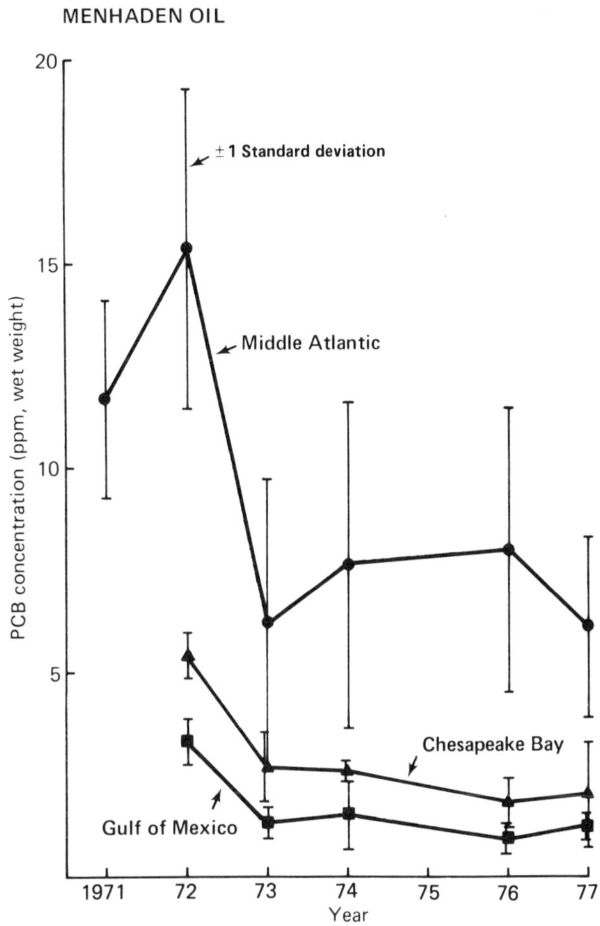

FIGURE 3. PCBs in menhaden *(Brevoortia tyrannus* and *B. patronus)* oil from three areas of the fishery. (From the data of Stout et al.[70])

PCB levels from fishes in Lake Simcoe, Ontario, Canada, decreased between 1970 and 1976. Ten species were sampled during two or more years. Because weights of the 367 specimens varied from year to year, a subset of 196 fish in comparable weight groups was selected for analysis of variance. PCBs declined significantly ($p < 0.05$ in the F- test) except in larger yellow perch *(Perca flavescens)* and smaller walleye *(Stizostedion vitreum vitreum)*. Significant decreases in fat content of several species over the time studied complicated interpretation of the data.[90] As will be seen again and again, the decline was greater when the initial chlorinated hydrocarbon level was higher. Levels change much less dramatically when the initial levels are low.

PCB levels decreased in the menhaden fishery over the years 1969 to 1977.[70] Whole fish and the commercially important products meal and oil were analyzed. Menhaden is the largest fishery in the U.S. in weight of fish. PCBs were measured in 314 samples collected from the Atlantic and Gulf Coasts and Chesapeake Bay. For statistical studies, 132 samples of menhaden oil were chosen. The null hypothesis of equality could be rejected for three areas of the fishery, but the differences between years were significant ($p < 0.05$) only in a few cases. Unequal numbers of samples and wide variations in PCB levels within years contributed to this situation. Coefficients of variation were as large as 86% (Figure 3).

Holden[91] found no change in PCBs in the blubber of gray seals *(Halichoerus grypus)* from the east coast of Scotland between 1968 and 1974. For a total of 108 specimens collected in 5 different years, the mean level in blubber was 36 ± 1.6 ppm.

Jones et al.[92] compared concentrations of PCBs in the liver, blubber, and brain of young harp seals *(Pagophidus groenlandicus)* collected from the ice in the Gulf of St. Lawrence in 1971 and 1973. Blubber contained 2.4 ± 1.4 ppm in the earlier year and 1.08 ± 0.53 ppm 2 years later. The level also decreased significantly ($p = 0.05$) in brain but not in liver. The 37% decrease in fat content in the liver may have masked any change in PCB content during this period. Neither the relationship between lipid and chlorinated hydrocarbon content nor the reason for the decline in lipid content has been established. PCBs as well as total DDT and dieldrin varied widely between individuals and without any correlation between compounds. The authors held that differences in lipid content of specimens and body burden in the mothers accounted for the variation. Not mentioned was the limited sample size, 16 animals in 1973 and 11 in 1971. With a sample of four adults, the PCB content in 1973 was higher in young than in older specimens. In a larger sample collected in 1970, the PCB concentration in females increased from birth to 5 to 6 years and then leveled off.[93]

Cod from the Gulf of St. Lawrence were analyzed in 1977, 1978, and 1979. Data from 122 specimens of liver were subjected to intensive statistical tests. Rectilinear MLR showed that PCBs were increasing while interactive MLR showed the opposite.[86] The authors felt that 3 years was a short interval for trend assessment.

Schneider[94] found no change in PCBs in cod from Kiel Bay, West Germany, between 1974/75 and 1977. He used a regression to calculate the value of 4.8 ppm in livers for the earlier time period, to compare with the 2.9 ± 1.0 ppm found in fish of the same length in 1977. Luckas and Lorenzen[95] reported that limitations in data made trend assessment in the North Sea and Ostsee (Kiel Bight) difficult. They cited earlier observations of a temporary increase, a period without change, or a slight decrease in PCBs in the area, as resulting from an inadequate data base.

In Finland, PCBs decreased in the marine ecosystem but stayed the same or increased in lakes. Paasivirta and Linko[96] found a significant decrease in PCB levels ($p < 0.001$) at two sites in the northern Baltic Sea. Between 1973 and 1978, the concentration in Baltic herring *(Clupea harengus)* fillets fell from 0.60 to 0.24 ppm and from 0.51 to 0.14 ppm at the two sites. Data for 1982 suggest that PCBs continued to disappear at the same rate.[69] From 1973 to 1978, PCB levels in perch *(Perca fluviatilis)* increased ($p < 0.01$) in one lake and showed no trend in another. The actual levels were low in both cases: reportedly increasing from 0.052 ± 0.020 to 0.078 ± 0.031 ppm in white lateral muscle in one case, and ranging between 0.056 and 0.098 ppm in the other.[96] The conclusions are based on stepwise multiple linear regression. Independent evaluation questioned the appropriateness of the statistical treatment.[97] Nonetheless, this small study may reflect as accurate a picture of PCBs as we are likely to obtain. In areas with high initial PCB contamination, reductions were seen, perhaps at the cost of low-level pollution spreading more widely. The reported decrease of PCBs in the muscle and liver of pike *(Esox lucius)* from the Turku Archipelago is based on an inadequate sample of 23 fish.[69] PCB levels in pike from Lake Päijänne have increased significantly ($p < 0.001$) from 1972 to 1980. The levels were some of the lowest in the world, 0.015 to 0.049 ppm in muscle. Because of the correlation between PCB level and fat content, the effect of fat content was eliminated by covariance analysis. The authors[98] did not believe local pollution was the cause, for PCBs had been banned in Sweden and were essentially never used in Finland. Instead they suggested that atmospheric transport apparently was averaging global pollution.

Although this chapter concerns animals in the wild, brief comment on beef and pork for human consumption may interest the reader. In these instances, the change in PCB levels reflects direct official action rather than environmental trends, as contamination levels in feedstuffs are regulated. In Ontario, Canada, the PCB concentrations of beef abdominal fat fell from 0.25 ± 0.058 ppm (fat basis) in 1969/70 to 0.0095 ± 0.022 ppm in 1981. In pork, the corresponding values were 0.33 ± 0.14 ppm and 0.003 ± 0.005 ppm.[99] Residues disappeared at a first order rate; the halflife was 3.2 years for beef and 1.8 years for pork. The levels in bovine milk decreased at about the same rate in southern Ontario. Between 1973 and 1977, the concentration in milk fat fell from 0.11 to 0.033 ppm.[100] Such a trend reflects the pattern of usage and disposal in an agricultural area. These data show the changes possible in agricultural commodities subject to direct intervention in contrast to the less direct control of environmental exposure of wildlife and fish.

These few studies from a number of sites in the Northern Hemisphere convey some of the challenge in assessing PCB trends. What is reported to be happening depends on location, type of sample, and adequacy of data.

B. Extensive Monitoring Programs

In some of the more intensive studies, I will cover the diverse species included in each regional program in order to evaluate contaminant trends in specific geographical areas. Then I shall describe integrated fish or wildlife monitoring programs which cover wider expanses.

1. Great Lakes — Overall Region

The Great Lakes of North America exemplify an aquatic environment surrounded by intense agricultural, industrial, urban, and recreational activities. The downstream course toward the sea provides a living model for studying the flow of contaminants through a complex system. Increasing population and pollution density[24] along Lake Huron and Lake Erie tend to obscure measurement of the PCB flux. Persistent chemicals remain for long periods, because outflow from the lakes is limited. Since 1978, the International Joint Commission[101] representing both Canada and the U.S. has coordinated monitoring of pollutants in the Great Lakes Basin ecosystem.

Lake Superior is farthest from the Atlantic Ocean and least densely populated. Between 1971 and 1975, PCBs in headless, eviscerated lake trout declined from 1.8 to 0.4 ppm.[102] For fish weighing 1.0 to 1.5 kg, the same trend was observed, from 1.9 to 0.49 ppm, although the authors[102] claimed PCBs had not decreased significantly. After stating that PCB values before 1970 were estimates, they apparently included both 1968 and 1969 data in concluding that there was no significant decline in PCB level in fish of a specific size. A smaller decrease, 0.8 to 0.3 ppm was observed for lake whitefish. The increase, 0.6 to 1.0 ppm for bloaters *(Coregonus hoyi)* may relate to the inclusion of one or more large fatty specimens in the data for 1975. All these conclusions are based on limited specimen numbers. According to another report,[27] PCB levels in lake trout were 2.3, 1.8, and 2.0 ppm in the edible portion in the years 1972, 1973, and 1977. The level fell to 0.85 ppm in *whole* trout near Thunder Bay in 1980. However, the exact status of Lake Superior has been clouded by the confounding of PCBs with toxaphene, mentioned earlier.

Only in three species in the main waters of Lake Huron did PCBs decrease between 1970 and 1976: alewives *(Alosa pseudoharengus)*, rainbow smelt *(Osmerus mordax)*, and walleye. PCBs in four other species increased at least at one location.[102] Not surprisingly the areas with least circulation are represented: cisco *(Coregonus artedii)* from Georgian Bay and yellow perch from North Channel. PCB levels in splake (hybrid between *Salvelinus fontinalis* and *S. namaycush)* and coho salmon *(Oncorhynchus kisutch)* (introduced in 1966) also increased in the main part of Lake Huron. For other locations of the above species and for other species, the PCB levels remained static. The reported increases were confounded by limited sample number and changes in size distribution or fat content from year to year. For instance, grouping splake by weight class showed that PCBs in headless, eviscerated fish actually decreased substantially from 1970 to 1974. Residues in the 1.0 to 1.5 kg weight class, for instance, fell from 1.5 to 0.3 ppm.

Lake St. Clair lies downstream from Lake Huron, upstream from Lake Erie. Frank et al.[103] reported that PCB levels in smallmouth bass *(Micropterus dolomieui)* in Lake St. Clair increased between 1968 and 1975. Before 1970, PCBs were only estimated, so this conclusion may be unwarranted. PCB levels in headless, eviscerated freshwater drum *(Aplodinotus grunniens)* were low and did not change substantially: 0.22 ppm for 0.25 to 0.50 kg specimens in 1971 and 0.17 ppm in 1976.

PCB levels in Lake Erie, the next downstream toward the Atlantic Ocean, appear to have declined somewhat over the period 1971 to 1975.[103] Data were limited, however, and reported increases between 1968 and later years may have been unwarranted. As stated above, the analytical technique in 1968 only allowed estimation of PCB content. Furthermore, in that procedure, TDE often obscures certain PCB components. Finally, the PCB standard (Aroclor® 1254 and 1260 in the ratio of 5:1 to 4:1) undoubtedly appropriate for later samples (for which it was designed presumably) may have undervalued the Aroclor® 1242 and 1254 that were often more typical in earlier years.

More recently PCB levels in spottail shiners *(Notropis hudsonius)* have reflected declining contamination in the nearshore environment of Lake Erie. Fish less than 1 year old have been sampled to monitor trends in this planktivorous species.[27,101] At Point Pelee, the most polluted site, PCB levels fell 82% from 840 ± 400 ppm in 1975 to 150 ± 37 ppm in 1980. Starting at a similar level in 1978, 157 ± 28 ppm, Lake Erie's Thunder Bay shiners reached 95 ± 29 ppm by 1980, i.e., a 40% decline in 2 years.

Although levels in coho salmon, rainbow smelt, and walleye (or yellow pickerel) were reported to have decreased in the main body of Lake Erie,[27] I found no trend in whole fish between 1977 and 1980. Only in the edible tissue of coho salmon did they decline from 3.2 ppm in 1972 and 2.8 ppm in 1973 to 0.91 ppm in 1977. The paucity of detail precludes real evaluation. Armstrong and Sloan[53] reported a 33% decline in six species of fishes after 1975.

The 1980 samples of top level predators from Lake Ontario showed a "minimal increase in PCB levels" over 1979.[27,101] Still, PCB pollution was not necessarily increasing. Instead the figures may reflect the usual fluctuations in contaminant data. For both lake trout and rainbow smelt, the levels were slightly lower than in 1977. In coho salmon the level declined 24% from 1977 to 1980, though the trend may not be statistically significant. These data may reflect the delayed response of top level predators to changing pollution levels. The data for the planktivorous spottail shiner reinforce this supposition. PCB levels decreased 58 to 82% at four locations in 2- to 5-year periods ending in 1980. The increase at one site from 153 ± 23 ppm in 1979 to 270 ± 51 ppm in 1980 is attributed to year-to-year variation, not increasing input. Both difficulties in obtaining representative samples and the complexities of the Great Lakes ecosystem contribute to such fluctuations.

All these studies underscore the variability of contaminant data and the hazards in attributing change with limited information. The above studies often lacked the intensity of sampling requisite for rigorous statistical evaluation. The Great Lakes Fisheries Laboratory designed a systematic assessment of three species of fish from Lake Michigan.[104] Large adult fish were collected in the fall at a specific location. The size was carefully regulated to give consistent samples suitable for yearly trend analysis.

Lake Michigan, which connects with Lake Huron, is the one Great Lake wholly within the U.S. Much of this lake is surrounded by intensive human activity. About 80 to 90% of the PCBs enter the region from the atmosphere vaporized either during incineration or directly from landfills.[101] The highly polluted sediments of Waukegan Harbor also contribute PCBs to the lake.

After 5 years of careful sampling, neither lake trout nor coho salmon showed a significant

Table 2
PCBs IN FISHES FROM LAKE MICHIGAN[104,107]

	PCBs (ppm)		
Year	**Lake trout**	**Coho salmon**	**Bloaters**
1972	13 (4.8)	11 (2.1)	5.7 (0.95)
1973	19 (2.1)	12 (0.8)	5.2 (0.37)
1974	23 (3.7)	10 (0.9)	5.6 (0.31)
1975	22 (2.9)	11 (0.6)	4.5 (0.36)
1976	19 (2.7)	9 (0.5)	4.1 (0.22)
1978			3.1 (0.28)
1980			2.2 (0.27)

Note: ppm: μg/g wet weight whole fish; arithmetic mean (95% confidence interval). Bloaters and lake trout were collected off Saugatuck, Mich., in southeastern Lake Michigan; Coho salmon were taken from east-central Lake Michigan. All fishes collected in the fall. Number of specimens: salmon and trout — 9 or 10 in 1972, 29 or 30 in other years, analyzed individually; bloaters — 100 to 170 analyzed as composites of 10 fish per sample.

PCB trend. The variability of PCB levels in 1972 (note the large 95% confidence intervals) remind the reader that the small number of specimens in 1972, nine or ten fish, do not comprise a representative sample (Table 2). Furthermore, the specimens were larger that year than later: 648 vs. 602 to 616 mm for trout and 693 vs. 620 to 665 mm for salmon. Unknown factors such as abundance of food, stress from adverse weather, or predation may have altered both the size and the contaminant distributions. The PCB level in lake trout showed no obvious trend, even when 1972 data were omitted. PCBs did decline 24% in coho salmon from 1973 to 1976. Lake trout live 10 to 20 years[105] and like dogfish in the marine environment,[48] they reflect long-term pollution in a top predator. Coho salmon spend about 18 months in Lake Michigan before returning to spawn.[106] They should respond more rapidly to declining contamination of the ecosystem because they are shorter lived.

The third species in the study, bloaters, was sampled more intensively. An important commercial species, bloaters are readily available and easy to catch. Seven samples of 100 to 170 fish each were collected between 1972 and 1980. Bloaters, like lake trout, are nonmigratory. Both reflect local conditions in contrast to coho salmon, which range widely and reflect lakewide pollution.[107] PCB levels in bloaters declined 27% by 1976 and 61% by 1980 (Table 2). Bloaters caught commercially at about 6 years of age[105] should respond to changing pollution more rapidly than long-living lake trout, but not as promptly as coho salmon.

Herring gull eggs monitored throughout the five Great Lakes provide an integrated approach to assessing a complex ecosystem. Herring gulls are advantageous as indicator organisms because they occur widely throughout the Northern Hemisphere, eat mainly fishes, bioaccumulate to high concentrations, and, as adults, are year round residents.[108] The PCB concentration in eggs from Lake Superior colonies did not change significantly from 1974 to 1979. Interference by toxaphene, found in increasing concentrations in Lake Superior specimens, may have obscured any trend. PCBs declined in a log-normal fashion on Lakes Ontario, Erie, and Huron. The "average population half-life" for the first-order model was 3.6 years for Lake Ontario, 8.4 years for Lake Erie, and 5.2 years for Lake Huron.[108] The average population half-life is a summation of dynamic factors in both the herring gull and its food. The actual half-life of PCBs in the gulls was assumed to be similar to the 250 days measured for ^{14}C-DDE.[108] Preliminary summaries of egg data for 1980[27] showed little change from the preceding year, except on Lake Ontario. In this area at the beginning of the study, levels were approximately twice as high as on Lakes Superior, Huron, and Erie. As the PCB levels for Lake Ontario herring gulls decline, the pattern may imitate the other lakes: initial dramatic declines followed by leveling off or at best gradual disappearance. The latter stage is difficult to discriminate from yearly variability.

Mineau et al.[108] were surprised that PCBs in herring gulls fell so rapidly following restriction in usage of these chlorinated hydrocarbons. They claim that reduced bioavailability stemmed from flushing, decreases in atmospheric contamination, or "irreversible" sedimentation. They predicted that contaminants would remain in the upper lakes, i.e., Superior and Huron, longer because of lower sedimentation rates, slower flushing, and greater input from the atmosphere onto the larger lake surface areas. Continued release of PCBs would explain the long average population half-life for Lake Erie (8.4 years). Extensive resuspension of sediments in the western part of Lake Erie may also play some role. As the diet of herring gulls is identified more exactly and the dynamics of PCB distribution are described, the relation between the food chain and the egg residue levels can be defined.

Herring gulls from Lake Michigan were monitored separately from birds on the lakes bordering both Canada and the U.S. PCB contamination in eggs decreased more than 50% from 1971 to 1980. The variability dropped at the same time. Specific data for one site are 129 ± 63 ppm in 1971, 151 ± 51 ppm in 1973, 110 ± 26 ppm in 1976, and 53 ± 10 ppm in 1980. Statistical interpretation of these data was not included.[27] The values are similar to those for Lake Ontario, which contained the highest levels of all.[27]

An intensive study of waterfowl nesting at the mouth of Green Bay in Lake Michigan shows the difficulty in detecting trends in environmental samples quickly. Mean PCB levels in red-breasted merganser *(Mergus serrator)* eggs did not change ($p > 0.05$) from 1977 to 1978, even though more than 200 eggs from a limited area were analyzed. PCB pollution in Green Bay still may have decreased, because the range dropped from 4.9 to 230 ppm in 1977 to 6.6 to 36 ppm in 1978.[79]

2. *Southern California Bight*

Perhaps the most intensely studied area in the world is the Southern California Bight. In 1969, Ventura County, the cities of San Diego and Los Angeles, and the county sanitation districts of Los Angeles and Orange County joined forces to found the Southern California Coastal Water Research Project (SCCWRP). They pooled scientific resources to acquire integrated regionwide data on the coastal waters from Point Conception to the Mexican border. All too frequently environmental research is short-term, narrow scope, point source problem oriented. From the beginning, SCCWRP chose instead "to understand the overall effects of man on the biology and chemistry of southern California's coastal waters".[109]

This more integrated approach to pollution assessment provides an opportunity to understand the flux of contaminants in a complex marine ecosystem adjacent to a metropolitan area. Eleven million persons, 5% of the total U.S. population, inhabit this coastal plain,[110] where intense agricultural and industrial activities present ample sources of pollutants. The Los Angeles city outfall into Santa Monica Bay and the county outfalls into the waters of the neighboring Palos Verdes Peninsula are the major "rivers" of Southern California. Formerly, they received industrial wastes directly. They continue to transport toxic substances concentrated by runoff from precipitation, aerial fallout, and agricultural drainage. They provide the main sources of fresh water entering the Southern California Bight. In 1979, the five largest dischargers of the region released 4×10^9 ℓ/day of municipal waste water, the predominant source of contaminants reaching Southern California coastal waters.[111] In 1979,for instance, waste water discharge contributed 0.8 t of PCBs compared to about 0.1

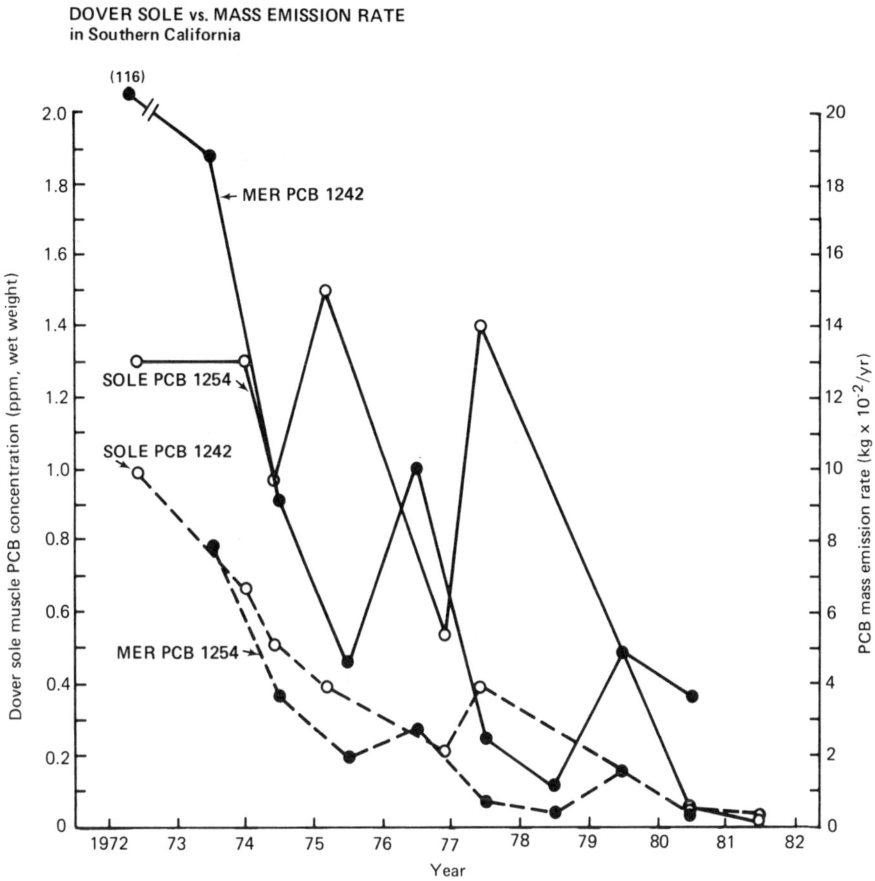

FIGURE 4. The relationship between PCB components in Southern California effluents and Dover sole *(Microstomus pacificus)* muscle. (From the data of Gosset;[116] see also Young and Heesen.[117])

t via storm runoff from the drainages of three major rivers of the Los Angeles Basin, the Los Angeles River, the San Gabriel River, and Ballona Creek.[112] As soon as the sewer outfalls were identified as the largest source of marine pollution in the bight, this influx was promptly restricted. Manufacturers and processors were prohibited from dumping industrial wastes into the sewer system.

The intensive studies by SCCWRP provide data for a comprehensive description of the many facets of pollution abatement. I am talking here about the complex combination of natural decomposition and dispersion and human efforts to limit use and release of toxic substances, and the observed responses of various components of the marine ecosystem.

SCCWRP has been monitoring chlorinated hydrocarbons off the coast of Southern California for the past decade. According to dated sediments from the Santa Barbara Basin (west of Santa Barbara and southeast of Point Conception), PCB deposition in the region began about 1945[113,114] the same time it occurred elsewhere in the U.S.[115] As the result of improved sewage treatment during the past decade, PCB emission from waste water in the Los Angeles County Sanitation District decreased 97% between 1972 and 1980 (Figure 4).[116] In the surface sediment near the outfall, PCB 1254 concentration decreased nearly 99% from 1972 to 1982; at the outfall site itself, the PCB 1242 and 1254 concentration decreased 94% from 1975 to 1981.[116]

What effects have these decreases in the amount of PCBs entering and residing in the marine ecosystem had on organisms? PCB levels in mollusks decreased much more rapidly than in flatfish. According to Bight-wide surveys of the open-coast mussel *(Mytilus californianus)* during 1971 and 1974, PCB contamination decreased significantly. The median decrease was 54%.[39] In contrast, PCB levels in the benthic flatfish, Dover sole *(Microstomus pacificus),* showed no statistically significant decrease over a similar period. In the regions of highest contamination, however, the maximum values did fall. Thus, at Palos Verdes, the highest value in the 1971/72 survey was 6.3 ppm in muscle compared to 2.5 ppm in the 1974/75 survey.[39] By 1981, PCBs had nearly disappeared from Dover sole muscle,[116,117] even at the most contaminated site. The PCB 1242 mean value had declined from 0.99 ± 0.15 ppm in 1972 to 0.028 ± 0.006 ppm, and PCB 1254 from 1.3 ± 0.2 ppm to 0.10 ± 0.02 ppm.

Although the level of PCB 1242 decreased in a fairly regular fashion, that of PCB 1254 fluctuated dramatically, as shown in Figure 4. The mass emission rate (MER) of PCB 1254 from the Los Angeles County Sanitation District outfalls at Palos Verdes showed only modest fluctuations, but the MER for PCB 1242 was more variable. The PCB 1254 level in Dover sole may be influenced by the MER for PCB 1242 because as the latter mixture weathers, many of the lower molecular weight components disappear, leaving behind residues resembling more closely PCB 1254. As early as 1975, McDermott et al.[81] noted that the proportions of PCB 1242 and 1254 in Dover sole did not bear a direct relationship to the proportions in municipal waste water discharges. They found 29% PCB 1254 in the discharges contrasted with 67% in Dover sole muscle. They suggested that factors such as volatility of individual components of the PCBs, sediment burden, and presence of other substances (e.g., metals, DDT, hydrogen sulfide) also influence the biological absorption processes, which are reflected in the ultimate levels in organisms. Furthermore, they found that proportions of PCB 1242 and 1254 in wild crab and mussels were similar to those in Dover sole.

In contrast, mussels from an uncontaminated source, exposed to the waste water plume at Palos Verdes for 3 months contained a decreased proportion of PCB 1254, namely 58%.[81] The marked decrease in PCB 1254 in mussels exposed higher above the bottom suggests that these PCBs derive from sediment and/or settling components of effluents. Closest to the bottom sediments, the PCB 1254 concentration approached 0.4 ppm, tenfold more than 35 m above at the surface.

Santa Monica Bay is a major sports fishing ground for the popular white croaker *(Genyonemus lineatus).* This species has been studied recently because one out of every three fish landed in Southern California is white croaker, and most are consumed by humans. PCB data were collected out of concern for the wholesomeness of these fish, which swim somewhat off the bottom, but feed on or in sandy bottom.[118] Between 1975 and 1980,[78,119] the mean PCB levels in edible tissue of white croaker collected at the nearby Los Angeles County Sewer outfall declined from 2.8 ppm total PCBs (PCB 1242 and PCB 1254) to 0.38 ppm. In 1980, PCB 1254 constituted 87% of these residues, a much higher proportion than found in Dover sole.

Waste water from sanitary treatment plants influences the behavior of PCBs in water and sediment. In the New York area, plankton blooms caused by the nutrient load absorb PCBs,[53] and remove them from the water. Moreover, particulate matter in sewage effluent buries PCB- and DDT-contaminated sediments. Such processes may serve to scavenge the area around the Los Angeles County sewer outfalls, since PCBs in fish decreased much more rapidly than originally predicted.[120] Ironically, improved waste water treatment can have a detrimental effect on buried PCB deposits. As improved treatment reduces the particulate content of the effluent, currents and wave action erode the overlying sediment. In fact, the Hendricks DDT model[121,122] suggests that zero discharge from the Los Angeles sewer outfalls would enable vast quantities of PCBs to be remobilized starting around 1990 (Figure 5). On the other hand, record storms in 1983 may have completely dissipated the contaminated sediment.

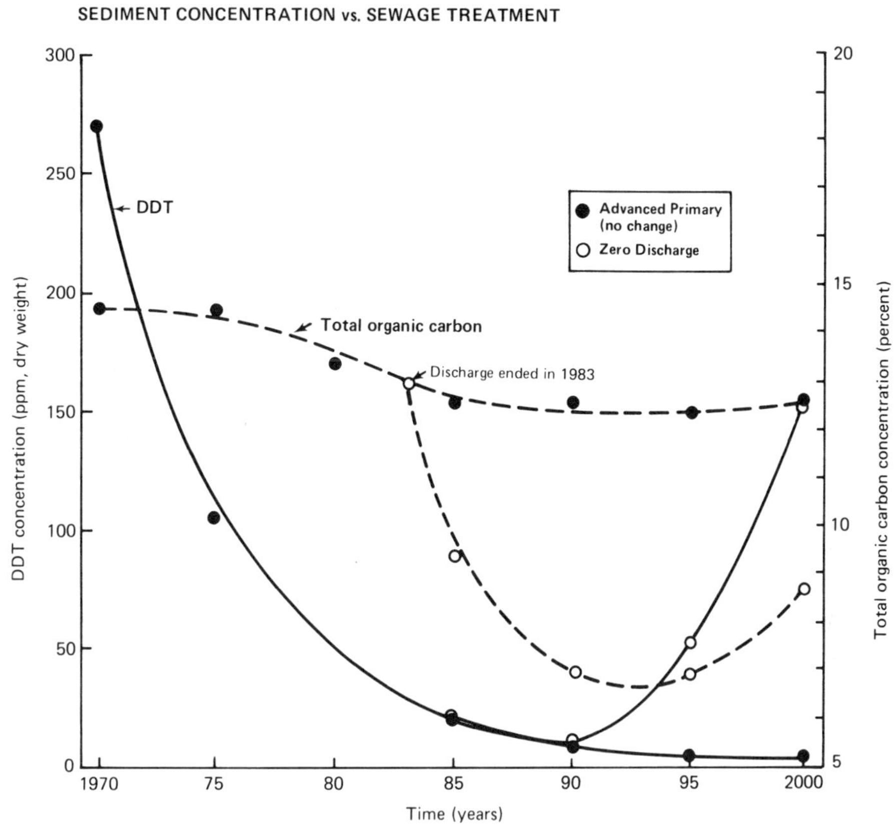

FIGURE 5. The influence of improved sewage treatment on the DDT concentration in sediment at the Los Angeles County sewer outfalls. Total organic carbon is related to the amount of particulate matter in sewage effluent. (After the model of T. Hendricks.[121,122])

3. Hudson River

Most of the studies described up to now have concerned areas with relatively diffuse sources of PCB contamination. The Los Angeles County sewer outfalls were considered point sources, but they discharged wastes from many sources. No direct disposal of pure PCBs was described. In contrast, the Hudson River in New York State was the site of massive intentional PCB dumping at a General Electric capacitor manufacturing plant. More than 200 t of PCBs were found in the river.[123] Dumping was reduced in 1976 and stopped in mid-1977. A monitoring program was established as part of the settlement agreement between New York State and General Electric after extensive pollution of the river was discovered. Monitoring involved both resident/fresh water and migratory/anadromous species (the authors' terminology).[52] For each species, specimens (30 per sample) were collected within 2 weeks of the target date each year.

PCB concentrations in mature resident/fresh water fishes from this heavily polluted area were directly proportional to lipid content from 1975 through 1980. This relationship applied to each collection site, and included specimens ranging over a number of age classes for sites downstream from the PCB source. No correlation was found between length or weight and PCB concentration.[52]

The behavior of PCBs depended on chlorine content. Overall, PCBs declined substantially in all species at all locations, the result of massive losses of Aroclor® 1016, the PCBs involved in the last years of discharge. Like Aroclor® 1242, Aroclor® 1016 contains 42% chlorine, but with reduced amounts of higher chlorinated components. While the components corresponding to 21 and 42% chlorination were decreasing, material resembling PCB 1254 generally remained constant or actually rose. The case of brown bullhead *(Ictalurus nebulosus)* shows an extreme of complexity in assessing PCB contamination. While Aroclor® 1016 decreased from 1900 ± 800 ppm lipid weight in skinless fillets to 700 ± 190 ppm between 1977 and 1980 and total PCBs from 2500 ± 1060 to 1480 ± 460 ppm, PCB 1254 increased from 390 ± 250 to 750 ± 290 ppm. (On a wet weight basis, the total PCBs dropped from 107 ± 49 to 12.3 ± 6.6 ppm.)

Aroclor® 1016 decreased significantly in yearling pumpkinseed *(Lepomis gibbosus)* from 1979 to 1980 at two sites. At a third site further downstream, it increased slightly but significantly. Since the correlation between PCB and lipid content was significant as well, outside sources of PCB apparently contributed to the PCB burden at this site.[52]

PCB 1254 levels in these fish varied much more. At Stillwater, 25 mi south of the industrial discharge, they decreased 51% from 1979 to 1980; farther south at Albany, they increased 157%, and farther downstream at Newburgh they remained constant. These results were attributed to resuspension of PCB-contaminated sediment during routine dredging. According to the ratios of PCB components in successive years, PCB 1254 had been more prevalent in previous years.

The mean annual decline of total PCB for all species was 34 ± 13%/year; Aroclor® 1016 decreased more rapidly, 47 ± 10%/year. Although not discussed in the paper,[52] PCBs disappeared more rapidly as the distance downstream from the source increased. In a 60-mi space, the rate of disappearance from all species increased 10%/year from 42 to 52%/year. Nonetheless, for three species taken individually, the rate increased 17.1%/year.

The declines in Aroclor® 1016 levels appear to be first order, i.e., disappearance rates in recent years were not different from previous years. The apparent "population" half-life for Aroclor® 1016 was 1.2 ± 0.4 years and for total PCBs, 2.3 ± 2.1 years. These are not true half lives for PCBs in a specific fish because they include both uptake and loss by all mechanisms. Instead, they compare lipid-based PCB levels in different years from similar populations at specific sites.

PCB levels in migratory/marine species were also studied to determine whether these species responded to the point source in the same way as resident/freshwater species. Aside from American shad *(Alosa sapidissima),* 83% of 30 sample sets showed correlations $p < 0.05$) between PCB content and lipid content. Only 24% of 50 data sets showed correlations of PCB content with fish length including one inverse relationship.

The lipid-based (see third footnote of chapter) PCB concentration of four species related inversely to body size. These species enter the river to spawn, but apparently do not feed there. Both large and small fishes (e.g., American shad averaging 1948 g and rainbow smelt averaging 9 g) followed this pattern. The smaller the fish the higher the PCB concentration, in quasiequilibrium with respect to PCB concentration in the local environment. The authors[3] suggest that PCBs accumulate passively showing a relationship to body size, i.e., the ratio of surface area to volume. Absorption may be through the gills, skin, and gut (only to the extent contaminated sediment was ingested). Smaller fish may accumulate more PCBs by filtering more water. If PCBs accumulate according to an equilibrium process, what is the maximum possible PCB concentration near the PCB source? In 1977, the lipid of resident goldfish *(Carassius auratus)* contained up to 10% PCBs, or 99,000 ppm.

PCBs in American shad dropped substantially between 1977 and 1979, and then leveled off or increased in 1980. The total PCB concentration in both sexes was 1.17 to 1.9 ppm in 1979/80. The increase of 0.3-0.4 ppm near the Tappan Zee Bridge was comprised of PCB 1221 which had been below the detectable limit in 1979. No explanation was available for the reappearance of the mainly monochlorinated PCB component, which had never had wide application and had virtually disappeared. Here is an example of the continuing problem of PCB pollution: often it goes undetected except in a chance case such as the shad data.[3]

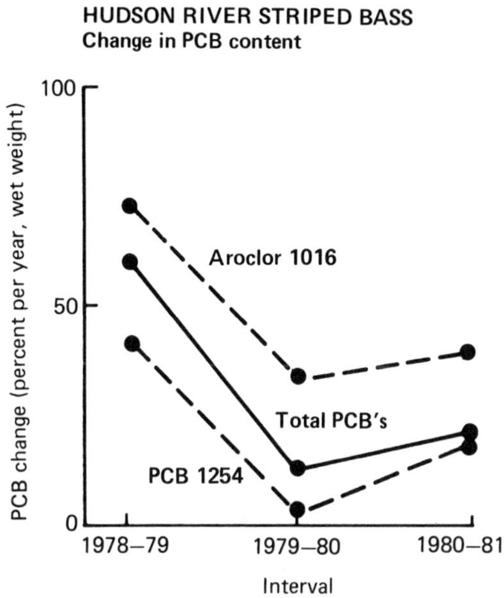

FIGURE 6. Rate of disappearance of PCBs from striped bass *(Morone saxatilis)* in the Hudson River after intentional dumping ceased. (From the data of Sloan and Armstrong.[3])

The 1978 PCB level in American eel *(Anguilla rostrata)* was so high (74 ± 67 ppm total PCBs in 42 specimens) that further monitoring did not seem worthwhile. Two years later, the level in six specimens had decreased to 9.1 ± 8.6 ppm PCBs, a value substantiated in 1981 by a mean of 10.8 ± 6.2 ppm in 30 eels. Even more dramatic was the virtual disappearance of Aroclor® 1016: from 40 ppm in 1978 to 0.46 to 0.49 ppm in 1980/81, a 76% annual decline, confirmed by similar findings at several other sites. Even closer to the dumpsite at the capacitor plant, Aroclor® 1016 only amounted to 0.93 ± 0.56 ppm and PCB 1254 to 12.2 ± 11.5 ppm in 1981. On a lipid basis, similar trends occurred. Thus, PCB 1254 became the predominant contaminant, more so than in any other species of fish.

Striped bass residues showed a marked decline from 18 ± 28 ppm total PCBs for 375 specimens in 1978 to 7.0 ± 6.1 ppm for 29 specimens in 1979 and 4.8 ± 5.0 ppm for 205 specimens in 1981. Aroclor® 1016 declined nearly 90% in that interval to 1.0 ± 2.2 ppm, but PCB 1254 declined only 55%. The average *annual* decline was 53% for Aroclor® 1016, but only 23% for PCB 1254. These average rates, however, obscure the true picture. Both components disappeared rapidly from 1978 to 1979 and then declined modestly at a more constant rate in the next 2 years (Figure 6).

Overall in the Hudson River between 1978 and 1981, Sloan and Armstrong[3] found the annual rate of decline of PCBs in migratory/anadromous fishes to be 42% for Aroclor® 1016 and 5% for PCB 1254. Elevated levels had been found before the study began, but the high values in resident/fresh water as well as migratory/anadromous species in 1978 may reflect elevated levels in the water column resulting from mobilization of sediments during the record spring floods of 1976 and 1977. Low levels in later years may result from stable water flow more recently. Under these conditions, burial can temporarily immobilize contaminated sediment. Further declines in PCB levels in the Hudson River will be less dramatic. Aroclor® 1016 levels are already quite low and residues resembling PCB 1254, now the preponderant component, seem to be much more persistent or are still being introduced into the river from diverse, seemingly stable sources.

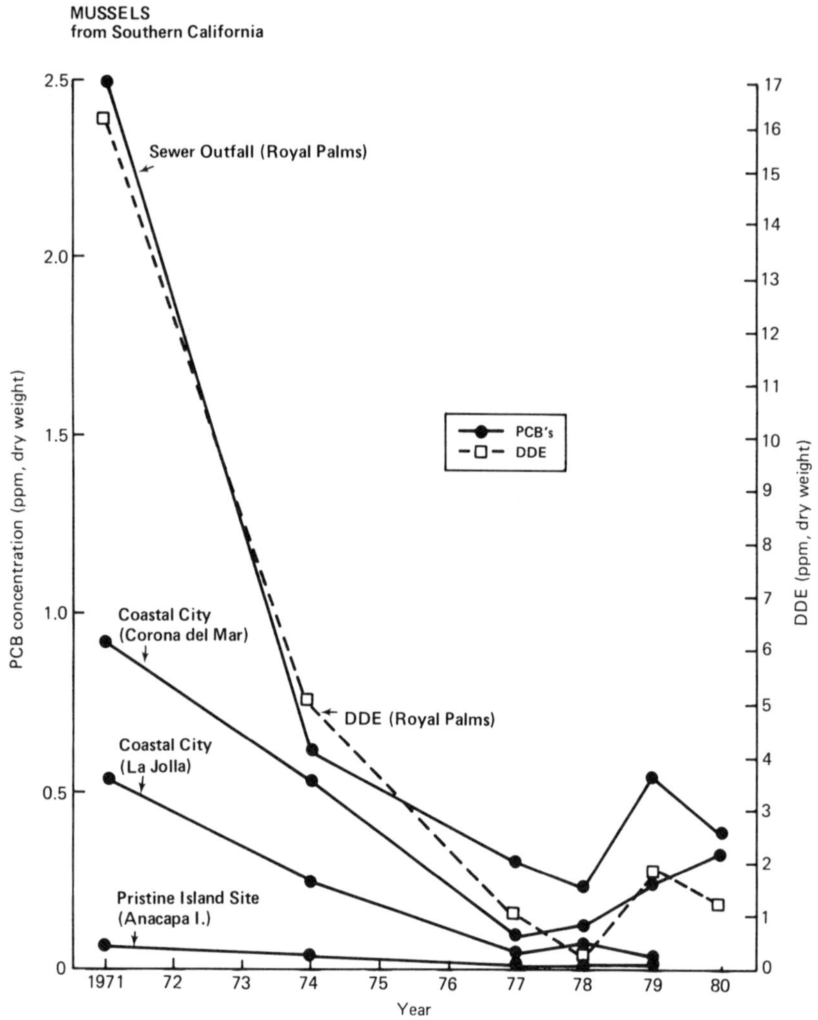

FIGURE 7. PCB and DDE levels in mussels *(Mytilus californianus)* at the Los Angeles County sewer outfall and PCB levels at three site on the coast of Southern California. (From the data of the Southern California Coastal Water Research Project,[151] Martin et al.,[34,124] and Risebrough et al.[35])

4. California Mussel Watch

Regional studies on the Hudson River provided an introduction to statewide and national monitoring programs. The California Mussel Watch, an outgrowth of the SCCWRP studies, was established for the purpose of monitoring pollutant levels in coastal waters throughout California. Mussels of the species *Mytilus edulis* or *M. californianus* were sampled depending on availability. Residue levels reflect the wide differences in the extent of pollution at individual sites. Mussels from Royal Palms State Beach, inshore from the sewer outfall of the Los Angeles County Sanitation Districts (LACSD), historically contained exceptionally high levels of both DDT and PCBs. They underwent a tenfold decrease in PCB content in the 7 years from 1971, when the focus of marine pollution was first identified, until 1978. The level fell from 2.5 ppm dry weight to 0.24 ppm. In 1979, however, the trend reversed and mussels contained 0.56 ppm PCBs.[124] In 1980, the content appeared to level off again at 0.39 ppm (Figure 7).[34]

What was the cause of the increase in PCB levels between 1978 and 1979? After a dramatic decline from 1971 to 1978, perhaps dumping regulations were being ignored. Or an old transformer may have fallen and spewed out PCBs, as occurred in the Duwamish River in the harbor of Seattle, Wash.[125] Perhaps the increase was only an artifact of the change in analytical techniques between 1978 and 1979. Other alternatives include residues mobilized by repairing or cleaning sewer lines or disturbing the sediment in the vicinity of the sewer outfall. In 1971, a 300 t reservoir of DDT[126] and 6 t of PCBs[127] surrounded the Los Angeles County sewer outfalls. Improved waste treatment, dredging, or unusual wind and wave action can increase the recirculation of pollutants normally buried progressively deeper by constant sedimentation. Examination of the concurrent trends for PCB and DDT suggests mobilization of previously deposited materials as a likely source of the slight increase in chlorinated hydrocarbon pollution levels noted above.

PCB levels in mussels from Corona del Mar, about 45 km southeast of Royal Palms, decreased ninefold from 0.92 ppm (dry weight) in 1971 to 0.102 ppm in 1977. Note that the level in 1977 was nine times as high as that found at the "pristine" site on Anacapa Island (see below).[124] What was the source of PCB pollution at Corona del Mar? If not a local spill, perhaps dispersion from the Los Angeles County sewer outfalls and surrounding sediments. The influence of this repository spread at least 100 km for DDT,[128] though mainly to the northwest. After 1977, the PCB level in California mussels from Corona del Mar increased each year to 0.25 ppm in 1979 and 0.33 ppm in 1980, a smaller but similar trend to that of Royal Palms (see Figure 7). This increase may be insignificant, because a five- to tenfold difference was considered minimal for significance[34] until a suitable data base for mussels was established and the statistical evaluation refined.

In 1980, resident mussels in Anaheim Bay south of Corona del Mar contained 2.4 ppm (dry weight) PCBs. This value equals the highest level found at Royal Palms in 1971. Elevated PCB levels in transplanted mussels at Port Hueneme (2.0 ppm) and Marina del Mar (1.8 ppm), a short distance north of Royal Palms, also suggest continuing widespread release of these compounds into the marine ecosystem of Southern California. The early dramatic decline in residue levels has not led to the complete disappearance of this pollutant.

In La Jolla, 100 km further southeast of Corona del Mar, the initial PCB level was only 20% of the level at Royal Palms. The disappearance of PCBs from this area has been even more profound, a 15-fold decline from 1971 to 1979. An early substantial decline was followed by a more gradual, but consistent decrease: 0.54 to 0.037 ppm dry weight over the 8-year period (Figure 7).

At Anacapa Island, a pristine site 110 km offshore and to the northwest of Royal Palms, PCB levels in California mussels decreased from 0.067 in 1971 to 0.010 ppm dry weight in 1979.[124] In contrast, the west part of San Miguel Island, another seemingly pristine site, twice as far from the sewer outfalls as Anacapa Island, contained an order of magnitude higher concentrations of PCBs (0.16 ppm) in 1978.[35] The levels at Ano Nuevo Island off the coast south of San Francisco were also higher (0.28 ppm) than at Royal Palms State Beach (0.24 ppm). The PCBs probably do not result from a spill or dumping since elevated levels of DDT also occur. Both San Miguel and Ano Nuevo Island contain extensive rookeries for marine mammals, such as California sea lions *(Zalophus californianus)* and northern elephant seals *(Mirounga angusterostris)*. Risebrough et al.[35] suggested PCBs and DDT are recycled into the food webs after excretion from marine mammals, which accumulate high levels of both substances. No data on fecal concentrations of DDT or PCBs are available. Flegal et al.[129] did find similar patterns of elevated mercury concentrations in California mussels adjacent to these pinniped and marine bird rookeries and in excrement from California sea lions on Ano Nuevo Island. Thus, PCBs ingested throughout the entire feeding range of the pinnipeds eventually are deposited at the remote seemingly pristine hauling out areas.

5. *National Mussel Watch*

Starting from Butler's oyster surveillance[32] and the SCCWRP mussel monitoring program,

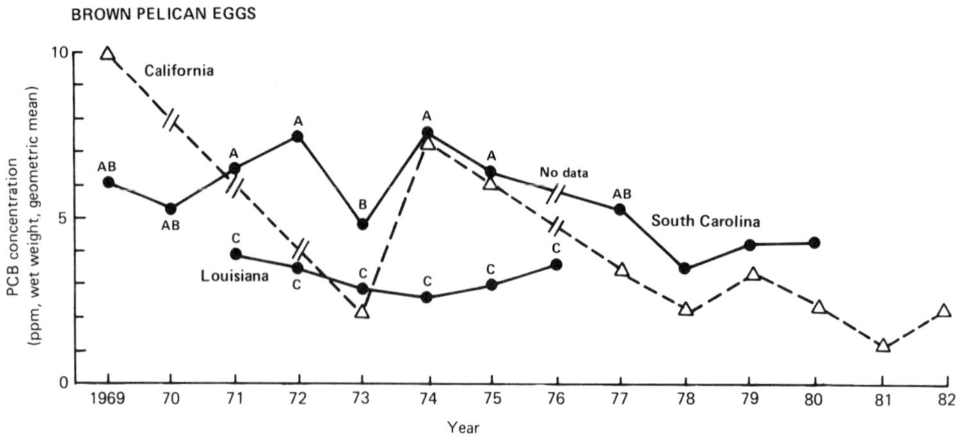

FIGURE 8. PCBs in brown pelicans *(Pelicanus occidentalis carolinensis* and *P. o. californicus).* Means which share a common letter do not differ significantly ($p > 0.05$). Data from California have not been evaluated statistically. (From the data of Blus,[83] Blus and Bunck,[132] Mendenhall and Prouty,[152] Blus et al.,[133] Anderson et al.,[59] and Gress.[135])

a national Mussel Watch was established in 1976 to monitor PCBs as well as DDT, heavy metals, radionuclides, and petroleum in coastal areas throughout the U.S. Part of an international program, Mussel Watch[20] employs mussels *(Mytilus* sp.) and oysters *(Ostrea* or *Crassostrea* sp.) to assay the marine environment. To eliminate differences in sampling, one scientist in a mobile laboratory collected all specimens at stations along the Atlantic, Gulf, and Pacific Coasts of the U.S. Interlaboratory comparisons with control samples were carried out in an attempt to improve consistency and accuracy of analyses.

Mussel Watch found no decrease in PCB levels in East Coast mussels during the years 1976 through 1978; these areas may be in the slowly changing phase of a decreasing exponential type curve,[28] such as has been noted for the DDT metabolite, DDE, in Southern California.[130] During the initial part of a seemingly bimodal loss process, about 85% of the original DDE disappeared with a half-life of 11 days. The remaining material had nearly a 200-day half-life. It is also possible that the initial restrictions on PCB usage had only limited effect on the primary sources of PCBs to the marine environment. Continuing study will help describe the results of the more recent near ban on PCBs. Assuming the changes are subtle, meticulous sampling design and statistical analysis will be needed to describe the process accurately.[28] A similar mussel watch has begun in Port Phillip Bay in Australia.[33]

6. Brown Pelicans

The U.S. Fish and Wildlife Service began monitoring PCBs in brown pelican eggs because of the massive decline in populations of this popular waterbird. In Louisiana, for instance, the estimated 50,000 to 85,000 birds before 1920 had disappeared by 1963,[58] apparently the result of DDE intoxication. Pelicans are an endangered species and people enjoy watching them. Their status is relevant to human beings in other ways, as well. Pelicans feed along the coasts where two thirds of all marine fishes and shellfishes spawn or spend their early lives.[131] Thus, pollutant levels in pelicans reflect the health of the nursery grounds of the seas. PCBs flow through estuaries and coastal waters on their way to the oceans. Controlling PCBs throughout the world helps protect 90 to 100 million t of foods potentially harvestable from the oceans each year.[131]

From 1969 to 1977, the levels of four chlorinated hydrocarbon insecticides declined significantly ($p < 0.05$) in brown pelican eggs in South Carolina, but PCB levels varied erratically without obvious trend (Figure 8).[83] Means for 1978 to 1980 were lower than in previous years;[132] the data await statistical evaluation. The geometric mean at two sites sampled extensively between 1969 and 1975 ranged from 4.8 to 7.63 ppm.[50] Interestingly, the level of PCBs in fresh eggs from nests which produced young successfully increased significantly from 4.3 ppm in 1971/72 to 12.0 ppm in 1975, whereas those from fresh eggs from nests which failed to produce viable young varied significantly but erratically. Taken together, fresh and embryonated pelican eggs, whether successful or unsuccessful, showed no temporal trend for PCBs that would suggest any effect from PCBs on the outcome of breeding.[83] Rather than pollution, anchovy *(Engraulis mordax)* biomass appears to be the main factor related to breeding success in California.[154]

In Florida, the trends in brown pelican eggs were not consistent. Along the Atlantic Coast, levels increased significantly ($p < 0.05$) from 2.7 ppm in 1969/70 to 6.1 ppm in 1974.[50] Specimens from Florida Bay and the Gulf Coast, however, showed little change, the former containing 0.75 ppm in 1969/70 and 0.62 ppm in 1974, the latter 0.70 and 1.18 ppm in the corresponding years. Characteristically, when chlorinated hydrocarbon levels were lower initially, significant changes occurred less frequently.

In Louisiana, PCB levels remained essentially unchanged from 1971 through 1976 (Figure 8). The geometric means varied only from 2.6 to 3.9 ppm,[133] while individual levels ranged from 0.69 to 8.9 ppm.

In Texas, King et al.[58] reported a significant ($p < 0.001$ t-test) decrease from 10 ± 5 ppm arithmetic (?) mean in 1970 to 3.0 ± 1.4 ppm in 1974. Because of the catastrophic decline in the brown pelican population in Texas, sampling was restricted to abandoned eggs. The number was quite limited; 11 eggs in 1970 and 5 eggs in 1974. Furthermore, the accuracy of analysis was vitiated by several factors. PCBs were not separated from other chlorinated hydrocarbons before analysis, and the analytical method became more precise in 1974. The individual values were not corrected for loss of moisture associated with incubation[134] or abandonment. For these reasons, Texas data are not conclusive. They were included only for general information.

PCB levels in brown pelican from Anacapa Island, Calif. probably decreased between 1969 and 1973 to 1975,[59] a trend confirmed by recent data.[135] Because the California pelicans are readily disturbed, eggs were collected only after the breeding season. In several years, only a few eggs were available. The limited sample sizes and changes in methodology precluded statistical evaluation of the data, which are shown in Figure 8.

7. *NPMP: Estuarine Organisms*

The National Pesticide Monitoring Program (NPMP) began evaluating organochlorine contaminants and toxic trace metals in 1967. A network of stations was established to sample fishes and wildlife throughout the U.S. Fresh water fishes, ducks, starlings, oysters, mollusks, and estuarine fishes have been analyzed on a regular basis for 15 years to determine present and potential future threats to the well-being of fishes and wildlife. PCBs were included in 1969, but accurate quantitation became practical only in 1971.

Butler's monumental report on estuarine mollusks, part of the National Pesticide Monitoring Program, noted the presence of PCBs in a few samples. Quantitation was not, however, routine in the early years of 1965 to 1972.[32] The follow-up survey in 1977 found no PCBs in oysters (<0.05) in the 87 estuaries sampled.[136]

The NPMP survey of estuarine fishes was designed to collect specimens less than 1 year old in both the spring and fall in 144 estuaries. To represent a broad spectrum of the food web, both carnivores and particle feeders were selected. Overall, 154 species representing 30% of the families of North American marine fishes were included. Every year the same species were chosen in each estuary to allow comparison between years. In all, from 1972 to 1976, 1524 samples composited from 38,000 whole fish were analyzed.

As detailed in Section I.D, this ambitious program has not been evaluated effectively.

Table 3
PCBs IN STARLINGS[71,138,139]

Year	No of pools	PCB incidence (%)	PCBs (ppm, wet weight) Geometric	Geometric (residue + 1)	Arithmetic	SE
1970	125	(100)	(0.36)		(0.66)	0.20
1972	130	(100)	(0.22)		(0.42)	0.15
1974	126	100	0.068	0.10/0.10	0.11	0.016
1976	125	21	0.24/0.003[a,b]	0.28/0.053	0.29/0.061	0.036
1979	112	83	0.082/0.039[c]	0.13/0.11	0.15/0.13	0.030
Stations Sampled in Both 1976 and 1979						
1976	106	23	0.24/0.003	0.28/0.058	0.29/[d]	0.039
1979	106	83	0.079/0.038	0.13/0.10	0.15/[d]	0.031

Note: ppm: µg/g wet weight; each sample consisted of 10 whole starlings with skin, wing tips, feet, and beak removed. Parentheses indicate the means in 1970 and 1972 cannot be compared to later data because the method of analysis changed in 1974.

[a] Mean of quantifiable values only/mean with 0.001 used arbitrarily for not detected values (quantitation limit was 0.01 ppm).
[b] Cain and Bunck[139] reported 0.01 ppm.
[c] Cain and Bunck[139] reported 0.05 ppm.
[d] Not calculated.

Butler and Schutzmann[47] concluded from statewide and grand total means that both maximum and average residue levels declined gradually. The means, however, are very sensitive to the extent of sampling at each site, because the residue levels varied widely between estuaries. For instance, whole fish contained less than 0.05 ppm PCBs in some estuaries and as much as 4.9 ppm in others. Butler,[32] in an earlier paper, had warned against comparison between states because of different sample sizes. Inspection is not an appropriate method of assessing trends for such a body of data. Two-way analyses of variance[97] or partially nested factorial analysis of variance[58] might be effective. The PCB incidence in estuarine fishes could not be compared with that in molluscs because PCBs are not found where shellfishes are commercially harvested.[47]

8. NPMP: Starlings

The starling survey covering the contiguous 48 states sampled 139 sites initially and after the inevitable falloff still included 112 sites in 1979. Starlings, introduced from Europe, range throughout the U.S. Common to the point of being pests, these birds are readily available for analysis. Their omnivorous diet includes insects, wild fruits, domestic crops, and other items. They reflect terrestrial levels in diverse foods.[137] Because starlings sometimes migrate, all data are considered on an integrated national basis.

Assessing the starling study is extremely difficult because of the statistical methods employed. Means included only quantifiable values without expressly noting the approach. Thus, the data exaggerated the actual levels, especially in 1976.[138] The levels are reported to have increased from 1976 to 1979, according to comparison of the geometric means.[139] My calculations[71] from the raw data did not confirm the means as calculated by Cain and Bunck.[139] PCB levels did increase when 0.001 ppm (one tenth the quantitation limit) was used for not detected values. On the other hand, the geometric mean, the geometric mean of PCB residue + 1, and the arithmetic mean of 1979 data were similar to the levels in 1974 (0.068 vs. 0.082 ppm, 0.103 vs. 0.130 ppm, and 0.113 vs. 0.150 ppm) (Table 3).

PCBs may have decreased from 1974 to 1976 and then increased again. Alternatively, these data may be an example of natural biological variation. In fact, I found that two-way analysis of variance rejected the null hypothesis between years ($p < 0.05$).[71] Scheffe's multiple comparison procedure indicated that both 1974 and 1979 differed from 1976. Beyond the issue of interpreting the data, it is still not clear that 1000 animals per year present an accurate picture of PCB contamination throughout the whole U.S.

9. NPMP: Ducks

The National Pesticide Monitoring Program selected ducks as representative of terrestrial pollution. Mallards *(Anas platyrhynchos)* occur throughout most of the U.S. The closely related black duck *(A. rubripes)* lives in certain areas of the Atlantic Flyway, where mallards are not found. Together these two species afford extensive coverage of a cross section of environmental exposure. Since black ducks feed more extensively on aquatic animals which concentrate chlorinated hydrocarbons, analysis of both species of ducks in the Atlantic Flyway expands the realm monitored.

A vast pool of samples otherwise to be discarded was available without effort.[140] Hunters throughout the U.S. were already submitting tens of thousands[141] of duck wings for an annual census of waterfowl productivity. Since DDT residues in mallard wings correlated ($p < 0.01$) with levels in breast muscle,[142] discarded wings could easily provide an indication of residue levels in the edible portion of prized waterfowl, which hunters would not readily donate for monitoring. The survey of duck wings aimed at detecting trends in continental populations not only of mallards and black ducks but also other waterfowls and the terrestrial environment more generally. As ducks are highly mobile, PCB levels reflect the status of the whole flyway in which they reside rather than in the specific state where they are shot.

Trends in PCB levels in duck wings are somewhat difficult to assess because of changes in methodology. It is only possible to compare the year 1969 to 1972 and separately 1976 to 1979. The early years cannot be compared to the later years. During the earlier period,[143] PCBs remained constant in the Atlantic Flyway and decreased 48 to 50% in the Central and Pacific Flyways ($p < 0.01$). In the Mississippi River Flyway, the apparent increase in 1972 resulted from skewed data. The mean level in four samples from Alabama was 6.3 ppm. Eliminating these data drops the flyway mean for 57 samples from 0.66 to 0.26 ppm, well below 0.44 ppm found in 1969. This example shows how a few data can bias the overall picture. The Alabama data may very well have been inaccurate. Gross industrial contamination near the Wheeler National Wildlife Refuge[144] caused high levels of DDE, TDE, and DDT. These may have interfered with PCB quantitation by the technique in use at that time. White[145] reported in his abstract that PCBs decreased significantly from 1972 to 1976 in the Atlantic Flyway (from 1.36 ± 0.15 to 0.52 ± 0.08 ppm). In the text, however, he stated that differences in quantitation precluded comparison between years.

From 1976 to 1979, the incidence of quantitatable levels of PCBs increased ($p < 0.05$) in all flyways but the Atlantic, where all samples contained PCBs in both years.[77] Incidences in earlier years were not reported. Declines in PCB levels were reported for mallards from all flyways, but $p > 0.05$. Furthermore, the comparison used only samples with detectable levels of residues. Data pairs were compared only if 50% of the pools had detectable PCB levels. Thus, all "zero" values were ignored.[77] The effect of this procedure depends on the distribution of zero values, but could be substantial. Means using 0.00 for all not-detected values[71] are shown in Table 4, along with published duck wing data. Partially nested factorial analysis[43] might be a better statistical approach than the *t*-test. The duck wing survey, containing results from 891 pools of 25 wings per sample, is one of the largest contaminant data sets in the world. Further statistical evaluation is certainly warranted at the earliest practical date.

Table 4
PCB LEVELS IN DUCK WINGS (1969—79)[77,143,171]

Species	Flyway	Year	No. of pools	PCBs mean (ppm) Incidence	Detect.[a]/all	SE
Black	Atlantic	1969	42		(1.4)	0.16
		1972	44		(1.4)	0.15
		1976	32	100a	0.52	0.08
		1979	24	100a	0.63	0.09
Mallard	Atlantic	1969	19		(1.3)	0.46
		1972	21		(1.2)	0.23
		1976	20	100a	0.52	0.18
		1979	29	100a	0.45	0.07
	Mississippi	1969	51		(0.44)	0.061
		1972	61		(0.66)	0.30
		1976	69	61a	0.23/0.14	0.03
		1979	64	98b	0.11/0.11	0.02
	Central	1969	49		(0.20)	0.039
		1972	56		(0.10)[b]	0.013
		1976	56	13a	0.15/0.02	0.01
		1979	54	90b	0.06/0.05	0.01
	Pacific	1969	51		(0.20)	0.014
		1972	55		(0.11)[b]	0.009
		1976	50	14a	0.16/0.02	0.04
		1979	44	93b	0.07/0.07	0.02

Note: ppm: μg/g wet weight; each sample consisted of 25 wings.

[a] Arithmetic means; for 1976 and 1979 "Detect." values include only samples with detectable PCB levels; "all" values are means using 0.00 for undetected values. Parentheses indicate means for 1969 and 1972 not comparable to later years because of differences in analysis. a,b: Incidences differ ($p < 0.05$) within a flyway when letters (a,b) differ.

[b] 1972 mean differed ($p < 0.05$) from 1969 mean in same flyway; means of detectable levels in 1976 and 1979 not significantly different ($p > 0.05$).

10. NPMP: Fresh Water Fishes

The fresh water fish program included 113 survey stations on major lakes and rivers. Three samples each composed of 3 to 5 whole fish were collected at each site. Adult specimens were to represent two bottom feeding and one predator species.[24] Five species were collected most frequently: common carp *(Cyprinus carpio),* channel catfish *(Ictalurus punctatus),* largemouth bass *(Micropterus salmoides),* white sucker *(Catostomus commersoni),* and yellow perch. They represent 45 to 50% of all the samples, and the incidence of each varied little over the years of study.

This monitoring effort considered both the incidence and the level of PCB contamination. PCBs seemed to be spreading, for by 1978/79, PCBs were found at 98% of the stations, up from 93% in 1974. Dispersal of contaminated sediments and atmospheric fallout seemed to be responsible. The characteristics of PCB residues shifted over the years. As the less chlorinated components disappeared, the more highly chlorinated ones came to predominate. Residues resembling PCB 1242 were missing after 1974. Intermediate components resembling PCB 1248 were present at about 40% of the stations from 1976 to 1979. The more highly chlorinated fraction resembling PCB 1254 was present most frequently at 84% of the stations in 1976/77 and at 98% in 1978/79. The most highly chlorinated fraction monitored (resembling PCB 1260) occurred at 84 to 88% of the stations in these years.

Comparisons of the 74 stations sampled in both early periods of the study, 1970 to 1972 vs. 1973 to 1974, suggested a decline of 26% on a wet basis and 48% on a lipid basis. The change may reflect major developments in methodology rather than environmental differences.

Mean PCB levels changed little between 1976 and 1979. Only the total PCBs based on lipid weight showed a significant difference between 1976/77 and 1978/79. The mean wet weight total PCBs and component PCBs on both wet and lipid bases did not differ significantly during this period. Data for 1974 contained analytical inconsistencies which clouded meaningful comparison. Data corrected for differences due to changed quantitative techniques suggested that PCB residue levels in fresh water fishes did not change from 1974 through 1979.

It is extremely difficult to portray adequately the findings of such a massive study. The maxima, means, and incidence of total PCBs and several components give only a brief review of the total effort. Data for the years 1973 through 1979 are summarized in Table 5.

The mixed nested-crossed model used for statistical calculations[43] allowed comparison of the trends at individual stations over time (i.e., year-station interactions). From this process, it was suggested that some stations still received fresh, unweathered PCBs like PCB 1242 or 1248. Most others were either no longer receiving any PCBs or alternatively were exposed only to PCB mixtures resembling the more chlorinated formulations, probably the products of weathering.

At individual stations, the concentrations of PCBs changed little. At 4 of the 78 sites, the lipid-weight levels increased, while at 5 of 78 stations, PCB levels declined ($p \leq 0.05$). Decreases occurred at sites where the initial levels were above 1 ppm. The components resembling PCB 1254 decreased at 14 stations and increased at 4, but did not exceed 1.5 ppm at any of these stations. The highest total PCB concentrations between 1973 and 1979 ranged from 71 to 93 ppm. At some highly contaminated stations, the levels may have decreased slightly but at the expense of spreading to natural sinks like the Great Lakes.[24]

A small subset of the National Pesticide Monitoring Program samples provided a more consistent data base for studying trends.[43] These cross-check samples were re-analyzed by the Columbia National Fisheries Research Laboratory for the quality assurance program. Methodology, though intended to be more consistent, nonetheless changed at intervals throughout the course of these studies. Furthermore, the cross-check program selected samples (1) known to have high residue levels or (2) caught at stations previously polluted with PCBs.[68] Obviously, they represent a biased sample from the point of view of trend assessment. Comparison of 92 cross-check samples from the early years, 1970 to 1972 vs. 1973/74, suggested differences ($p < 0.01$). Values declined more than 65%, part of which may have been an artifact of a rapidly changing quantitation technique. Again at these point sources, the pattern of pollution contrasted with that observed in areas distant from direct PCB contamination. Data based on lipid weight exhibited greater relative precision than those based on wet weight, similar to findings for the Hudson River. The overall picture at the 15 polluted sites between 1970 and 1976/77 is shown in Figure 9.

III. CONCLUSIONS

Despite years of concern about PCB pollution, PCB levels in terrestrial and aquatic organisms have declined mainly in the most polluted areas. PCB levels have remained stable or actually increased where they were previously low or absent. PCB levels have dropped substantially in the Hudson River, the site of intentional massive dumping of Aroclor® 1016. Total PCBs in striped bass dropped from 18.1 ppm in 1978 to 4.8 ppm in 1981. Aroclor® 1016, containing 42% chlorine declined 90% and PCB 1254 55%. By comparison, levels in fresh water fishes sampled throughout the U.S. for the National Pesticide Monitoring Program followed a pattern typical of lower levels of pollution. PCB levels ranged up and

Table 5
INCIDENCE AND CONCENTRATION OF PCBs IN NATIONAL PESTICIDE MONITORING PROGRAM SAMPLES OF FRESH WATER FISHES[24,68]

	Total PCBs (ppm)				PCB 1242 (ppm)			PCB 1248 (ppm)		PCB 1254 (ppm)			
	73*	74	76—77	78—79	73	74	76—77	76—77	78—79	73	74	76—77	78—79
No. of stations	**	87	106	108									
No. of samples	**	335	305	315									
Lipid content (%)	5.5	5.8	6.8	6.5									
Length (cm)	**	**	33.6	33.0									
Residue incidence (%)	70	93	92	98	14	5.4	0	39	42	62	86	84	98
Residue content (ppm)													
Wet weight basis													
Mean (all)	0.78	0.95	0.87	0.84	0.11	0.01	0	0.15	0.14	0.58	0.82	0.47	0.46
Mean (78)		1.1	0.97	0.97							0.96	0.53	0.50
Maximum	25.0	75	71	93	15	4.5	0	52	67	25	75	16	22
Lipid weight basis													
Mean	5.7	9.7	8.0	8.8	0.3	0.1	0	0.64	0.73	3.7	7.0	4.0	4.7
Mean (78)		10	9.7	11							7.8	5.0	5.4
Maximum	812	1060	740	480	180	95.5	0	470	350	710	1060	280	110

Note: ppm: μg/g whole fish; mean: ''geometric'' mean using $\log_{10}$ (PCB residue + 1). Mean (78): geometric mean for 78 stations represented in all years 1974 to 1979; no data for PCB 1242 or PCB 1248. *First year PCB 1242, 1254, and 1260 were determined separately. No data for PCB 1242 in 78—79. No data for PCB 1248 in 73 or 74, **Unknown.

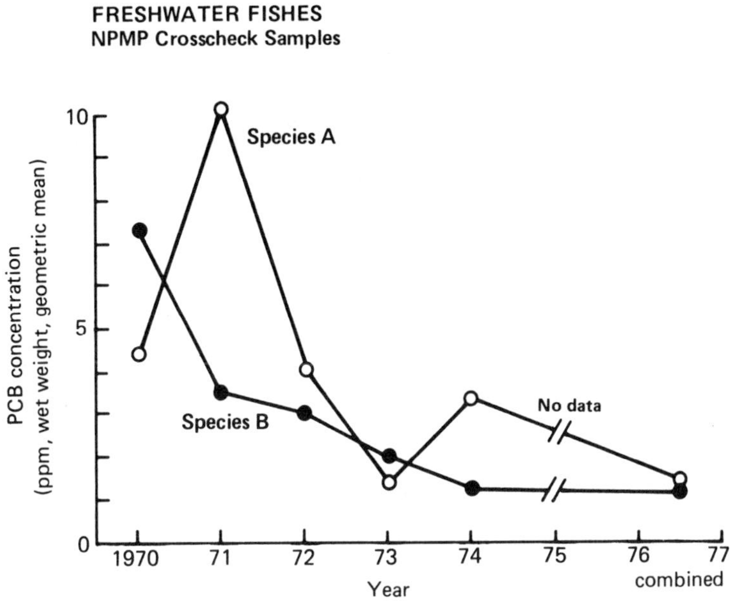

FIGURE 9. PCBs in cross check samples from the National Pesticide Monitoring Program survey of freshwater fishes. Specimens from 15 polluted sites were analyzed by a single laboratory in the years 1970—1977. (From the data of Schmitt.[43])

down from year to year without showing distinct trends: from 0.78 to 0.95 to 0.84 ppm total PCBs in the years 1973, 1974, and 1978/79.

Other data are similarly variable. Declines in PCB levels were noted in ten species of fishes from Lake Simcoe, Canada, in young harp seals from the Gulf of St. Lawrence, and in herring from the Baltic Sea. On the other hand, no significant change in PCB levels was observed in gray seals from the east coast of Scotland, brown pelicans from South Carolina (U.S.) or mallard ducks throughout the U.S. Thus, the specific trend of PCB contamination depends on the exact location and species.

Why have we seen no clear worldwide PCB trend? Biological variation and environmental fluctuation contribute to the variability of PCB data, but more important still, a vast reservoir remains from the estimated 680,000 t of PCBs produced worldwide starting in 1929.[33] Recent human poisonings in Taiwan witness the continuing use of PCBs in open systems.[146] Even in the U.S. a ruptured transformer contaminated poultry and eggs in several states in 1979.[147]

Beyond these episodes of essentially direct contamination of human and animal feed, polluted landfills and sediment continue to release PCBs, which are recycled starting with worms[148] and detritivores,[149] and continuing for decades in at least marine mammals[64] and sharks.[48]

Once PCBs cease to be released, they will still continue to dissipate slowly. Their disappearance from organisms presumably follows an exponential curve. After an initial rapid decline, further loss will be much more gradual,[28,130] until PCBs degrade or reach the ocean depths,[150] where remobilization occurs more slowly.

IV. OUTLOOK

The purpose of this paper has been to assess the status of PCB pollution. Readers undoubtedly hoped to see some change, preferably a substantial decrease, in environmental levels of PCBs. They assumed that heightened awareness of the problem and subsequent restrictions in use would limit the amount dispersing throughout the world. They may have assumed that with all the developments in analytical techniques, trends could be determined easily, rapidly, and accurately.

As we have seen, these hopes have not yet been fulfilled. Measuring trends of these complex, refractory mixtures is extremely complicated. Gradual dissipation and degradation of PCB residues can be discerned only after an extended period. Even then the evolution of analytical techniques has introduced elements of doubt.

With all these limitations, why bother to monitor at all? If we cannot detect small changes rapidly, we can at least see gradual shifts downward or upward after extended periods of surveillance. Such changes serve as an early warning that desired trends may not be achieved, that feared increases in pollution may actually occur, that further action may be necessary. Even more important, we have gained some insights. While we cannot quantitatively compare specific data, we can still apply our knowledge in general terms to predicting the future of pollution.

V. NEEDS

This chapter would not be complete without a brief statement on future needs. We must accept the intrinsic imprecision of environmental data until discrete individual compounds are being monitored. As long as analytical techniques for complex mixtures continue to be refined, we must recognize the uncertainty of comparing data over time. We must also recognize the need for patience, for even in this era of microprocessors and space travel, trend assesesment will never be an instantaneous process.

A more concrete issue is appropriate statistical methods for assessment of PCB (and other pollutant) data. Chemists and biologists need to join with statisticians in an intensive examination of the nature of our data and effective statistical approaches. Chemists and biologists seldom have the expertise to devise statistically valid analyses, while statisticians need the experience and insights of biologists and chemists.

We also need to continue sampling on a regular basis. As we have seen, residue levels vary widely. Occasional samples at long intervals will not provide the substantive information only available from intensive monitoring.

Finally, we must realize that ultimately all the inhabitants of planet Earth pay for quick "free" disposal of toxic wastes. "Disposal" in landfills and sanitary sewers is only the beginning of long-term pollution. Real control of toxic substances requires confinement and destruction at the source. Toxic substances already in landfills and sediment require special attention, including new methods for immobilization or removal. Since new processes such as advanced sanitary waste water treatment can be detrimental rather than beneficial, we need to evaluate the long-range impact of techniques before introducing them in polluted areas.

ACKNOWLEDGMENTS

Many of the scientists noted in this chapter generously shared their insights and recent research even before publication. I wish to thank specifically T. F. Bidleman, L. J. Blus, C. M. Bunck, J. W. Farrington, R. Frank, F. Gress, S. D. Haseltine, A. V. Holden, E. G. Horn, A. Jensen, A. J. Mearns, H. A. Schafer, C. J. Schmitt, W. R. Swain, J. F. Uthe, W. A. Willford, and D. R. Young.

J. A. Dassow, E. J. Gauglitz, Jr., C. R. Houle, M. M. Krahn, A. J. Mearns, R. Miller, G. H. Stout, and especially W. D. MacLeod, Jr. provided critical review of the manuscript. R. F. Kappenman provided guidance for the statistical evaluation. D. A. DeWitt prepared the manuscript with great skill and endless patience. G. H. Stout proofed the final copy.

I wish to thank J. Spinelli and W. Aron for their continuing support throughout the writing of this chapter.

REFERENCES

1. **Maugh, T. H., II,** Polychlorinated biphenyls: still prevalent, but less of a problem, *Science,* 178, 388, 1972.
2. **Stratton, C. L., Whitlock, S. A., and Allan, J. M.,** A method for the sampling and analysis of polychlorinated biphenyls (PCBs) in ambient air, EPA-600/4-78-048, U.S. Environmental Protection Agency, Research Triangle Park, North Carolina, 1978, as quoted in **Billings, W. N. and Bidleman, T. F.,** High volume collection of chlorinated hydrocarbons in urban air using three solid adsorbents, *Atmos. Environ.,* 17, 383, 1983.
3. **Sloan, R. J. and Armstrong, R. W.,** PCB patterns in Hudson River fish. II. Migrant/marine species, *Proc. Hudson River Environ. Soc.,* Norrie Point, N.Y., September 1981, in press, 1985.
4. **Lawler, Matusky, and Skelly Engineers,** Upper Hudson River PCB no action alternative study: final report, Pearl River, N.Y., 1978, as quoted in **Horn, E. G., Hetling, L. J., and Tofflemire, T. J.,** The problem of PCBs in the Hudson River system, in *Health Effects of Halogenated Aromatic Hydrocarbons,* Nicholson, W. J. and Moore, J. A., Eds., *Ann. N.Y. Acad. Sci.,* 320, 591, 1979.
5. **Jensen, S.,** Report of a new chemical hazard, *New Sci.,* 32, 612, 1966.
6. **Risebrough, R. W., Rieche, P., Peakall, D. B., Herman, S. G., and Kirven, M. N.,** Polychlorinated biphenyls in the global ecosystem, *Nature (London),* 220, 1098, 1968.
7. **Jensen, S., Johnels, A. G., Olsson, M., and Otterlind, G.,** DDT and PCB in marine animals from Swedish waters, *Nature (London),* 224, 247, 1969.
8. **Risebrough, R. W., Walker, W., II, Schmidt, T. T., de Lappe, B. W., and Connors, C. W.,** Transfer of chlorinated biphenyls to Antarctica, *Nature (London),* 264, 738, 1976.
9. **Holmes, D. C., Simmons, J. H., and Tatton, J. O'G.,** Chlorinated hydrocarbons in British wildlife, *Nature (London),* 216, 227, 1967.
10. **Stout, V. F.,** Pesticide levels in fish of the Northeast Pacific, *Bull. Environ. Contam. Toxicol.,* 3, 240, 1968.
11. **Beezhold, F. L. and Stout, V. F.,** The use and effect of mixed standards on the quantitation of polychlorinated biphenyls, *Bull. Environ. Contam. Toxicol.,* 10, 10, 1973.
12. **Stout, V. F. and Beezhold, F. L.,** Chlorinated hydrocarbon levels in fishes and shellfishes of the northeastern Pacific Ocean, including the Hawaiian Islands, *Mar. Fish. Rev.,* 43(1), 1, 1981.
13. **Holden, A. V.,** International cooperative study of organochlorine and mercury residues in wildlife, 1969—71, *Pest. Monit. J.,* 7, 37, 1973.
14. **Veith, G. D., Kuehl, D. W., Leonard, E. N., Welch, K., and Pratt, G.,** Polychlorinated biphenyls and other organic chemical residues in fish from major United States watersheds near the Great Lakes, 1978, *Pest. Monit. J.,* 15, 1, 1981.
15. **Uthe, J. F., Musial, C. J., and Palmork, K. H.,** A report on an intercomparison study of the determination of chlorinated biphenyl (PCB) isomers in fish oil, *Int. Council Expl. Sea Mar. Environ. Qual. Comm.,* C. M., E 49, 1981.
16. **Stout, V. F. and Beezhold, F. L.,** Analysis of chlorinated hydrocarbon pollutants: a simplified extraction and cleanup procedure for fishery products, *Fish. Bull.,* 76, 880, 1979.
17. **Holden, A. V.,** International cooperative study of organochlorine pesticide residues in terrestrial and aquatic wildlife, 1967/1968, *Pest. Monit. J.,* 4, 117, 1970.
18. **Villeneuve, J. P., Marchand, M., Elder, D., Fowler, S. W., LaRosa, J., Duursma, E. K., Vas, D., and Parsi, P.,** Intercalibration measurements for non-nuclear pollutants. Organochlorine compounds: progress and results to date, in *Activities of the International Laboratory of Marine Radioactivity 1976 Report,* IAEA-187, International Atomic Energy Agency, Vienna, 1976, 99.
19. **Holden, A. V.,** personal communication, 1982.
20. **Goldberg, E. D., Bowen, V. T., Farrington, J. W., Harvey, G., Martin, J. H., Parker, P. L., Risebrough, R. W., Robertson, W., Schneider, E., and Gamble, E.,** The mussel watch, *Environ. Conserv.,* 5, 101, 1978.
21. **Richardson, B. J. and Waid, J. S.,** PCBs in the Port Phillip Region. Survey of Polychlorinated Biphenyls (PCBs) in Port Phillip Bay, Environmental Studies Program Task No. PO8-802, Ministry for Conservation, Victoria, Australia, 1980.
22. **Bidleman, T. F.,** Interlaboratory analysis of high molecular weight organochlorines in ambient air, *Atmos. Environ.,* 15, 619, 1981.
23. **Horwitz, W., Kamps, L. R., and Boyer, K. W.,** Qualtiy assurance in the analysis of foods for trace constituents, *J. Assoc. Off. Anal. Chem.,* 63, 1344, 1980.
24. **Schmitt, C. J., Ribick, M. A., Ludke, J. L., and May, T. W.,** National Pesticide Monitoring Program: Organochlorine Residues in Freshwater Fish, 1976—79, Resource Publ. 152, U.S. Department of the Interior, Fish and Wildlife Service, Washington, D.C., 1983.
25. **Swain, W. R.,** Chlorinated organic residues in fish, water, and precipitation from the vicinity of Isle Royale, Lake Superior, *J. Great Lakes Res.,* 4, 398, 1978.

26. **Swain, W. R. and Mullin, M. D.,** Refined Analysis of Residue Forming Organic Substances in Lake Trout from the Vicinity of Isle Royale, Lake Superior, Abstract, 25th Annu. Meet., Int. Assoc. Great Lakes Res., *Sault Ste. Marie, Ontario, 1982.*
27. Great Lakes Water Quality Board, 1981 Report on Great Lakes Water Quality, Appendix, Great Lakes Surveillance, Report to the International Joint Commission, November 1981.
28. **Farrington, J. W., Goldberg, E. D., Risebrough, R. W., Martin, J. H., and Bowen, V. T.,** U.S. "Mussel watch" 1976—1978: an overview of the trace metal, DDE, PCB, hydrocarbon and artificial radionuclide data, *Environ. Sci. Technol.,* 17, 490, 1983.
29. **Willford, W. A.,** The Role of the Great Lakes Fishery Laboratory (USFWS) in Surveillance of Contaminants in Great Lakes Fishes, Agenda Item 20, Interim Meeting, Great Lakes Fishery Commission, Ann Arbor, Mich., November 29, 1978.
30. **Addison, R. F. and Brodie, P. F.,** Organochlorine residues in maternal blubber, milk, and pup blubber from grey seals *(Halichoerus grypus)* from Sable Island, Nova Scotia, *J. Fish. Res. Bd. Can.,* 34, 937, 1977.
31. **Frank, R., Lumsden, H., Barr, J. F., and Braun, H. E.,** Residues of organochlorine insecticides, industrial chemicals, and mercury in eggs and in tissues taken from healthy and emaciated common loons Ontario, Canada, 1968—1980, *Arch. Environ. Contam. Toxicol.,* 12, 641, 1983.
32. **Butler, P. A.,** Organochlorine residues in estuarine mollusks, 1965—72 — National Pesticide Monitoring Program, *Pest. Monit. J.,* 6, 238, 1973.
33. **Richardson, B. J. and Waid, J. S.,** Polychlorinated biphenyls (PCBs): an Australian viewpoint on a global problem, *Search,* 13, 17, 1982.
34. **Martin, M., Crane, D., Lew, T., and Seto, W.,** Synthetic organic compound in mussels, *M. californianus* and *M. edulis,* from California's coast, bays, and estuaries, in *California Mussel Watch: 1980—1981,* Part III, State Water Resources Control Board, Sacramento, 1982.
35. **Risebrough, R. W., de Lappe, B. W., Letterman, E. F., Lane, J. L., Firestone-Gillis, M., Springer, A. M., and Walker, W., II,** Organic pollutants in mussels, *Mytilus californianus* and *M. edulis,* along the California coast, in *California Mussel Watch: 1977—1978,* Vol. 3, Water Quality Monitoring Report No. 79—22, State Water Resources Control Board, Sacramento, 1980.
36. **Ballschmiter, K., Buchert, H., Bihler, S., and Zell, M.,** Baseline studies of the global pollution. IV. The pattern of pollution by organo-chlorine compounds in the North Atlantic as accumulated by fish, *Fresenius' Z. Anal. Chem.,* 306, 323, 1981.
37. **Parrish, P. R., Lowe, J. I., Wilson, A. J., Jr., and Patrick, J. M., Jr.,** Effects of Aroclor® 1254, a PCB, on oysters, *Crassostrea virginica* (bivalvia: protobranchia: Ostreidae), *Assoc. Southeast Biol. Bull.,* 19, 90, 1972.
38. **Parrish, P. R., Hansen, D. J., Couch, J. N., Patrick, J. M., Jr., and Cook, G. H.,** Effects of the polychlorinated biphenyl Aroclor® 1016, on estuarine animals, *Assoc. Southeast Biol. Bull.,* 21, 74, 1974.
39. **McDermott, D. J., Young, D. R., and Heesen, T. C.,** PCB contamination of southern California marine organisms, in Proc. Natl. Conf. on Polychlorinated Biphenyls, Buckley, J. L., Ed., U.S. Environmental Protection Agency, Washington, D.C., 1976, 209.
40. **Stout, V. F. and Lewis, L. G.,** Role of disposal of PCB-contaminated sediment in the accumulation of PCB's by marine animals, in Aquatic Disposal Field Investigations Duwamish Waterway Disposal Site Puget Sound, Washington, Tech. Rep. D-77-24, Appendix, B., U.S. Army Engineer Waterways Experiment Station, Vicksburg, November, 1977.
41. **Phillips, D. J. H.,** *Quantitative Aquatic Biological Indicators, Their Use to Monitor Trace Metal and Organochlorine Pollution,* Applied Science Publishers, London, 1980, 290.
42. **Young, D. R., Heesen, T. C., and McDermott, D. J.,** An offshore biomonitoring system for chlorinated hydrocarbons, *Mar. Pollut. Bull.,* 7, 156, 1976.
43. **Schmitt, C. J.,** Analysis of variance as a method for examining contaminant residues in fish: National Pesticide Monitoring Program, in *Aquatic Toxicology and Hazard Assessment, Fourth Conference,* American Society for Testing and Materials Spec. Tech. Publ. No. 737, Branson, D. R. and Dickson, K. L., Eds., ASTM, Philadelphia, 1981, 270.
44. **Skea, J. C., Simonin, H. A., Dean, H. J., Colquhoun, J. R., Spagnoli, J. J., and Veith, G. D.,** Bioaccumulation of Aroclor 1016 in Hudson River fish, *Bull. Environ. Contam. Toxicol.,* 22, 332, 1979.
45. **Zitko, V.,** Uptake of chlorinated paraffins and PCB from suspended solids and food by juvenile Atlantic salmon, *Bull. Environ. Contam. Toxicol.,* 12, 406, 1974.
46. **Lieb, A. J., Bills, D. D., and Sinnhuber, R. O.,** Accumulation of dietary polychlorinated biphenyls (Aroclor 1254) by rainbow trout *(Salmo gairdneri), J. Agric. Food. Chem.,* 22, 638, 1974.
47. **Butler, P. A. and Schutzmann, R. L.,** Residues of pesticides and PCBs in estuarine fish, 1972—76 National Pesticide Monitoring Program, *Pest. Monit. J.,* 12, 51, 1978.
48. **Butler, P. A. and Schutzmann, R. L.,** Bioaccumulation of DDT and PCB in tissues of marine fishes, *Aquatic Toxicology,* American Society for Testing and Materials Spec. Tech. Publ. No. 667, Marking, L. L. and Kimerle, R. A., Eds., ASTM, Philadelphia, 1979, 212.

49. **Gilman, A. P., Fox, G. A., Peakall, D. B., Teeple, S. M., Carroll, T. R., and Haymes, G. T.,** Reproductive parameters and egg contaminant levels of Great Lakes herring gulls, *J. Wildl. Manage.*, 41, 458, 1977.
50. **Blus, L. J., Lamont, T. G., and Neely, B. S., Jr.,** Effects of organochlorine residues on eggshell thickness, reproduction, and population status of brown pelicans *(Pelecanus occidentalis)* in South Carolina and Florida, 1969—76, *Pest. Monit. J.*, 12, 172, 1979.
51. **Bache, C. A., Serum, J. W., Youngs, W. D., and Lisk, D. J.,** Polychlorinated biphenyl residues: accumulation in Cayuga Lake trout with age, *Science,* 177, 1191, 1972.
52. **Armstrong, R. W. and Sloan, R. J.,** PCB patterns in Hudson River fish. I. Resident/freshwater species, *Proc. Hudson River Environ. Soc.*, Norrie Point, N.Y., September 1981, in press, 1985.
53. **Armstrong, R. W. and Sloan, R. J.,** *Trends in Levels of Several Known Chemical Contaminants in Fish from New York State Waters,* Tech. Rep. 80-2, New York State Department of Environmental Conservation, Albany, 1980.
54. **Hansen, M. M., Jensen, A., and Lassen, H.,** Trend monitoring. I. Problems in the interpretation of contaminant concentrations using fish as indicator organisms, *Int. Council Expl. Sea Mar. Environ. Qual. Comm.*, C. M., E 25, 1982.
55. **Olsson, M., Jensen, S., and Reutergard, L.,** Seasonal variation of PCB levels in fish — An important factor in planning aquatic monitoring programs, *Ambio,* 7, 66, 1978.
56. **Wilson, A. J. and Forester, J.,** Persistence of Aroclor 1254 in a contaminated estuary, *Bull. Environ. Contam. Toxicol.*, 19, 637, 1978.
57. **Nisbet, I. C. T. and Reynolds, L. M.,** Organochlorine residues in common terns and associated estuarine organisms, Massachusetts, 1971—81, *Mar. Environ. Res.*, 11, 33, 1984.
58. **King, K. A., Flickinger, E. L., and Hildebrand, H. H.,** The decline of brown pelicans on the Louisiana and Texas Gulf Coast, *Southwest Nat.*, 21, 417, 1977.
59. **Anderson, D. W., Jurek, R. M., and Keith, J. O.,** The status of brown pelicans at Anacapa Island in 1975, *Calif. Fish Game,* 63, 4, 1977.
60. **Holden, A. V.,** Organochlorine residues in blubber of grey seals *(Halichoerus grypus)* from the Farne Islands, *Int. Council Expl. Sea Mar. Environ. Qual. Comm. Ref. Mar. Mammals Comm.*, C.M., E 41, 1978.
61. **Addison, R. F. and Smith, R. G.,** Organochlorine residue levels in Arctic ringed seals: variation with age and sex, *Oikos,* 25, 335, 1974, as cited in **Holden, A. V.,** Pollutants and seals — a review, *Mammal Rev.*, 8, 53, 1978.
62. **Albright, L. J., Northcote, T. G., Oloffs, P. C., and Szeto, S. Y.,** Chlorinated hydrocarbon residues in fish, crabs, and shellfish of the lower Fraser River, its estuary, and selected locations in Georgia Strait, British Columbia — 1972—3, *Pest. Monit. J.*, 9, 134, 1975.
63. **Gahler, A. R., Cummins, J. M., Blazevich, J. N., Rieck, R. H., Arp, R. L., Gangmark, C. E., Pope, S. V. W., and Filip, S.,** Chemical Contaminants in Edible, Non-Salmonid Fish and Crabs from Commencement Bay, Washington, EPA-910/9-82-093, U.S. Environmental Protection Agency, Seattle, 1982.
64. **Buckley, E. H.,** Accumulation of airborne polychlorinated biphenyls in foliage, *Science,* 216, 520, 1982.
65. **Mearns, A. J.,** Responses of coastal fishes and invertebrates to wastewater discharges, *Prog. Water Technol.*, 4, 19, 1979.
66. **Butler, P. A.,** Biological problems in estuarine monitoring, Paper EPA-600/4-74-004, 1974, 126, Proc. Seminar on Methodology for Monitoring the Marine Environment, U.S. Environmental Protection Agency, Washington, D.C., 1974.
67. **Bimbo, A.,** Personal communication, 1982.
68. **Schmitt, C. J., Ludke, J. L., and Walsh, D. F.,** Organochlorine residues in fish: National Pesticide Monitoring program, 1970—74, *Pest. Monit. J.*, 14, 136, 1981.
69. **Moilanen, R., Pyysalo, H., Wickstrom, K., and Linko, R.,** Time trends of chlordane, DDT, and PCB concentrations in pike *(Esox lucius)* and Baltic herring *(Clupea harengus)* in the Turku Archipelago, Northern Baltic Sea for the period 1971—1982, *Bull. Environ. Contam. Toxicol.*, 29, 334, 1982.
70. **Stout, V. F., Houle, C. R., and Beezhold, F. L.,** A survey of chlorinated hydrocarbon residues in menhaden fishery products, *Mar. Fish. Rev.*, 43(3), 1, 1981.
71. **Stout, V. F.,** Unpublished data.
72. **Keck, G., Paubel, P., and Monneret, R. J.,** Organochlorine and mercury residues in peregrine falcon eggs in France, *Bull. Environ. Contam. Toxicol.*, 28, 705, 1982.
73. **Gordon, M., Knauer, G. A., and Martin, J. H.,** *Mytilus californianus* as a bioindicator of trace metal pollution: variability and statistical considerations, *Mar. Pollut. Bull.*, 11, 195, 1980.
74. **Jensen, A. and Lassen, H.,** Trend monitoring of contaminants. Sampling design considerations, *Int. Council Expl. Sea Mar. Environ. Qual. Comm.*, C.M., E 63, 1981.

75. **Kaiser, K. L. E.,** Early trend determination of organochlorine contamination from residue ratios in the sea lamprey *(Petromyzon marinus)* and its lake whitefish *(Coregonus clupeaformis)* host, *Can. J. Fish. Aquat. Sci.*, 39, 571, 1982.
76. **Lassen, H.** Trend monitoring. II. Statistical theory, *Int. Council Expl. Sea Mar. Environ. Qual. Comm.*, C.M., E 26, 1982.
77. **Cain, B. W.,** Nationwide residues of organochlorine compounds in wings of adult mallards and black ducks, 1979—80, *Pest. Monit. J.*, 15, 128, 1981.
78. **Gossett, R. W., Puffer, H. W., Arthur, R. H., Jr., and Young, D. R.,** DDT, PCB, and benzo(*a*)pyrene levels in white croaker *(Genyonemus lineatus)* from southern California, *Mar. Pollut. Bull.*, 14, 60, 1983.
79. **Haseltine, S. D., Heinz, G. H., Reichel, W. L., and Moore, J. F.,** Organochlorine and metal residues in eggs of waterfowl nesting on islands in Lake Michigan off Door County, Wisconsin, 1977—78, *Pest. Monit. J.*, 15, 90, 1981.
80. **Young, D. R., McDermott-Ehrlich, D., and Heesen, T. C.,** Sediments as sources of DDT and PCB, *Mar. Pollut. Bull.*, 8, 254, 1977.
81. **McDermott, D. J., Young, D. R., and Heesen, T. C.,** Polychlorinated Biphenyls in Marine Organisms off Southern California, TM 223, NTIS # PB274464/AS, Southern California Coastal Water Research Project, El Segundo, 1975.
82. **Sarkka, J., Hattula, M.-L., Janatuinen, J., Paasivirta, J., and Palokangas, R.,** Chlorinated hydrocarbons and mercury in birds of Lake Paijanne, Finland — 1972-74, *Pest. Monit. J.*, 12, 26, 1978.
83. **Blus, L. J.,** Further interpretation of the relation of organochlorine residues in brown pelican eggs to reproductive success, *Environ. Pollut.*, 28(A), 15, 1982.
84. **Hansen, M. M.,** Trend monitoring. III. Statistical analysis of factors affecting the concentrations of cadmium, copper, and zinc in three species of fish *Int. Council Expl. Sea Mar. Environ. Qual. Comm.*, C.M., E 27, 1982.
85. **Jensen, A.,** Trend monitoring. IV. Statistical analysis of factors affecting the concentrations of mercury in flounder, *Int. Council Expl. Sea Mar. Environ. Qual. Comm.*, C.M., E 28, 1982.
86. **Scott, D. P., Uthe, J. F., and Chou, C. L.,** Further considerations of time trend determinations of contaminant levels in Canadian cod, *Int. Council Expl. Sea Mar. Environ. Qual. Comm.*, C.M., E 47, 1981.
87. **Klaas, E. E. and Belisle, A. A.,** Organochlorine pesticide and polychlorinated biphenyl residues in selected fauna from a New Jersey salt marsh — 1967 vs. 1973, *Pest. Monit. J.*, 10, 149, 1977.
88. **Pearce, P. A., Peakall, D. B., and Reynolds, L. M.,** Shell thinning and residues of organochlorines and mercury in seabird eggs, Eastern Canada, 1970—76, *Pest. Monit. J.*, 13, 61, 1979.
89. **Deubert, K. H., Rule, P., and Corte-Real, I.,** PCB residues in *Mercenaria mercenaria* from New Bedford harbor, 1978, *Bull. Environ. Contam. Toxicol.*, 27, 683, 1981.
90. **Frank, R., Holdrinet, M. V. H., Desjardine, R. L., and Dodge, D. P.,** Organochlorine and mercury residues in fish from Lake Simcoe, Ontario, 1970—76, *Environ. Biol. Fish.*, 3, 275, 1978.
91. **Holden, A. V.,** Pollutants and seals — a review, *Mammal Rev.*, 8, 53, 1978.
92. **Jones, D., Ronald, K., Lavigne, D. M., Frank, R., Holdrinet, M., and Uthe, J. F.,** Organochlorine and mercury residues in the harp seal *(Pagophilus groenlandicus)*, *Sci. Total Environ.*, 5, 181, 1976.
93. **Frank, R., Ronald, K., and Braun, H. E.,** Organochlorine residues in harp seals *(Pagophilus groenlandicus)* caught in eastern Canadian waters, *J. Fish. Res. Bd. Can.*, 30, 1053, 1973.
94. **Schneider, R.,** Polychlorinated biphenyls (PCBs) in cod tissues from the Western Baltic: significance of equilibrium partitioning and lipid composition in the bioaccumulation of lipophilic pollutants in gill-breathing animals, *Meeresforschung*, 29, 69, 1982.
95. **Luckas, B. and Lorenzen, W.,** Zum Vorkommen von chloroganischen Pestiziden und polychlorierten Biphenylen in Meerestieren der Küsten Schleswig-Holsteins, *Dtsch. Lebensm. Rundsch.*, 77, 437, 1981.
96. **Paasivirta, J. and Linko, R.,** Environmental toxins in Finnish wildlife. A study on time trends of residue contents in fish during 1973—1978, *Chemosphere*, 9, 643, 1980.
97. **Kappenman, R. F.,** Personal Communication, 1982.
98. **Paasivirta, J., Sarkka, J., Aho, M., Surma-Aho, K., Tarhanen, J., and Roos, A.,** Recent trends of biocides in pikes of the Lake Paijanne, *Chemosphere*, 10, 405, 1981.
99. **Frank, R., Braun, H. E., and Fleming, G.,** Organochlorine and organophosphorus residues in fat of bovine and porcine carcasses marketed in Ontario 1969—81, *J. Food Prot.*, 46, 893, 1983.
100. **Frank, R., Braun, H. E., Holdrinet, M., Sirons, G. J., Smith, E. H., and Dixon, D. W.,** Organochlorine insecticides and industrial pollutants in the milk supply of southern Ontario, Canada — 1977, *J. Food Prot.*, 42, 31, 1979.
101. Great Lakes Water Quality Board, 1981 Report on Great Lakes Water Quality, Report to the International Joint Commission, November 1981.
102. **Frank, R., Holdrinet, M., Braun, H. E., Dodge, D. P., and Sprangler, G. E.,** Residues of organochlorine insecticides and polychlorinated biphenyls in fish from Lakes Huron and Superior, Canada — 1968—76, *Pest. Monit. J.*, 12, 60, 1978.

103. **Frank, R., Braun, H. E., Holdrinet, M., Dodge, D. P., and Nepszy, S. J.,** Residues of organochlorine insecticides and polychlorinated biphenyls in fish from Lakes Saint Clair and Erie, Canada — 1968—76, *Pest. Monit. J.,* 12, 69, 1978.
104. **Willford, W.,** Results of fish-contaminant monitoring, personal communication, 1982.
105. **Scott, W. B. and Crossman, E. J.,** *Freshwater Fishes of Canada,* Bulletin 184, Fisheries Research Board of Canada, Ottawa, 1973, 220.
106. **Jensen, A. L., Spigarelli, S. A., and Thommes, M. M.,** PCB uptake by five species of fish in Lake Michigan, Green Bay of Lake Michigan, and Cayuga Lake, New York, *Can. J. Fish. Aquat. Sci.,* 39, 700, 1982.
107. **Willford, W. A., Hesselberg, R. J., and Nicholson, L. W.,** Trends of polychlorinated biphenyls in three Lake Michigan fishes, in Proc. Natl. Conf. on Polychlorinated Biphenyls, Buckley, J. L., Ed., U.S. Environmental Protection Agency, Washington, D.C., 1976, 177.
108. **Mineau, P., Fox, G. A., Norstrom, R. J., Weseloh, D. V., Hallett, D. J., and Ellenton, J. A.,** Using the herring gull to monitor levels and effects of organochlorine contamination in the Canadian Great Lakes, in *Toxic Contaminants in the Great Lakes,* Nriagu, J. O. and Simmons, M. S., Eds., John Wiley & Sons, New York, 1984, 425.
109. **Bascom, W.,** Director's statement, in *Coastal Water Research Project Biennial Report for the Years 1979—1980,* NTIS # PB81-158271, Bascom, W., Ed., Southern California Coastal Water Research Project, Long Beach, 1980, 1.
110. **Young, D. R., McDermott, D. J., Hessen, T. C., and Jan, T. K.,** Pollutant inputs and distributions off sourthern California, in *Marine Chemistry in the Coastal Environment,* ACS Symp. Ser. 18, Church, T. M., Ed., American Chemical Society, Washington, D.C., 1975, 424.
111. **Schafer, H. A.,** Characteristics of municipal wastewater, in *Coastal Water Research Project Biennial Report for the Years 1979—1980,* NTIS #PB81-158271, Bascom, W., Ed., Southern California Coastal Water Research Project, Long Beach, 1980, 235.
112. **Young, D. R., Jan, T.-K., Gossett, R. W., and Hershelman, G. P.,** Trace pollutants in surface runoff, in *Coastal Water Research Project Biennial Report for the Years 1979—1980,* NTIS # PB81-158271, Bascom, W., Ed., Southern California Coastal Water Research Project, Long Beach, 1980, 163.
113. **Hom, W., Risebrough, R. W., Soutar, A., and Young, D. R.,** Deposition of DDE and polychlorinated biphenyls in dated sediments of the Santa Barbara Basin, *Science,* 184, 1197, 1974.
114. **MacGregor, J. S.,** Changes in the amount and proportions of DDT and its metabolites, DDE and DDD, in the marine environment off southern California, 1949—1972, *Fish. Bull.,* 72, 275, 1974.
115. **Neidermyer, W. J. and Hickey, J. J.,** Chronology of organochlorine compounds in Lake Michigan fish, 1929—66, *Pest. Monit. J.,* 10, 92, 1976.
116. **Gossett, R. W.,** Personal communication, 1982.
117. **Young, D. R. and Heesen, T. C.,** DDT, PCB and chlorinated benzenes in the marine ecosystem off southern California, in *Water Chlorination — Environmental Impact and Health Effects,* Vol. 2, Jolley, R. L., Gorchev, H., and Hamilton, D. H., Jr., Eds., Ann Arbor Science, Ann Arbor, Mich., 1978, 267.
118. **Skogsberg, T.,** *The Fishes of the Family Sciaenidae (Croakers) of California,* Fish Bull. No. 54, California Division of Fish and Game, Bureau of Marine Fisheries, Sacramento, 1939, 48.
119. **Puffer, H. W. and Gossett, R. W.,** PCB, DDT and benzo(a)pyrene in raw and pan fried white croaker *(Genyonemus lineatus), Bull. Environ. Contam. Toxicol.,* 30, 65, 1983.
120. **Young, D.,** Changes in chlorinated hydrocarbons and metals, in March 1978 report to LA/OMA on task order No. 3 — forecast effects (under the conditions of the baseline alternative), Bascom, W., Ed., p. 59, in *Modification of Secondary Treatment Requirements for Discharges into Marine Waters,* Hearings before the Subcommittee of Water Resources of the Committee on Public Works and Transportation, HR 95-54, U.S. House of Representatives, 95th Congress, 2nd Session, May 24 and 25, 1978, 133.
121. **Hendricks, T.,** Changes in the bottom near Whites Point and Orange County as determined by a mathematical model using data on organic carbon, DDT, and copper, in March 1978 report to LA/MOA on task order No. 3 — forecast effects (under the conditions of the baseline alternative), Bascom, W., Ed., p. 9, in *Modification of Secondary Treatment Requirements for Discharges into Marine Waters,* Hearings before the Subcommittee on Water Resources of the Committee on Public Works and Transportation, HR 95-54, U.S. House of Representatives, 95th Congress, 2nd Session, May 24 and 25, 1978, 133.
122. **Hendricks, T.,** Areal distribution of sewage particulate flux to Palos Verdes shelf; fate of particulates subsequent to deposition; resultant sediment chemical characteristics, in *Joint Water Pollution Control Plant Application for Modification of Secondary Treatment Requirements for Publicly Owned Treatment Works Which Discharge into Marine Waters, Vol. 1, Appendix IX,* Garrison, W. E., Ed., Los Angeles County Sanitation Districts, Whittier, 1979.
123. **Horn, E. G., Hetling, L. J., and Tofflemire, T. J.,** The problem of PCBs in the Hudson River system, in *Health Effects of Halogenated Aromatic Hydrocarbons,* Nicholson, W. J. and Moore, J. A., Eds., *Ann. N.Y. Acad. Sci.,* 320, 591, 1979.

124. **Martin, M., Crane, D., Lew, T., and Seto, W.,** Synthetic organic compounds in mussels, *Mytilus californianus,* and *M. edulis,* along the California coast and selected harbors and bays, in *California Mussel Watch: 1979—1980,* reprinted 1982, part II, State Water Resources Control Board, Sacramento, 1980.
125. **De Yonge, J.,** Poisonous chemical spills into Duwamish, *Seattle Post-Intelligencer,* September 21, 1974.
126. **MacGregor, J. S.,** DDT and its metabolites in the sediments off southern California, *Fish. Bull.,* 74, 27, 1976.
127. **Young, D. R., McDermott, D. J., and Heesen, T. C.,** Marine inputs of polychlorinated biphenyls off Southern California, in Proc. Natl. Conf. on Polychlorinated Biphenyls, Buckley, J. L., Ed., U.S. Environmental Protection Agency, Washington, D.C., 1976, 199.
128. **Southern California Coastal Water Research Project,** Chlorinated hydrocarbons in intertidal organisms of the bight, in *The Ecology of the Southern California Bight: Implications for Water Quality Management,* Section 8.5.2, Southern California Coastal Water Research Project, El Segundo, 1973, 307.
129. **Flegal, A. R., Stephenson, M., Martin, M., and Martin, J. H.,** Elevated concentrations of mercury in mussels *(Mytilus californianus)* associated with pinniped colonies, *Mar. Biol.,* 65, 45, 1981.
130. **Berkihiser, E.,** Decrease of DDT in mussels, in *Coastal Water Research Project, NTIS # PB274468/AS, Annual Report for the Year Ending June 30, 1974,* Bascom, W., Ed., Southern California Coastal Water Research Project, El Segundo, 1974, 101.
131. National Marine Fisheries Service, Strategic Plan, U.S. Department of Commerce, National Oceanic and Atmospheric Administration, Washington, D.C., 1983, 5.
132. **Blus, L. J. and Bunck, C. M.,** Personal communication, 1982.
133. **Blus, L., Cromartie, E., McNease, L., and Joanen, T.,** Brown pelican: population status, reproductive success, and organochlorine residues in Louisiana, 1971—1976, *Bull. Environ. Contam. Toxicol.,* 22, 128, 1979.
134. **Peakall, D. B. and Gilman, A. P.,** Limitations of expressing organochlorine levels in eggs on a lipid-weight basis, *Environ. Contam. Toxicol.,* 23, 287, 1979.
135. **Gress, F.,** Personal communication, 1983.
136. **Butler, P. A., Kennedy, C. D., and Schutzmann, R. L.,** Pesticide residues in estuarine mollusks, 1977 versus 1972 — National Pesticide Monitoring Program, *Pest. Monit. J.,* 12, 99, 1978.
137. **White, D. H.,** Nationwide residues of organochlorines in starlings, 1974, *Pest. Monit. J.,* 10, 10, 1976.
138. **White, D. H.,** Nationwide residues of organochlorine compounds in starlings *(Sturnus vulgaris),* 1976, *Pest. Monit. J.,* 12, 193, 1979.
139. **Cain, B. W. and Bunck, C. M.,** Residues of organochlorine compounds in starlings *(Sturnus vulgaris),* 1979, *Environ. Monit. Assess.,* 3, 161, 1983.
140. **Heath, R. G. and Prouty, R. M.,** Trial monitoring of pesticides in wings of mallards and black ducks, *Bull. Environ. Contam. Toxicol.,* 2, 101, 1967.
141. **Heath, R. G. and Hill, S. A.,** Nationwide organochlorine and mercury residues in wings of adult mallards and black ducks during the 1969—70 hunting season, *Pest. Monit. J.,* 7, 153, 1974.
142. **Dindal, D. L. and Peterle, T. J.,** Wing and body tissue relationships of DDT and metabolite residues in mallard and lesser scaup ducks, *Bull. Environ. Contam. Toxicol.,* 3, 37, 1968.
143. **White, D. H. and Heath, R. G.,** Nationwide residues of organochlorines in wings of adult mallards and black ducks, 1972—73, *Pest. Monit. J.,* 9, 176, 1976.
144. **O'Shea, T. J., Fleming, W. J., III, and Cromartie, E.,** DDT contamination at Wheeler National Wildlife Refuge, *Science,* 209, 509, 1980.
145. **White, D. H.,** Nationwide residues of organochlorine compounds in wings of adult mallards and black ducks, 1976—77, *Pest. Monit. J.,* 13, 12, 1979.
146. **Anon.,** PCB widespread poisoning incident in Taiwan reported, *Food Chem. News,* 21(43) 23, 1980.
147. **King, W.,** Federal Agencies didn't move fast enought in PCB case, say critics, *Seattle Times,* November 11, 1979, B13.
148. **Elder D. L., Fowler, S. W., and Polikarpov, G. G.,** Remobilization of sediment-associated PCBs by the worm *Nereis diversicolor, Bull. Environ. Contam. Toxicol.,* 21, 448, 1979.
149. **Marinucci, A. C. and Bartha, R.,** Accumulation of the polychlorinated biphenyl Aroclor 1242 from contaminated detritus and water by the saltmarsh detritivore, *Uca pugnax, Bull. Environ. Contam. Toxicol.,* 29, 326, 1982.
150. **Elder, D. L. and Fowler, S. W.,** Polychlorinated biphenyls: penetration into the deep ocean by zooplankton fecal pellet transport, *Science,* 197, 459, 1977.
151. Southern California Coastal Water Research Project, as quoted in Reference 35.
152. **Mendenhall, V. M. and Prouty, R. M.,** Recovery of breeding success in a population of brown pelicans, *Proc. Colon. Waterbird Group,* 1978, 65.
153. American Chemical Society Subcommittee on Environmental Analytical Chemistry, Crummett, W. B., Chairman, Guidelines for data acquisition and data quality evaluation in environmental chemistry, *Anal. Chem.,* 52, 2242, 1980.

154. **Methot, R. and MacCall, A.,** U.S. Department of Commerce NOAA, NMFS, Southwest Fisheries Center, Lajolla, Calif., 1983, unpublished data.

Chapter 10

NONMETABOLIC ALTERATION OF PCBs

John A. G. Roach

TABLE OF CONTENTS

I. INTRODUCTION

Predictions about the nonmetabolic alteration of PCBs in the environment must address the fact the PCBs do not behave as a homogenous substance. The physical and chemical properties of PCBs vary greatly, depending on the degree and position of chlorine substitution on the biphenyl ring system. Differences in volatility and water solubility of the individual chlorinated biphenyls tend to fractionate the residue so that the actual composition of a thin film, a vapor, or a solution may differ markedly from the original material. It has even been argued that monochlorobiphenyls should be regulated differently from more highly chlorinated biphenyls because of the differences in their chemical and physical properties.[1]

PCBs are stable to chemical alteration as evidenced by their use in a wide variety of applications that require inert behavior. However, nonmetabolic alteration of these compounds does occur in the environment. Processes which may be involved include aqueous hydrolysis, photo alteration, air oxidation, and thermal alterations. The study of the effects of any one of these processes on PCBs in the environment is a complex task. Thus, most of the information for this discussion has been obtained under the more controlled conditions of the laboratory.

II. HYDROLYSIS AND ALCOHOLYSIS

A chlorine on a chlorobiphenyl molecule is not easily displaced in a nucleophilic substitution reaction. Most PCBs are stable to hydrolysis under moderate conditions. The 4- and 4,4′- positions of decachlorobiphenyl are the positions most susceptible to chlorine displacement by hydroxide and alkoxide ions. Decachlorobiphenyl can be hydrolyzed to octachloro-4,4′-biphenylol by heating with aqueous alkali in an autoclave.[2,3] Refluxing with 2% potassium hydroxide in ethanol has been used to remove pesticides that interfere with PCB analysis[4] and to extract PCBs from paper products.[5] Under these conditions, decachlorobiphenyl is readily converted to diethoxyoctachlorobiphenyl. In contrast to decachlorobiphenyl, 2,5-dichlorobiphenyl must be heated with sodium methoxide to produce 2-chloro-5-biphenylol.[6] The nonmetabolic hydrolysis or alcoholysis of PCBs in the environment is unlikely because of the vigorous conditions required.

III. PHOTOLYSIS

PCBs are easily photodegraded. UV light can provide ring activation for both nucleophilic and radical reactions.[7] The photoreactivity of PCBs has been used to confirm PCB residues[8] and to remove PCB interferences in the analysis of chlorinated paraffin residues.[9] A pilot plant has even been constructed to test the feasibility of disposing of PCBs in industrial waste effluent with a combination of irradiation and ozone.[10]

The photoreactivity of a chlorinated biphenyl depends on the number and positions of chlorine atoms in the molecule. When chlorinated biphenyls are photolyzed in organic solvents, those molecules containing more chlorine dechlorinate more readily than the less chlorinated congeners.[8-11] Examination of the photoproducts of isomeric chlorinated biphenyls demonstrates that the ease of dechlorination is *ortho* > *meta* > *para*.[12]

Safe and Hutzinger[13] reported stepwise loss of chlorine, rearrangement, condensation, and polar products when 2,2′,4,4′,6,6-hexachlorobiphenyl was irradiated at 310 nm in organic solvents. These processes have since been observed with a number of pure chlorobiphenyls and PCB preparations using laboratory UV sources and sunlight. Reductive dechlorination is the predominant photochemical reaction of PCBs in hydrocarbon solvents.[11] Dechlorination proceeds at a faster rate in alcoholic solvents than in hydrocarbon solvents. Photolysis in fluorocarbon solvents and thin films results in chlorination and rearrangements in addition to dechlorination.[14] Hydroxylic organic solvents result in polar oxygenated PCB photoproducts as well as dechlorination of PCBs.[13,14]

Water reacts with PCBs during photolysis. Irradiation of a thin film without water results in dechlorination of the PCBs and the formation of PCB condensation products such as terphenyls and quaterphenyls.[15] Irradiation of a thin film in the presence of water yields hydroxylated products.[11] Irradiation of aqueous suspensions of PCBs with *ortho* chlorines has resulted in the formation of chlorinated dibenzofurans in very low yields.[16]

There are differences in the chemistry that occur in a thin film, a vapor, or a solution. Therefore, the location of a PCB residue determines which photoreactions will take place. Interactions between chlorinated biphenyls such as condensation and chlorination can be photoinduced in a thin film. In a solution, the effect of the solvent predominates. In the vapor state, PCBs photolyze rapidly or slowly depending on the presence of other substances. The location of a residue determines the amount of incident light that it receives. However, the result of this insolation is affected by sensitizers, quenchers, and reactants present with the residue. Thus, the diversity of the microenvironments in which PCB residues may be found limits the certainty with which laboratory data may be used to assess photo alteration of PCBs in the environment.

If incidents of accidental contamination are disregarded, PCBs enter our diet primarily from aquatic sources.[7] Thus, the aquatic chemistry of PCBs is highly relevant to our dietary intake of PCBs. Bunce et al.[18] estimated several environmental factors affecting the photo degradation of PCBs in the shallow waters of Lake Erie. They concluded that up to 5% of the lightly chlorinated PCBs might lose a chlorine each year, but that at least one chlorine should be lost from every highly chlorinated molecule annually. The less-chlorinated PCBs are more reactive metabolically than the highly chlorinated PCBs. This suggests that the most important effect of solar radiation on PCBs in the environment is replenishment of metabolically active PCBs from more highly chlorinated congeners.

IV. OXIDATION AND THERMAL ALTERATION

Good thermal stability and excellent dielectric properties led to the use of PCBs in applications where they would be subject to oxidation and considerable thermal and electrical stress. Analyses of PCBs have identified contaminants and products of use that are potentially more hazardous than the PCBs.[19,20]

Chlorinated dibenzofurans are contaminants at the parts per million level in PCB mixtures and other industrial chemicals.[21,22] Vos et al.[23] separated PCBs into fractions by column chromatography and found that PCB fractions containing chlorinated dibenzofurans were more toxic in a chick-embryo bioassay than the other PCB fractions. The possible presence of highly toxic contaminants in PCBs led to the examination of commercial PCB mixtures by several laboratories.[24-26] The results demonstrated the presence of chlorinated dibenzofurans containing two or more chlorines in all PCB mixtures examined except Aroclor® 1221.[27-29] Methodology to detect alteration products of PCBs must distinguish between these trace impurities and any products of use.

The alteration of PCBs in electrical components has not been satisfactorily demonstrated because of the difficulty in obtaining used and unused portions of the same production batch of PCBs. Phillipson et al.[30] carefully devised a procedure to quantitate chlorinated dibenzofurans in Aroclor® 1260-based insulating fluids (askarels) which had been used in high-voltage electrical network transformers for periods ranging from 13 to 20 years. After separation from PCBs and other askarel components by alumina and Florisil column chromatography, chlorinated dibenzofurans were identified by combined gas chromatography-mass spectrometry (GC/MS) and quantitated by electron capture gas-liquid chromatography. Recoveries of 9 di- to octachlorodibenzofurans from spiked askarel matrices ranged from

72 to 103%. The behavior of 17 additional chlorinated dibenzofuran isomers in the procedure indicated that the method detected a large majority of the 131 possible di- to octachloro-dibenzofurans. The total amount of chlorinated dibenzofuran in each of the used askarel samples was found to be not more than three times the chlorinated dibenzofuran content of an unused Aroclor® reference material. When asked about the significance of these findings, Phillipson concluded that the levels of chlorinated dibenzofurans in the used askarels were not sufficiently greater than the reference material and might simply reflect reduced levels of contaminants in Aroclors® of later manufacture.

Anodic oxidation of chlorinated biphenyls provides an insight into the events that occur when the electrified matrix is conducting rather than insulating. Fenn et al.[31] report 3-hydroxy-2,2′,5,5′-tetrachlorobiphenyl is the major hydroxychlorobiphenyl product in the electrochemical oxidation of 2,2′,5,5′-tetrachlorobiphenyl in an aqueous system. Any hydroxychlorobiphenyl formed in the large-scale electrolysis of PCBs is further oxidized at the potentials required for the oxidation of the PCBs. The principal electrolysis products of PCBs at controlled potentials are chlorophenylbenzoquinones. Chlorinated dibenzofurans are not formed during anodic oxidation of PCBs in this aqueous systems. Operation of electrical equipment containing PCB insulating fluids does not provide these aqueous reaction conditions. However, these products do suggest the fate of PCBs subjected to oxidative electrochemical disposal.[32]

The tragic Yusho incident provided the impetus to study the alteration of PCBs in heat exchange fluids. This accidental contamination of rice oil with PCB heat transfer fluid in February 1968, affected some 1696 people.[33] The manufacturer had used a PCB, Kanechlor® 400, to heat rice oil at reduced pressure in order to remove odorous matter from the oil. Small holes in the heating pipes permitted the PCB to leak into the oil.[34] Comparison of the contaminated oil with Kanechlor® 400 showed that the oil contained an inordinately high level of chlorinated dibenzofurans based on the level of PCBs in the oil.[26,35] Kanechlor® 400 was put into a sealed glass ampoule and heated at 300°C for 15 days, but no increase in chlorinated dibenzofurans was detected.[36] The chlorinated dibenzofuran level in the oil was about 250 times higher than the level found in Kanechlor® 400.[37] This large difference in levels made it difficult to attribute the chlorinated dibenzofuran content of the oil only to a simple concentration of these impurities during processing. Morita et al.[38] examined in detail several PCB preparations and contaminated rice oil samples. A PCB heat transfer fluid that had been used for 2 years appeared to contain about four times the amount of chlorinated dibenzofurans found in comparable unused fluids. They reported that the Yusho oil contained levels of chlorinated dibenzofurans about 500 times that of Kanechlor® 400. It was suggested that oxygen catalyzed by the metal heat transfer tubing would react with *ortho* substituents in the PCBs during processing to form chlorinated dibenzofurans in the heat transfer fluid. This led to a series of heating experiments with Aroclor® 1248. Samples were sealed in glass tubes with air, oxygen, or nitrogen and heated in an oven. Chlorinated dibenzofurans were formed with a maximum yield of about 0.2%. Temperatures over 270°C were necessary for the transformation but above 330°C the chlorinated dibenzofurans decomposed. The best yields were obtained with oxygen in about 1 week at 300°C. Prolonged heating at 300°C decreased the chlorinated dibenzofuran levels indicating that once formed, the chlorinated dibenzofurans were decomposed at this temperature.[39]

The U.S Food and Drug Administration, in evaluating the data from the Yusho incident to assess the health hazard associated with the ingestion of PCBs, noted an inconsistency in the Yusho data. The 1969 report of Tsukamoto et al.[40] said that the oil contained approximately 2000 to 3000 ppm PCBs. The 1976 report of Kuratsune et al.[26] indicated that oil contained 1000 ppm PCBs. In order to examine the oil more closely, the FDA requested a sample of the oil from Dr. Kuratsune of Kyushu University, Fukuoka, Japan. GC analysis showed that the sample contained 1000 ppm PCB. Neutron activation analyses showed that the oil contained twice the amount of chlorine that could be expected in 1000 ppm PCB. It was surmised that the sample contained more chlorinated compounds, but their nature was not evident until the samples were examined with a shorter GC column and higher operating temperatures than are normally used for PCB analysis. The resultant GC/MS analyses identified chlorinated quaterphenyls as the other major chlorinated contaminants in the oil.[41] Miyata et al.[42] published similar results concurrently. This group then simulated the oil deodorization step of rice oil processing in the laboratory to note the effect on the oil contaminants. Over 90% of the PCBs were removed from the oil during deodorization. The remaining PCBs were congeners of high chlorine content. The chlorinated quaterphenyls were unaffected by deodorization. The evaporation rate of the chlorinated dibenzofurans was lower than that of the PCBs so that the relative abundance of the chlorinated dibenzofurans was increased in the oil and the composition of this contaminant shifted to the more highly chlorinated dibenzofuran congeners. The relative amounts of these contaminants still did not coincide with those in the Yusho oil.[43] This suggested that the differences might be attributed to the composition of the heat transfer fluid involved in the Yusho incident. This brought the investigators back to the examination of alterations that occurred with use. Analyses of three used heat transfer fluids revealed chlorinated quaterphenyls at levels as high as 0.03%.[44] But these results did not demonstrate the formation of quaterphenyls with use. Experiments such as those of Morita et al.[39] were required. Accordingly, Kanechlor® 400 was heated in glass ampoules under varying conditions including the presence of stainless steel and water. The levels of chlorinated quaterphenyls and dibenzofurans in the Kanechlor® 400 rose markedly. Both water and steel accelerated the formation of these substances at 360°C.[44]

The number of variables involved in the alteration of Kanechlor® 400 in the Yusho incident was relatively small compared to that available in the general environment. However, it took 10 years to build a reasonably complete picture of the Yusho incident. The search for analogous substances[45] and processes in the environment has turned up one other significant source of toxic thermal alteration products. Chlorinated dibenzofurans have been found in the combustion products of waste incinerators.[46-49] These chlorinated dibenzofurans have been attributed to the incomplete incineration of PCBs.[50,51] Buser et al.[52] reported that the yield of chlorinated dibenzofurans at a combustion temperature of 550°C ranged from 3 to 25%. Thus, use and disposal are the two most important factors in the thermal formation of more toxic substances from PCBs. Fortunately, these are factors we should be able to control.

REFERENCES

1. **Neely, W. B.,** Complex problems — simple solutions, *Chemteck,* 249, 1981.
2. **Smith, F.,** Octachlorobiphenylquinone, U.S. Patent 2, 449, 088, 1984.
3. Societe d'Electrochime, d'Electrometallurgie et des Acieries Electriques d'Ugine, purification of Octachlorodihydroxybiphenyl, Belgian Patent 613, 066, 1962.
4. **Trotter, W.,** Removing the interference of DDT and it analogs in the analysis for residues of polychlorinated biphenyls, *J. Assoc. Off. Anal. Chem.,* 58, 461, 1975.
5. **Young, S. J. V., Finsterwalker, C., and Burke, J. A.,** Extraction and cleanup for determination of polychlorinated biphenyls in paper and paperboard, *J. Assoc. Off. Anal. Chem.,* 56, 957, 1973.
6. **deCrauw, R.,** The principle of induced alternating polarity in connection with the reactions of derivatives of *p*-dichlorobenzene and other compounds with sodium methylate, *Rec. Trav. Chim.,* 50, 753, 1931.
7. **Crosby, D. G., Moilanen, K. W., and Wong, A. S.,** Environmental generation and degradation of dibenzodioxins and dibenzofurans, *Environ. Health Perspect.,* 5, 259, 1973.
8. **Hannan, E. J., Bills, D. D., and Herring, J. L.,** Analysis of polychorinated biphenyls by gas chromatography and ultraviolet irradiation, *J. Agric. Food Chem.,* 21, 87, 1973.

9. **Friedman, D. and Lombardo, P.,** Photochemical technique for the elimination of chlorinated aromatic interferences in the gas-liquid chromatographic analysis for chlorinated paraffins, *J. Assoc. Off. Anal. Chem.*, 58, 703, 1975.
10. **Arisman, R. K., Musick, R. C., Zeff, J. D., and Crase, T. C.,** Experience in operation of a ultraviolet-ozone (Ultrox®) pilot plant for destroying polychlorinated biphenyls in industrial waste influent, *Proc. 35th Ind. Waste Conf.*, Ann Arbor Science, Ann Arbor, Mich., 1981, 802.
11. **Hutzinger, O., Safe, S., and Zitko, R.,** Photochemical degradation of chlorobiphenyls (PCBs), *Environ. Health Perspect.*, 1, 15, 1972.
12. **Ruzo, L. O., Zabik, M. J., and Schuetz, R. D.,** Photochemistry of bioactive compounds: photoproducts and kinetics of polychlorinated biphenyls, *J. Agric. Food. Chem.*, 22, 199, 1974.
13. **Safe, S. and Hutzinger, O.,** Polychlorinated biphenyls: photolysis of 2,4,6,2′,4′,6′,-hexachlorobiphenyl, *Nature (London)*, 232, 641, 1971.
14. **Hustert, K. and Korte, F.,** Ecological chemistry. XXXVIII. Synthesis of polychlorinated biphenyls and their reactions under UV irradiation, *Chemosphere*, 1, 7, 1972.
15. **Hutzinger, O., Safe, S., and Zitko, V.,** Photodegradation of chlorobiphenyls, *The Chemistry of PCBs*, CRC Press, Boca Raton, Fla., 1974, 126.
16. **Crosby, D. G. and Moilanen, K. W.,** Photodecomposition of chlorinated biphenyls and dibenzofurans, *Bull. Environ. Contam. Toxicol.*, 10, 372, 1973.
17. **Corlde, F., Corneliussen, P., Jelinek, C., Hackley, B., Lehman, R., McLaughlin, J., Rhoden, R., and Shapiro, R.,** Human exposure to polychlorinated biphenyls and polybrominated biphenyls, *Environ. Health Perspect.*, 24, 157, 1978.
18. **Bunce, N. J., Kumar, Y., and Brownlee, B. G.,** An assessment of the impact of solar degradation of polychlorinated biphenyls in the aquatic environment, *Chemosphere*, 2, 155, 1978.
19. **Pomerantz, I., Burke, J., Firestone, D., McKinney, J., Roach, J., and Trotter, W.,** Chemistry of PCBs and PBBs, *Environ. Health Perspect.*, 24, 133, 1978.
20. **Kashimoto, T., Miyata, H., and Kunita, N.,** The presence of polychlorinated quaterphenyls in the tissues of Yusho victims, *Food Cosmet. Toxicol.*, 19, 335, 1981.
21. **Huff, J. E., Moore, J. A., Saracci, R., and Tomatis, L.,** Long-term hazards of polychlorinated dibenzodioxins and polychlorinated dibenzofurans, *Environ. Health Perspect.*, 36, 221, 1980.
22. **Buser, H. R.,** Identification of polychlorinated dibenzo-*p*-dioxins and dibenzofurans in environmental and industrial samples using high-resolution gas chromatography and mass spectrometry, in *Recent Developments in Mass Spectrometry in Biochemistry and Medicine*, Vol. 6, Frigerio, A. and McCamish, M., Eds. Elsevier, Amsterdam, 1980, 515.
23. **Vos, J. G., Koeman, J. H., van der Maas, H. L., ten Noever de Brauw, and de Vos, R. H.,** Identification and toxicological evaluation of chlorinated dibenzofuran and chlorinated naphthalene in two commercial polychlorinated biphenyls, *Food Cosmet. Toxicol.*, 8, 625, 1970.
24. **Roach, J. A. G. and Pomerantz, I. H.,** The finding of chlorinated dibenzofurans in a Japanese polychlorinated biphenyl sample, *Bull. Environ Contam. Toxicol.*, 12, 338, 1974.
25. **Bowes, G. W., Simoneit, B. R. T., Burlingame, A. L., and Rosebrough, R. W.,** Identification of chlorinated dibenzofurans in American polychlorinated biphenyls, *Nature (London)*, 256, 305, 1975.
26. **Nagayama, J., Masuda, T., and Kuratsune, M.,** Chlorinated dibenzofurans in Kanechlors and rice oils used by patients with Yusho, *Fukuoka Acta Med.*, 66, 593, 1975.
27. **Roach, J. A. G. and Pomerantz, I. H.,** The Finding of Chlorinated Dibenzofurans in Aroclor PCBs of Recent Manufacture, paper 58, 88th Annu. Meet. Assoc. Off. Anal. Chem., Washington, D.C., October 14, 1974.
28. **Bowes, G. W., Mulvihill, M. J., DeCamp, M. R., and Kende, A. S.,** Gas chromatographic characteristics of authentic chlorinated dibenzofurans; identification of two isomers in American and Japanese polychlorinated biphenyls, *J. Agric. Food Chem.*, 23, 1222, 1975.
29. **Roach, J. A. G., Sphon, J. A., and Dusold, L. R.,** Application of selected ion recording techniques to the qualitative and quantitative analysis of impurities and metabolites of chlorinated biphenyl, 10th Middle Atlantic Regional Meet. American Chem. Soc., Philadelphia, February 23, 1976.
30. **Phillipson, D. W., Farrell, T. J., and Puma, B. J.,** Chlorinated dibenzofurans in Aroclor 1260 and Aroclor 1260-based transformer fluids. paper 76, 92nd Annu. Meet. Assoc. Off. Anal. Chem., Washington, D.C., October 16, 1978.
31. **Fenn, R. J., Krantz, K. W., and Stuart, J. D.,** Anodic oxidation of two polychlorinated biphenyls, *J. Electrochem. Soc.*, 123, 1643, 1976.
32. **Miller, R. A., Johnson, J. D., Helsel, R. W., and Pitts, D. M.,** Destruction of toxic chemicals by catalyzed wet oxidation, in *Proc. 35th Ind. Waste Conf.*, Ann Arbor Science, Ann Arbor, Mich., 1981, 425.
33. **Taki, I.,** Foreword, The eighth reports of the study on Yusho and PCB, *Fukuoka Acta Med.*, 72, 131, 1981.

34. **Higuchi, K.,** *PCB Poisoning and Pollution,* Academic Press, New York, 1976, 17.
35. **Miyata, H., Kashimoto, T., and Kunita, N.,** Detection and determination of polychlorodibenzofurans in normal human tissues and Kanemi rice oils caused "Kanemi Yusho," *J. Food Hyg. Soc.,* 18, 260, 1977.
36. **Miyata, H., Nakamura, A., and Kashimoto, T.,** Separation and detection of polychlorodibenzofurans (PCDFs) on Japanese commercial PCBs (Kanechlors) and their heated preparation, *J. Food Hyg. Soc.,* 17, 227, 1976.
37. **Rappe, C. and Gara, A.,** Analysis of polychlorinated dibenzofurans in Yusho oil using high resolution gas chromatography-mass spectrometry, *Chromosphere,* 5, 231, 1977.
38. **Morita, M., Nakagawa, J., Akiyama, K., Mimura, S., and Isono, N.,** Detailed examination of polychlorinated dibenzofurans in PCB preparations and Kanemi Yusho oil, *Bull. Environ. Contam. Toxicol.,* 18, 67, 1977.
39. **Morita, M., Nakagawa, J., and Rappe, C.,** Polychlorinated dibenzofuran (PCDF) formation from PCB mixture by heat and oxygen, *Bull. Environ. Contam. Toxicol.,* 19, 665, 1978.
40. **Tsukamoto, H., Makisumi, S., Hirose, H., et al.,** Detection of toxic compounds in rice bran oils, *Fukuoka-Igaku-Zasshi,* 60, 496, 1969.
41. **Kamps, L. R., Trotter, W. J., Young, S. J., Carson, L. J., Roach, J. A. G., Sphon, J. A., Tanner, J. T., and McMahon, B.,** Polychlorinated quaterphenyls identified in rice oil associated with Japanese "Yusho" poisoning, *Bull. Environ. Contam. Toxicol.,* 20, 589, 1978.
42. **Miyata, H., Kashimoto, T., and Kunita, N.,** Studies on the compounds related to PCB (V) detection and determination of unknown organochlorinated compounds in Kanemi rice oil caused the "Yusho", *J. Food Hyg. Soc.,* 19, 364, 1978.
43. **Miyata, H. and Kashimoto, T.,** Investigation on organochlorinated compound formed in Kanemi rice oil that caused the "Yusho", *J. Food Hyg. Soc.,* 20, 1, 1979.
44. **Miyata, H., Murakami, Y., and Kashimoto, T.,** Studies on the compounds related to PCB (VI) determination of polychlorinated quaterphenyl (PCQ) in Kanemi rice oil caused the "Yusho" and investigation on the PCQ formation, *J. Food Hyg. Soc.,* 19, 417, 1978.
45. **Bowes, G. W., Simoneit, B. R., Burlingame, A. L., deLappe, B. W., and Risebrough, R. W.,** The search for chlorinated dibenzofurans and chlorinated dibenzodioxins in wildlife populations showing elevated levels of embryonic death, *Environ. Health Perspect.,* 5, 191, 1973.
46. **Olie, K., Vermeulen, P. L., and Hutzinger, O.,** Chlorodibenzo *p*-dioxins and chloridibenzofurans are trace components of fly ash and flue gas of some municipal incinerators in the Netherlands, *Chemosphere,* 8, 455, 1977.
47. **Buser, H. R. and Bosshardt, H. P.,** Identification of polychlorinated dibenzofuran isomers in fly ash and PCB pyrolyses, *Chemosphere,* 5, 419, 1978.
48. **Cavallaro, A., Bandi, G., Invernizzi, G., Luciani, L., Mongini, E., and Gorni, A.,** Sampling, occurrence and evaluation of PCDDs and PCDFs from incinterated solid urban waste, *Chemosphere,* 9, 611, 1980.
49. **Liberti, A., Brocco, D., Cecinato, A., and Possanzini, M.,** Determination of polychlorinated dibenzo-*p*- dioxins and dibenzofurans in the emission from urban incinerators, *Mikrochim. Acta,* 1, 271, 1981.
50. **Buser, H. R. and Rappe, C.,** Formation of polychlorinated dibenzofurans (PCDFs) from the pyrolysis of individual PCB isomers, *Chemosphere,* 3, 157, 1979.
51. **Duvall, D. S., Rubey, W. A., and Mescher, J. A.,** Applications of the thermal decomposition analytical system (TDAS), *Proc. Annu. Meet. Air Pollut. Control Assoc.,* 73, 80, 1980.
52. **Buser, H. R., Bosshardt, H. P., and Rappe, C.,** Formation of polychlorinated dibenzofurans (PCDFs) from the pyrolysis of PCBs, *Chemosphere,* 1, 109, 1978.

INDEX

A

B

C

D

E

F

G

H

I

J

K

L

M

N

O

P

Q

R

T

U

V

W

X

Y

Z